Political Ecology

Enrique Leff

Political Ecology

Deconstructing Capital and Territorializing Life

palgrave
macmillan

Enrique Leff
Instituto de Investigaciones Sociales
National Autonomous University of Mexico
Mexico City, Mexico

ISBN 978-3-030-63324-0 ISBN 978-3-030-63325-7 (eBook)
https://doi.org/10.1007/978-3-030-63325-7

Cover illustration: Maram_shutterstock.com

This Palgrave Macmillan imprint is published by the registered company Springer Nature Switzerland AG.
The registered company address is: Gewerbestrasse 11, 6330 Cham, Switzerland

To Jacquie

...always

Preface

This book was written in no less than 40 years. To be sure, human thinking, much like one's own, is an endless odyssey in which each port becomes an excuse to recharge batteries and prepare for novel adventures that lead to new horizons. Notwithstanding this recognition, it came as a surprise to me to see that the waves and winds of time had not swept away the treasures I had left behind. Indeed, rather than fossils of the past, they were milestones that could be reconvened in an organized body of thinking, a body whose bones and flesh had been waiting to be reassembled into a book. Of course, over time, some words, concepts, and phrases had become rusty and needed polishing. Tissues and joints were also in need of new blood to be integrated into a more coherent argumentative line that would provide an evolving subject to the structure of the book. However, instead of trying to update the older texts into my current discursive style, I decided to blend old and new into a progressive concatenation of concepts, as one would do with the chronological exhibition of a painter's images where the spectator appreciates the evolution of abstract modes that emerge like the chromatic harmonic progression of a symphony, rather than a static show of pictures.

This book has gained its present consistency after revising, linking, and blending dated texts that came to light in four decades. "Twenty years is nothing," says a famous tango. In this time span, *two nothings aspired to something*. Not only has the style of the author changed: the topic at stake—the nature of the environmental question—has evolved, and thinking about it has matured. This book reflects this evolution of subject and thought, a shifting of environmental thinking caused by the development of eco-territorial conflicts and the social responses given to the environmental question. The chapters in this book reveal the emergence of new concepts in the transition from ecological economics and historical materialism to ecological Marxism: the forging of political ecology from the magma of significance that was no longer thinkable in the schemes of Marxism—the transition from value theory in political economy, to *ecological distribution* and *ecologies of difference*; a transition that would further shift to an environmental rationality grounded in the ontology of diversity, of difference and of otherness. This transformation would give consistency to a theoretical discourse able to respond to the questions generated by the radicalization of the environmental question.

In welding together the different moments captured in the sequence of my dated publications, I have tried not to erase the inscriptions and footprints left in my original texts. I have revised all of the writings included as chapters in this book, altering only parts of the originals, to give them a more "evolutionary consistency" for this new composition. Thus, their different styles and vocabulary, yet visible and readable, appear as new modulations of a larger symphony. Although, perhaps, the book should be read beyond the idea of chromatic harmonies in a musical piece, and approached as one of those animations by William Kentridge that show the progression of history and life, movements that erase past events and design the future. As in Kentridge's *Film Flip Book* or *Second-hand Reading*, I can identify myself with the author, his running back and forth over the written text, unbinding and rebinding pages, overpassing and erasing new signs, yet maintaining the original text where printed letters remain as solid as inscriptions in rock, where the footprints and stepping stones over which the author reflects are an eternal return to the original breakthrough idea in the environmental crisis of civilization of modern times.

Beyond the vicissitudes involved in this adventure of writing about what is not right, about the faulty ideas forged by philosophical categories and scientific concepts in the history of metaphysics and in the oblivion of life, beyond the task of polishing words and semantics, to weave and fit the knots of conceptual textures, beyond binding together the resulting chapters, this book's endeavor is to share the core of Latin America's thinking in political ecology. Underlying the text beats the heart of an emancipatory human experience. Beyond the will to deconstruct the legacy of philosophical thinking and *logocentric science* as the historical reason of the environmental crisis, this book highlights the responsibility of giving words to processes of dispossession that continue to impact people and to the ecological degradation of their life-territories. This is the conflict of life where political ecology has emerged from; not only as a new discipline in the social sciences but as a process of *re-existence* that reorients the civilization process towards sustainability from the conditions of life in the planet and the rights of existence of the peoples of the Earth.

Mexico City, Mexico
May 24, 2020

Enrique Leff

Prologue

I am very happy and honored to have been asked to write the Prologue to this important book. It gives me the opportunity to convey my intellectual and political enthusiasm for the vision of political ecology crafted by its author over the span of four decades of dedicated theoretico-political work. Its importance stems not only from Leff's enormous contributions to a particular interdisciplinary field of scholarship, political ecology, of which he is one of the most influential practitioners in the world, but from what it says about two of the most fundamental processes with which we are confronted in the world at present: the impact of capitalism on all living beings and ecosystems on Earth, including human social relations, on the one hand; and, on the other, the necessary and actual responses to the ensuing social and ecological devastation by groups of people all over the world committed to the defense of life and their territories from the developmentalist and capitalist onslaught. This dual character of the present collection is fittingly condensed in the book's subtitle: "Deconstructing Capital/ Territorializing Life." From the outset, then, we are invited to consider the essays that follow as having a twofold goal, at once deconstructive of discourses and processes of exploitation and domination and reconstructive of people's defense of life and territory.

It would be difficult, if not impossible, to find a Latin American scholar more prepared to carry out this complex mandate that Enrique Leff. From his early days, particularly the publication of his first landmark book, *Ecología y Capital* (Leff 1986, translated into English as *Green Production* in 1995), which made a splash in Latin America when it first appeared, all the way to his last three, genuinely major, works, tellingly focused on the very concept of "life" (*La Apuesta por la Vida—The Wager for Life*, 2014; *El Fuego de la Vida—The Fire of Life*, 2018; and the forthcoming *El Conflicto de la Vida—The Conflict of Life*), Leff has vastly demonstrated his profound understanding of the intertwined dynamics of biological, social, economic, cultural and political life in the contemporary world. In doing so, he has cemented his reputation, widely shared in Latin America and the Caribbean, as the continent's leading scholar in the environmental studies field, broadly speaking. By "broadly speaking" here I mean those fields dealing with the manifold interrelations between Nature, culture, and politics, including political ecology and environmental philosophy but extending to areas of knowledge such as ecological anthropology, environmental sociology, human geography, and ecological economics. Enrique Leff is, simply put, Latin America's premier environmental thinker.

It is unusual to find an author who feels as much at home with European philosophy and theory (from Heraclitus to Marx, Heidegger, Sartre, Lacan, Levinas, Foucault, and the Frankfurt School) as with Latin American approaches (his knowledge of Latin American scholarship on "the environmental question" is likely the most comprehensive in the continent). He is as conversant with Marxism, psychoanalysis, hermeneutics, and phenomenology as he is in the newer lands of post-structuralism and post-dualist social theory. Also unusual is his detailed knowledge of the environmental knowledge produced by social movements, from Zapatismo in Mexico to environmental, indigenous, and Black movements in South America. His long trajectory in these fields, from *Ecología y Capital* to *El Fuego de la Vida*, and the many books and edited volumes in between, are an indubitable demonstration of the richness and depth of the work of this complex thinker. I would like to emphasize that Leff is also a bridge-builder: between disciplines and fields of studies; between European and Latin American social and ecological theory; and between academic theorizing and activist knowledge production. Rather than dissolving the differences harbored inside these various strands, however, he carefully treats them, in rigorous Foucauldian fashion, as belonging to diverse onto-epistemic configurations which nevertheless have a lot to contribute to each other and to the larger goal of conceptualizing environmental problems in all of their biological, social, political, and philosophical complexity. Through his work, as attested by the selection of texts included in this volume, he achieves a veritable reterritorializing of knowledge, resulting in a unique epistemic tapestry with its own textures, nuances, and possibilities.

Political Ecology: Deconstructing Capital/ Territorializing Life constitutes an excellent introduction to Leff's thought and a well-chosen anthology of his entire oeuvre. The essays comprising it reflect most of his main theoretical-political preoccupations (e.g., with epistemology, deconstruction, Marxism, critiques of modernity, the question of value, sustainable development, socio-environmental conflicts, entropy, difference, ontology, social movements, de-growth, and so forth); they involve tangible issues concerning environmental degradation, territories, and identities; and they enlighten us on the cultural-political processes that unfold as various social groups engage in what he terms the social reappropriation of nature. Concepts such as environmental rationality, *saber ambiental*, environmental complexity, epistemologies of nature, ecological episteme, negentropic production, eco-technological productivity, and so forth, are indelibly associated with his name, as are discussions on the transdisciplinarity of environmental knowledges, environmental sustainability, and the ecological turn in social movements, among others. While important topics such as patriarchy, racism, or the trendy anthropology and geography of the more-than-human do not receive substantial explicit attention in these essays, they are by no means completely absent from consideration.

Little of this vast body of work is known in English, despite Leff's highly regarded reputation in the Anglo-American academic context. His contributions to the prestigious and pioneering journal *Capitalism, Nature, Socialism*, founded in 1988 (on whose Editorial Board he served for many years), along with his sole volume in English to date, already mentioned (Leff 1995), gave him a certain visibility in the United States and the United Kingdom. It was high time, however, for a more comprehensive volume of his work, and I am really glad that this has finally happened.

I believe the author's editorial strategy for this collection, poetically expressed in the Foreword ("instead of trying to update the older texts into my current discursive style, I decided to blend old and new into a progressive concatenation of concepts, as one would do with the chronological exhibition of a painter's images where the spectator appreciates the evolution of abstract modes that emerge like the chromatic harmonic progression of a symphony, rather than a static show of pictures"), has handsomely paid off. To resort to another musical metaphor, one might say that the works selected, taken as a whole, function as so many variations on the two major themes identified in the book's subtitle— the analysis of capital and the concern with life. While the former originates in his long-standing interest in Marxism, and his commitment to the collective elaboration of a cogent ecological Marxism, the latter has taken him fully into the domain of ontology, and to posit the historical struggle between an ontology of capital and an ontology of life, or between the social and biological death produced by capital and the fire of life he rightly sees reverberating underneath the struggles for territorial defense by indigenous and Black peoples but also in a wide array of rural and urban struggles at present.

Along the way, we find eye-opening discussions of modernity's domineering rationality and, conversely, about the all-to-important goal of the emancipation of life; original deliberations about the demands of a negentropic world founded on ontological difference in contrast to the imposition of a single, globalized world designed according to the dictates of markets, competitiveness, productivity, and never-ending growth; and clarifying analyses of the contradictory character of sustainable development. An important thread of inquiry concerns the rethinking of value in light of the increasing capitalization of Nature through technology. Above all, many of chapters gravitate around what is perhaps Leff's most cherished notion, that of a new environmental rationality, one capable of bringing economics and society into the fold of negentropic production and within the geographical, thermodynamic and ecological limits of the planet. This is not all, because to this rich conceptual apparatus he adds the emphasis, arising from social movements, on rexistence and the reconstitution of worlds through the cultural reappropriation of nature. The reason for his appeal to social movements is none other than his insistence on an ethics of difference as the *sine qua non* for fulfilling the ontological mandate of the reconstitution of worlds—whether referring to the difference harbored at the heart of biological life itself (biodiversity) or to that enacted by indigenous peoples and many others through their struggles. His wager for ontological pluralism is explicitly mindful of not falling back into forms of ontological monism (his critique of Murray Bookchin), and careful of not reinscribing his synthesis into the traps of modernist dualisms or systems theory reductionism.

To sum up, one could say that against the "forgetfulness of life", Leff counterposes the steady reconstruction of the conditions for life on the planet; he does so by envisioning a careful and caring re-weaving of the webs of interdependence that make up the bodies, places, territories, and regions that we *are* and inhabit. Straddling and productively blurring the boundary between Anglo-American, continental European, and Latin-American political ecologies, always with a decolonizing thrust that permits him to expose effectively the unavoidable fact that most scholarly knowledge has practiced, and resulted in, a fatal bracketing of Life from the space of thought—what he calls *the ignorance of knowledge*—this prolific and accomplished author has given us, throughout his lucid, decades-long work, a framework for thinking collectively about the civilizational transitions to which we are all summoned by the current multifaceted crisis of climate, energy, inequality, poverty, and meaning.

While Leff's is not always an easy prose, in the pages that follow the reader will find a generous invitation to seriously consider thought's ability—indeed, ineluctable duty—to reimagine our modes of being in the world, so that humanity as a whole can finally recommence its journey towards becoming what he aptly calls "peoples of the Earth." What he imagines, in the last instance, is the possibility for the emancipation of life from the shackles of the particular mode of thought and existence underlying extractivist global capitalism, those modes of being-in-the-world that have brought about the civilizational crisis in which we are engulfed at present. This journey finds in his capacious concept of environmental rationality a key to possible transitions to an era in which humans finally relearn to coexist with the planet in a mutually enhancing manner.

Arturo Escobar
Professor of Anthropology Emeritus
University of North Carolina, Chapel Hill
Adjunct Professor, PhD Program
in Environmental Sciences, Universidad
del Valle, Cali, Colombia
Adjunct Professor, PhD Program in Design
and Creation, Universidad de Caldas
Manizales, Colombia

Acknowledgements

I would like to express my recognition and gratitude to Maria Alessandra Woolson for her valuable help in revising the English translation of my book.

Contents

1

Unbinding the Sustainability of Life on Earth

Half a century ago, the alarm sounded worldwide announcing an unprecedented event in human history: a planetary environmental crisis. This crisis was no ecological catastrophe drawn by the adjustment of the metabolism of life on the planet as in past geological history of the Earth. This emergent crisis was rightly perceived as a *crisis of civilization*, of the limits of economic growth and of the unleashing of Promethean technological forces that had overcome the resilience of the fragile web of life in the biosphere. To control the disintegration and collapse of the life-support system of the planet and to counterbalance the destructive effects generated by the momentum of the techno-economic drive of the globalization process, a new keyword was forged: *sustainability*. Unbinding sustainability on Earth was the call to save the planet from its unintended and unforeseen downfall, in order to respond to its catastrophic destiny.

Thus, humanity was convened to ground the development process on a sustainable basis. This imperative has led to a series of questions: What are the abyssal foundations of the economy that drives the entropic

This is a revised version of the unpublished text of a contribution offered at the Conference on Sustainability Unbound, 2011–2012, the Saul O. Sidore Memorial Lecture Series, University of New Hampshire, 21–22 March 2012.

E. Leff, *Political Ecology*, https://doi.org/10.1007/978-3-030-63325-7_1

degradation of the biosphere? What is repressing the sustainability of life on the planet? From what constraints of a domineering reason do we need to unbind sustainability in order to liberate its potentialities for life? Is this just an occasion to cut the ribbon in a glamorous celebration of a simulacrum, an unexpected surprise that pops out of the cage of rationality in modernity like a Jack-in-the-box, a Joker suddenly appearing at the turn of the corner hidden from our eyes for as long as civilization has existed? From what intentions and power strategies has such a word emerged? What is the real and profound meaning of *sustainability*? What has been locking it, quieting its revealing message, restraining its potentialities, blocking its transgressional purpose? From what malevolent witchcraft do we need to exorcise it, to let it free? Is it bound by the hegemonic powers that dominate and rule the world order; or by fear of liberating a revolutionary process that subverts the ontological security of an Establishment shaken by environmental risks? There seemed to be something enigmatic and suspicious in the term *sustainability*: a false promise, a dangerous risk; unrevealed powers would like to bind it to a constrained expression, to limit and deviate its meaning to a color-blind vision of the *greening of the world*.

In fact, sustainability was an unspoken word, a practically nonexistent concept in the vocabulary of economic theory, public policy and international affairs before April 22, 1970. On that day, the moment Earth Day was founded, at the height of counterculture in the United States; the word sustainability gave voice to an emerging consciousness that channeled the energy of the anti-war protest to the environmental movement. Previously, the gerund *sustaining* and the adjective *sustainable* appeared in dictionaries as "the ability to sustain something" or as "something capable of being sustained." But the concept of sustainability as a social norm, as a condition of the economy and of life itself, was forged later on as a neologism in the Brundtland Report, *Our Common Future* (WCED 1987), where it was introduced in the discourse on development by the International Union for the Conservation of Nature (IUCN) through the decade. Before then, we lived in a world governed by the ideology of progress, of unlimited growth and Promethean technology. The irruption of the environmental crisis, the idea of an imminent ecological collapse, came as a cultural shock. One day we woke up and were confronted with a crisis of civilization that we had not anticipated. How could this outcome of history have remained unforeseeable to modern culture, blind to the Enlightenment of reason, occluded by the *truth* of science? How could the principles of progress based on the right and duty of human beings to exploit Nature—an ideology embedded in Western Judeo-Christian civilization and the modern principle of reason—prove to be so wrong? What had gone astray and out of control of our keenly sought transparent and objective knowledge in the age of *Enlightenment of Reason*?

And yet, something was profoundly mistaken in the modern conception of human life and development, in the dominant forms of knowing and modes of production, in our ways of understanding life, of inhabiting the planet and *being in the world*. The "civilized" world, the society of knowledge, woke up one day alienated by reason, unable to breathe its polluted air, to bathe in its contaminated waters, smeared by the murky world that was the product of a history blind to life, living Jose Saramago's filthy metaphor of an insane scatological world.

Indubitably, there had been premonitory insights of uneasiness caused by living in a world governed by technology: the Nietzschean spreading wasteland that accompanied the nihilism of reason and Heidegger's world of *Gestell*; the *unheimlich* Freudian malaise in culture; or more recently, Italo Calvino's *cloud of smog*, Rachel Carlson's *Silent Spring* and Fernando Pessoa's phantasmagorical vision of the polluted world to come, of "Fog and Smoke," impossible to grasp either by reason or by the senses.[1] But it was *The Limits of Growth*, the MIT/Club of Rome study published in 1972 at the time of the UN World Conference on Human Environment held in Stockholm, that sounded the ecologic alarm. If the economy, if population and if technology were to continue their growth trends, there would be an ecological collapse by the turn of the century (Meadows et al. 1972).

This news caught the whole of humanity unprepared. The first reactions and responses appeared immediately as an antidote to the world's sickness: birth control, zero growth and the human harnessing of technology. While the "underdeveloped countries" responded by claiming their right to development, in seeing the halt to growth as a

[1] Fernando Pessoa, *O libro do desassossego*, São Paulo, Companhia das Letras, 2002, pp. 385/346–347.

condemnation to poverty, the economic system resisted the challenge in its paradigmatic terms and powerful means. Unable to assimilate the radical critique that the environmental crisis posed to the rationality of modernity—and to the metaphysical history of Western civilization—the economic establishment tried to resist it by minimizing its consequences while simultaneously, in unacknowledged contradiction, seeking to reabsorb it within the principles of economic rationality.

The response from established mainstream economics was to internalize the ignored ecological externalities by extending the tentacles of the economic system to those—until then—free and abundant elements of the environment: air and water first; then forests and biodiversity; to finally end up codifying life support systems, environmental goods and services, and climate change in terms of economic values. Therefore, all biological, ecological and environmental processes have been absorbed by a totalitarian and omnivorous techno-economic rationality. The *great transformation* (Polanyi 1944) driven by capitalism converted Nature into raw materials and natural resources, and human beings into labor force. In the present phase of economic globalization, Nature's organizing systems in the biosphere are disintegrating into mass and energy, molecules and genes, manipulated by technology. *Homo sapiens sapiens* has been converted to *Homo economicus*; he no longer needs to think to establish his place on Earth; he has to adapt to global change by exercising his "rational choice" in response to market signals.

From the emergent discipline of ecological economics, the response was in a way radical, and yet profoundly naïve. In its more optimistic vein, it postulated the economic system restrain itself according to conditions for planetary sustainability, to reach a "steady state" and ecological balance, in order to allow depleted resources to be regenerated and degraded environments to be restored (Daly 1991). Its radical current, initiated by Nicholas Georgescu-Roegen (1971), saw the economic system as a mega-machine that fed on Nature, which is then transformed by the production process according to the second law of thermodynamics. Therefore, the promise of unlimited growth and the self-complacency of the economic system resulted in an unsustainable economic process: like the Goyescan image of Saturn sinking in his swampy feet and feeding on his offspring, the economic process became the furnace that consumes scarce Nature; according to the entropy law it transforms and degrades all matter and energy into non-recyclable pollution and heat. This is particularly so when fossil energy sources fuel the economy. As a result, the economic system triggers, unleashes and exacerbates the *entropic degradation of the planet.*

Thus, sustainability appeared on the horizon of a world convulsed by an environmental crisis, in which the basic structures of life support systems have been broken and degraded. Sustainability became the watchword to readjust the economic process, social practices and personal behavior to bring the world back into ecological balance. However, to unbind sustainability much more was needed than just oiling the rusted economic mechanisms to internalize environmental externalities and generate efficient economic instruments for environmental management. Building sustainability into the economic process implied something more than merely setting boundaries to limit the economic system. Beyond the claim for economic de-growth (see Chap. 8 below), liberating sustainability calls for the deconstruction of economic reason (Gorz 1989). But strong epistemological obstacles within economic rationality prevent this paradigm change internalizing the thermodynamic and ecological conditions necessary to launch a sustainable economic process.

Sustainability became an impossible Quixotic dream. Its purpose was to resist the momentum of an economic growth unbound by modernity's Promethean promise that was embodied in the core roots and soul of economic rationality. Ecological economics was right in postulating that a sustainable economy could not violate the limits and conditions of Nature. But it did not provide the psychic perspective that would cure the economy of its psychotic "growth-mania"; nor did it facilitate the biotechnologist's efforts to modify the unsustainable gene that resists all attempts to control its potency, or the surgeon's scalpel that could cut out the malignant tumor in the economic organism and enable it to adapt to the ecological conditions of our living planet.

Sustainability was seen as a messianic signifier to save humanity, and restore the technological world that had degraded and perverted life. But the profound wound that was opened by the human *logos* imposed its ways over the emergence and becoming of *physis*, and its aftermath in the Cartesian separation of subject and object, of reason and feeling, of culture and Nature; it needed more than a keyword or a concept to heal. The simulation strategies of discourse and *geopolitics of sustainable development* can easily tamper with such concepts (Leff 2002a). The polysemy of the word lends itself to this manipulation. While in Spanish we can distinguish *sustentabilidad* as the need to base something on a solid support—in this case, ecological sustainability—and *sostenibilidad* as a process that is maintained over time, in English as well as in other languages (French, Portuguese, etc.) this linguistic distinction does not occur. Thus, the need to internalize the ecological basis of sustainability was quickly transformed into a willful expression of *sustainable development*. This in turn, meant a will to reset the boundaries of the economic process with no restrictions other than the "amount of ecology" that could be assimilated by the economy without compromising its expansive process. As a result, a "green economy" emerged, and the concept of sustainability became corrupted. But the dominant discourse on sustainability acted as a boomerang: it went in circles and came back to its initial point, maybe a little more greenish, but reinforced in its sameness, trapped in the enclosure of "reflexive modernization" (Beck et al. 1994).

Indeed, the discourse declares to care for humanity, for a sustainable future. Thus, a slogan has been disseminated worldwide to define sustainability as "development that satisfies present needs without compromising the needs of the future" (WCED 1987). Sustainable development pretends to be a holistic paradigm and to stand solidly on three feet: economy, society, and environment. It pretends to be open to epistemological democracy and participative management. And yet, in practice, the geopolitics of sustainable development exerts a wider and deeper subjection of the world to the functioning and expansion of the global economic system, to the empire of the market, to the will to unbind a "sustainable development" process through the economic valuation of the natural goods and environmental services of the planet.

This explains the rise of carbon sinks and carbon bonds, while the most common greenhouse-effect gas has already surpassed the threshold of 400 ppm in an already saturated atmosphere, increasing the risk of environmental catastrophes.[2] And what we get from the world leaders is nothing more than rhetorical responses, like those of the G-8 or G-20 about their willingness to reduce greenhouse gases emissions in 2050, a sufficiently long time-span to allow them to be unaccountable for not having taken actions consistent with such declarations, reminiscent of Doctor Dulcamara, that marvelous character in Donizetti's opera buffa, *L'Elisir d'Amore*. The doctor promotes his magic potion (actually cheap Bordeaux wine) telling Nemorino that he will notice the effect only the day after, "*tanto tempo sufficiente per cavarmela e fugir*."[3]

In the meantime, China's emissions have already surpassed those of the United States; India and the "emergent countries" are not far behind. The technological illusion promoted by the Wuppertal Institute in Germany in its "Material Input per Unit of Service (MIPS)" and associated Factor X project to decrease material flows (i.e. dematerialize production) in the industrialized countries by a factor of ten has failed; and will continue to fail because one thing that technology cannot reverse is the irreversible degradation of matter and energy entering the economic process, that is, the ineluctable entropic law. Together with the failure of the Kyoto Protocol to control climate change, the Clean Development Mechanism has become the Trojan Horse that has infiltrated the tropical regions and the global South to facilitate the neo-economic colonization of its territories. By promoting *Reduction of Emissions from Deforestation and Desertification* (known by the acronym REDD, but don't be alarmed, it is not a red–green solution coming from the politburo of an eco-Marxist regime), the international financial and environmental organisms, together with governments of developed countries and transnational enterprises, pretend to reduce the emissions from deforestation and desertification by assigning economic value to the conservation of forests

[2] Steadily rising CO_2 concentration exceeded 400 ppm in air for the first time in human history in 2013. Prior to the onset of the Industrial Revolution, CO_2 levels had fluctuated over the millennia but had never exceeded 300 ppm at any point in the last 800,000 years. The April monthly average exceeded 410 parts per million for the first time in recorded history according to the Keeling Curve measurement series made at the Mauna Loa Observatory in Hawaii. This represents a 30 percent increase in carbon dioxide concentration in the global atmosphere since the Keeling Curve began in 1958 (cf. Scripps Institution of Oceanography at the University of California San Diego: http://scripps.ucsd.edu/programs/keelingcurve/).

[3] "Enough time to get by and flee," my translation.

and biodiversity. Thus, the biosphere is being refunctionalized to serve the purposes of the expanding economy.

Like a wild horse that bucks to avoid being harnessed, the economy rejects sustainability as the norm that binds its unrestrained desire to expand. *Sustainable development* has been capital's syntagmatic disguise in its ecological stage, to serve this *will to power*. But, is this the only way we can think about sustainability? Or, is there a deeper meaning and potentiality to be revealed, to be discovered, to be liberated from the restrictive rational cage in which ecological, cultural and human potentiality has been oppressed and silenced, by the "enlightened" diffusion of global hegemonic economic reason?

Unbinding sustainability is a project that goes beyond expanding the boundaries of the economy to embrace its environing hinterland. It goes beyond an interdisciplinary project to open the frontiers of knowledge to neighboring disciplines, as if pretending that the environment is nothing more than a new dimension to be internalized in the normal paradigms of science, just as economics pretends to internalize ecological externalities. The environment emerges in a different dimension, in another ontological regime, grounding its roots in new territories, projecting toward the horizons of a new world that rises from the conditions of life. The environment is not a discipline to complete the perfect circle of sciences. The environment is, in fact, exterior to the *logocentrism of sciences* (Derrida 1978, 1982) and the enclosure of the economic system. Here, economists are right; but it is an externality that resists being reabsorbed in that rationality that has ignored it. The environment emerges externally to economic rationality, bounded by the principles of a universal, homogeneous technological world. Unbinding sustainability is the act of liberating life from the dominance of theoretical-instrumental-economic rationality; that rationality which tightened its boundaries, and engulfed itself in a totalitarian will to power, exterminating life, constraining diversity, and ignoring the original principle of *differance* (Derrida 1982) from which human thinking diverted the creative evolution of life from its natural course toward an abysmal unsustainable state (Leff 2018).

Sustainability appears on the horizon of those other invisible but possible worlds, if only we were able to unleash the potentialities of life that have been constrained by force of a rationality *anti-natura*. The process of economic rationalization that has deviated and degraded the courses of life, leading to the *entropic decay of the planet* has been trapped in the discourse of sustainable development. What forces and potentialities have to be released in order for sustainability to be unbound, and travel in its odyssey toward the horizon of sustainable futures? This question leads to an underlying question. What is life? This inquiry opens up the most fundamental, radical, and abyssal question necessary to liberate the *life* that animates the sustainability that we seek. To liberate life, we have to deconstruct the obscure reasons that, throughout the history of metaphysics, have led humanity to oblivion of life, to the insensibility for life itself, leaving us thoughtless and speechless to answer this question.[4]

And yet, according to the existential ontology proclaimed and sung by the Beatles, “Let it be… there will be an answer… let it be.” Deep ecology followed the call to *let Nature be*: not only to preserve biodiversity and cultural diversity but to allow all life forms, present, and potential, to come into being. But, how to understand the Heideggerian call to *let Being be*, when biodiversity is being eroded by economic uniformity; when life is being distorted by biotechnology; when letting things be implies giving way to anything that technology can create and destroy in its way?

The question of sustainability becomes an abstract query when we do not define what we want to sustain. If we are going to make the economy sustainable, because we think of it as the ultimate purpose of human

[4] I have tried to shed some light into this question in my book: *El Fuego de la Vida. Heidegger ante la Cuestión Ambiental* [*The Fire of Life. Heidegger in the Face of the Environmental Question*] (Leff 2018). The Yiddish poet Avrom Sutzkever expressed the astonishment, the wordless response to the most extreme and uncanny human experience: the Holocaust. His poem *Vi azoi*, says the unsayable that calls for an answer to break the chains of the past, to offer new words to silence and feeling, to awake numbness, to crawl up and liberate life, making it humanly and ecologically sustainable:

> How? How, and with what will you fill your cup after your liberation? In your joy, are you ready to feel all of yesterday's dark lamentation? Where the days have congealed into skulls in a bottomless, endless abyss? You will search for the keys to your doors whose locks are all shattered and dead. You'll think: it was better before as you chew on the sidewalks like bread and the time gnaws you silent and numb like a cricket held inside a fist. And your memories will all be compared to a buried, forgotten old town and your outsider eyes they will stare like a mole crawling down, crawling down…. (Kahn 2014)

civilization, then we must explore the economic rationality constructed by modernity, whether it can become "sustainable," and if so, what are its consequences for life in the planet. We should also address its capacity to internalize ecological norms in order to orient economic processes toward the sustainability of growth. The world has been steered toward the *entropic death of the planet*, but this destination is not an inescapable fatality; it is the biosphere's metabolic tendency once the *anti-natura* rationality of modernity has intervened in life. Emancipating life implies the deconstruction of the history of metaphysics that set off this trend. However, if the economic process is not the means to make life sustainable, then we must first question what kind of life we are seeking and why the capitalist mode of production cannot make it possible. For life to be sustainable, we must explore which are the *conditions for the sustainability of life*. What we cannot do is merely take life for granted. Yes, life was given to us, but not forever and not under any condition. Humanity cannot waive its responsibility for the destiny of life: and that life on the planet depends on the ways that we chose to inhabit it and on the meaning we ascribe to life for human existence.

Because the rationality instituted in the world has intervened in the destiny of life in the planet by opposing an antagonistic techno-economic process to Nature, the ultimate question for unbinding of sustainability is the conditions for the sustainability of life. The world is governed by economic rationality. And the open economic process that grows without limits feeds on Nature, on matter and energy, on biodiversity and life forms reduced to mere objects, as raw materials for the economy, not for the value and meaning of life. Objectified Nature is fed to the megamachine of the global economy, and following the law of entropy, it is transformed into commodities, pollution, and heat. This degradation process is manifest in the increasing deforestation, desertification, biodiversity loss, and climate change that are leading to the entropic degradation of the planet. This is indeed not the "sustainable development" we want to unbind.

The question then arises: is there a possibility of deriving a sustainable economy, an economy that works with and through the creative forces of Nature and not against them, an economy built on the ecological potentials of the planet, an ecological and human economy that is grounded on principles and conditions of life, that is intertwined with the web of life in the biosphere, and recognizes the conditions of human life and cultural diversity? My answer is yes, but this implies the deconstruction of *economic rationality* and *logocentrism of science* to open the way for the construction of a new paradigm of production founded on the ontology of life: a paradigm of eco-technological-cultural productivity guided by the principles of the *negentropic potentials*[5] of the planet and environmental rationality (Leff 1986, 1995, 2004, 2019).[6]

If capitalist-induced *entropic degradation* is what is driving ecological destruction of life-support systems and cultural resiliency, then sustainability will depend on the possibility of envisioning and constructing a sustainable mode of production based on the *negentropic potentials of life*: a mode of production based on the process of photosynthesis, on managing the ecological arrangements of the biosphere that create life, and on the ethical responsibility, the creative action and the symbolic condition of human existence. This is a fundamental reorientation of the civilization process that is to be grounded on a new rationality to preserve and sustain, enhance and magnify life in the planet. Photosynthesis is the main negentropic process in the creation of life on Earth: it is the transformation of radiant solar energy into biomass, upon which all the complex forms of life depend. It is true that to maintain this evolving complex organization of life, entropy works in self-organizing of living organisms, in the matter and energy cycles of the food chains, in the metabolism of the biosphere, and in the economic process. Notwithstanding the entropic decay in all transformation processes of Nature, negentropy is the

[5] I adopt Erwin Schrödinger's concept of *negative entropy* as the principle feature of life (Schrödinger 1944) by transforming solar energy into biochemical compounds via photosynthesis. Political ecology faces the challenge of extending this thermodynamic process to the construction of a social order based on the immanence of life: the ecosystemic productivity of the biosphere and cultural creativity with Nature. Thus, political ecology encompasses the power strategies adopted through the social rationalities that direct the flows of matter and energy either to increase the entropic degradation of the planet or to build *negentropic societies*.

[6] *Environmental rationality* is the fundamental category that articulates my thinking on the environmental question from beginning to end, from my books *Ecología y Capital* (1986, 1994) and *Racionalidad Ambiental* (2004) to *La Apuesta por la Vida* (2014), *El Fuego de la Vida* (2018) and *El Conflicto de la Vida* (2020). Although environmental rationality is referred to throughout the chapters of this book, only in the works mentioned above will the interested reader find an elaboration of the concept.

thermodynamic and material principle that creates life and maintains the life support systems of the planet. Sustainability becomes a complex negentropic/entropic organization processes.

The sustainable *negentropic paradigm of production* that I have proposed (Leff 1975, 1986, 1995, 2019) is built on the complex integration of three orders of productivity: ecological, technological, and cultural. Ecological productivity is based on the ecological potential of different ecosystems. Ecological research has shown that the most productive ecosystems, those of the humid tropics, produce biomass at natural yearly rates up to about 8 to 10 percent (Rodin et al. 1975). This ecological potential can be enhanced by scientific research, ecological technologies, and productive practices that include high-efficiency photosynthesis, management of ecological processes and selective regeneration of valuable species, as well as multiple associated cropping, agroforestry arrangements and agro-ecological techniques, all of them guided by cultural practices that define the cultural-economic value of the eco-technological output of different productive units.

This alternative paradigm of production is forged in spatial and time frames of *cultural imaginaries*[7] and ecological practices of the peoples of the Earth. Thus, a rich diversity of peoples and cultures are grounding new theoretical perspectives in the reconstruction of their life territories. Otherwise than the *a priori* conditions of knowledge and abandoning the linear Eurocentric conception of time, territoriality appears as the manifestation of the "unequal accumulation of times" that cohabit in geographical space, as expressed by the Brazilian geographer Milton Santos (1996). *Negentropic production* is embedded in environmental rationality, in a phenomenological-hermeneutical-existential understanding of space-time from the ontology of life, as a re-engaging of historical becoming of diverse ontological regimes and modes of *living-in-the-world* (Deleuze and Guattari 1987; Leff 2018).

[7] I adopt and use throughout the book the sociological category of *social imaginaries*, borrowed from Cornelius Castoriadis (1998), to refer to the institution of the diverse ways of *living in the world* of *cultural beings* within an *ontology of life*. Social imaginaries do not only reflect and express imagination as acts of consciousness, but as *magmas of significations* embodied in habitus and practices, connecting social behavior with the metabolism of life territories.

In the culture of original peoples, as the Inca tradition that prevails in the imaginaries and practices of the Andean people today, ecological floors were integrated by means of principles of complementarity and reciprocity that commanded the organization of geographical space (Murra 1956, 1975). Coevolution of culture with Nature and the relations to their territories gain visibility, opening new perspectives for the construction of alternative sustainable societies. This conception of history, of territory, and the cultural heritage of natural resources has important political implications for the reinvention of the identities of traditional peoples, such as the actuality of ancestry invoked by the afro-Colombians of the South Pacific and the principle of "living well" (*Sumak Kawsay* or *Suma Qamaña*) of the Quechua and Aymara peoples of Bolivia and Ecuador today.

The privileged spaces for deploying the strategy of *negentropic production* are the rural areas of the planet, especially those territories inhabited by indigenous peasant peoples. This new paradigm can be contested and contrasted in the academic field with the established economic rationality; but even if it could prove consistent and triumph in the intellectual arena, it will only be able to verify its possible historical truth by grounding its principles in the life territories of the peoples of the Earth, through their imaginaries and practices. It will only confront the established economic world order in the field of political ecology, through social movements, in their struggles to reappropriate their bio-cultural patrimony. Thus, new socio-environmental indigenous and peasant movements are emerging in this "postmodern" perspective in the *endowment* and *enownment*[8] of their cultural beings embedded in the immanence of life.

[8] I adopt here the neologism *enowning*, a term with which Parvis Emad and Kenneth Maly rightly translated the Heideggerian concept of *Ereignis* in the title of his *Contributions to Philosophy (From Enowning)* (Heidegger 1999, pp. xx–xxi) to highlight the fundamental fact that Being does not appropriate entities as one were to extract something from elsewhere to make it your own, or as capitalism expropriates Nature, but rather *enowns*, makes something its own, because it properly belongs to itself or emerges from itself. I will use the term *enownment of Nature* to envision the emergent ways of reinvention of the identities and reinscription of the imaginaries and practices of the people in the immanence of life, the ways in which a culture embodies its bio-thermodynamic living conditions into its cultural being, and the "self-management" of its bio-cultural patrimony under the conditions of life. The utopian *enownment* of Nature will be distinguished from traditional modes of appropriation of Nature and the capitalist expropriation of Nature.

An emblematic example is the struggle of the *Seringueiros* in the Brazilian Amazon region who, from their trade-union claims to the land where they worked as agrarian proletarians of the rubber industry in the late nineteenth century, have come to establish their *extractive reserves* as a strategy for sustainable production. Seringueiros have reinvented their identity by adopting the name of the tree (*Siringa*) from which they derive their livelihood. Extractive reserves are just one alternative to the strategy of the *Green Economy*: to the Clean Development Mechanism and REDD Programs; to oil fracking and transgenic latifundia. Facing the colonizing and exploitative character of the geopolitics of globalization and "sustainable development," a series of critical and creative responses are emerging in the conflictive field of political ecology lead by Latin American and Third World peoples.

The imaginaries that root ecosystems dwellers to their territories are becoming manifest as strategies of *resistance–rexistence*, as expressed in the political discourse of the Afro-Colombian movement of the *Process of Black Communities* of the Colombian Pacific coast in the reinvention of their identities and enownment of their rich heritage of biodiversity. These principles and targets are spreading in the claims of indigenous peoples for autonomy and dignity in the emergence of socio-environmental cultural organizations and emancipation movements in defense of their cultural rights to their territories and productive practices in many Third World countries and all latitudes of the Latin American region: from the COICA Amazonian cultures and the Aymaras and Quechuas of the Andean region to the Zapatista movement in Mexico; from the Mapuche in the south of the continent to the Seris or Comcaac in the arid north of Mexico.

The peoples of Earth are *territorializing sustainability* in continuous resistance to capitalist expansion and their struggles with the dominant *geopolitics of sustainable development.* One recent expression of this confrontation is the *Tipnis* movement in Bolivia in defense of their ecological and cultural heritage against the construction of a highway over their traditional biologically diverse territory to give Brazil access to the Pacific. Likewise, today the indigenous and peasant peoples, together with large groups of intellectuals and civil society of the southeast of Mexico are opposing the project of the Maya Train and development infrastructure in the Tehuantepec Isthmus promoted by the new "progressivist" government of Lopez Obrador.

The thermodynamic, ecological, symbolical, and cultural principles that give support to an alternative mode of production embodied in the *social imaginaries of sustainability of the peoples of the Earth* (Leff 2010) mobilize the emancipatory processes of peasant and indigenous communities in their creative innovation processes for the construction of *negentropic societies*. These are the points of anchorage to root environmental rationality and unbind the sustainability of human life within the ecological conditions of the planet. There is not one strategy defined in advancing to build sustainability, and there cannot be any triumphant optimism that indigenous and peasant peoples' persistence will prevail and not be absorbed by modernity or swept away by the momentum of entropic degradation triggered, driven, and fueled by the world economic order. The task arising from the environmental crisis for political ecology and environmental sociology is not merely to analyze the efficacy of these resistance and *rexistence* movements, of the adaptation or extermination of "traditional peoples" throughout the socioenvironmental conflicts under the hegemony of power in the global world order. There must be accompanying sociological imagination to their struggles for emancipation mobilized by their *social imaginaries of sustainability*, rooted in their ancestral memories and their life territories, as those being expressed today by many indigenous peoples' movements in Latin America, to give support to the trust that other worlds are still possible, that the sustainability of life has not been engulfed by the unsustainable progress that drives the world toward the entropic degradation of the Earth.

I envision sustainability as a world made of many worlds; as the diversity of life itself, where the future emerges from the heterogenesis of the coevolution of biological and cultural diversity; an economy where production is based on the ecological conditions and potentialities of our living planet developing from the negentropic principle of life, together with cultural inventiveness of the peoples of the Earth: a human world constrained by the entropic condition of life, by the limit law of entropy, but triggered to the *negentropic productivity of life*. If life has been able to survive and to continue its creative evolution under this entropic condition, why should the emergence of the symbolic order derived from an unsustainable economy lead humanity to the end of history? The sustainability of the world order demands the human capacity to harness technology under the conditions for the sustainability of life, where the will to power of life prevails over the will to power embedded in the techno-economic rationality that drives the entropic death of our living planet.

The limit-law of entropy condemns the hegemonic economic order. Even though Heidegger understood our present world, derived from the history of metaphysical thinking, as being trapped and molded by technology (Heidegger 1935/1977), we cannot succumb to such fatalistic views of the outcome of man, of the future of humankind, of the deathly destiny of life. We must resist the forces that would have us live in a wasteland and turn our atmosphere into a non-breathable gas chamber as our inevitable future. However it may be, a sustainable world will not emerge spontaneously from the *noosphere*, as Teilhard de Chardin (1961) thought it would. It will take a long historical transformation to deconstruct the dominant rationality of modernity and to transcend to other possible worlds built on the principles of environmental rationality. The mirror of representation that reflects the universality of Being, the forced unification of the World, and our selfish Ego must be broken. This historical transition will be a process guided by a politics of diversity, difference, and otherness; through a dialogue of knowledge/*savoirs* open to other languages and other voices, to other feelings and other reasons.

The construction of a sustainable world built on the ontology of diversity demands more than a culture of tolerance and intercultural politics. Living together in a world ruled by *ethics of otherness* (Levinas 1977, 1999) means living in uncertainty, accepting the other as an absolute other, a being that cannot be translated into my own codes of thinking, absorbed by sameness, dominated by rational powers. A sustainable future will not be the result of a consensus derived by *rational communication* (Habermas 1984, 1987) but will imply opening our rational beings to other sensibilities and imaginaries, to other practices and *habitus* not amenable to the arguments or reduced to the norms of rational discourse. A sustainable future will be the outcome of a *dialogue of knowledge/savoirs*, beyond transdisciplinary bounds, in the fertilizing encounter of different cultural beings (Leff 2004). This process will not emerge as the confluence of different visions and interests for the appropriation of Nature free of conflict. At the bottom of political ecology lies the question of conflictive interests in the appropriation of Nature, and the tremendous challenge that for humanity means to learn to live under the conditions of life on the planet.

To transcend this conflict and lo liberate sustainability from the chains of technology we must hold back our will to dominate others. We must risk coming closer to Nature, caressing the world and re-erotizing life. "The caress does not know what it seeks," wrote Emmanuel Levinas (1993). This caress is driven by a desire of life to liberate its creative potentials taking care of the conditions of human existence. The caress is a *wager for life*, the drive to release the power of *physis* and to enlighten the *fire of life*, harnessing our *human logos* and unconscious desires. This is the sustainability I envision, the force that must erotize human thinking and guide human actions to build a sustainable living world: unleashing the potency of life; allowing life to be, conducted by the meanings of life; devoting life to life, under the conditions of life.

Bibliography

Beck, U., Giddens, A. & S. Lash (1994), *Reflexive modernization: politics, tradition and aesthetics in the modern social order*, Stanford: Stanford University Press.

Castoriadis, C. (1998), *The imaginary institution of society*, Cambridge: MIT Press.

Chardin, P. Teilhard de (1961), *The phenomenon of man*, San Francisco: Harper & Row Publishers.

Daly, H. (1991), *Steady-state economics*, Washington: Island Press.

Deleuze, G. & Guattari, F. (1987), *A thousand plateaus. Capitalism & schizophrenia*, Minneapolis/London: University of Minnesota Press.

Derrida, J. (1978), *Writing and difference*, Chicago: The University of Chicago Press.

———. (1982), *Margins of philosophy*, Chicago: Chicago University Press.

Georgescu-Roegen, N. (1971), *The entropy law and the economic process*, Cambridge, Mass.: Harvard University Press.

Gorz, A. (1989), *Critique of economic reason*, London/New York: Verso.

Habermas, J. (1984), *Theory of communicative action, Vol. I: Reason and the rationalization of society*, Boston, Mass.: Beacon Press.

Habermas, J. (1987), *Theory of communicative action, Vol. II: Live-world and system: a critique of functionalist reason*, Boston, Mass.: Beacon Press.

Heidegger, M. (1935/1977), "The question concerning technology", in *Heidegger. Basic writings*, trans. D. Farrell Krell, New York: HarperCollins, pp. 283–317.

———. (1999), *Contributions to philosophy (from enowning),* Translated by Parvis Amad & Kenneth Maly, Bloomimngton & Indianapolis: Indiana University Press.

Kahn, D. (2014), "The chain of the Past" (Daniel Kahn on Avrom Sutzkever), Azymptote Blog, April, 2014.

Leff, E. (1975), "Hacia un proyecto de ecodesarrollo", *Comercio Exterior*, Vol. XXV, No. 1, México, pp. 88–94.

———. (1986), *Ecología y capital: hacia una perspectiva ambiental del desarrollo*, México: UNAM.

———. (1994), *Ecología y capital: racionalidad ambiental, democracia participativa y desarrollo sustentable*, México: Siglo XXI Editores.

———. (1995), *Green production: towards an environmental rationality*, New York: Guilford.

———. (2002a), "La geopolítica de la biodiversidad y del desarrollo sustentable: economización del mundo, racionalidad ambiental y reapropiación social de la naturaleza", in Ceceña, A.E. & Sader, E., *La guerra infinita: hegemonía y terror mundial*, Buenos Aires: CLACSO-ASDI, pp. 191–216.

———. (2004), *Racionalidad ambiental: la apropiación social de la naturaleza.* México: Siglo XXI Editores.

———. (2010), "Imaginarios sociales y sustentabilidad", in *Cultura y representaciones sociales.* Num. 9, México, pp. 42–121.

———. (2014), *La apuesta por la vida: imaginarios sociales e imaginación sociológica en los territorios del Sur*, México: Siglo XXI Editores.

———. (2018), *El fuego de la vida: Heidegger ante la cuestión ambiental*, Mexico: Siglo XXI Editores.

———. (2019), "Negentropic production", in Kothari, Ashish, Demaria, F., Acosta, A., Salleh, A. and Escobar, A. Eds., *Pluriverse: a postdevelopment lexicon.* London: Zed Books.

Levinas, E. (1977/1997), *Totalidad e infinito: ensayo sobre la exterioridad*, Salamanca: Sígueme, 4th edition [*Totality and infinity: an essay on exteriority*, Pittsburgh: Duquesne University Press, 1969].

———. (1993), *El tiempo y el otro*, Barcelona: Paidós [*Time and the Other*, Pittsburgh: Duquesne University Press, 1987].

———. (1999), *De otro modo que ser, o más allá de la esencia*, Salamanca: Ediciones Sígueme [Otherwise than Being or beyond essence, Springer Science + Public Media, B.V., 1991].
Meadows, D.H., D.L. Meadows, J. Randers & W.W. Behrens III (1972), *The limits to growth*, New York: Universe Books.
Murra, J. (1956), *The economic organization of the Inca state*, Chicago: University of Chicago.
———. (1975), *Formaciones económicas y políticas del mundo andino*, Lima: IEP
Polanyi, K. (1944) *The great transformation: the political and economic origins of our time*, Boston: Beacon Press.
Rodin, L.E., N.I. Bazilevich e N.N. Rozov (1975), "Primary productivity of the main world ecosystem", *Ecology First International Congress*, The Hague, September 1974, in W.H. van Dobben e Lowe-McConell, *Unifying concepts in ecology*, The Hague, W. Jung B. V. Publishers and Wageningen: Centre for Agricultural Publishing and Documentation.
Santos, M. (1996), *A natureza do espaço: técnica e tempo; razão e emoção*, São Paulo: Hucitec.
Schrödinger, E. (1944/1969) *¿What is life? The physical aspect of the living cell*, London/New York: Cambridge University Press.
WCED (1987), *Our common future*, Oxford: Oxford University Press.

2

The Social Reappropriation of Nature

Valuing Nature in Ecological Economics

Sustainability is not the state that life has reached in the Anthropocene era,[1] manifesting from the natural emergency as *physis* through the intervention of *logos* in the creative evolution of life. Conversely, sustainability is a call from life arising from the unsustainable state of the planet, from the environmental crisis driven by the techno-economic rationality that governs the world that has intervened in Nature and caused the natural courses of life to deviate. The over-exploitation of natural resources and environmental degradation is the outcome of economic rationality that has alienated culture from the sources of life, externalizing Nature from

Published originally in Leff, E. (1995), ¿De Quién es la Naturaleza? Sobre la Reapropiación Social de los Recursos Naturales, *Gaceta Ecológica*, No. 37, INE-SEMARNAP, México, pp. 58–64. An English version was published in *Capitalism, Nature, Socialism*, 10(3), No. 38, September 1999, pp. 89–104.

[1] The *Anthropocene* is the proposed name of a new era dating from the commencement of significant human impact on the Earth's geology and ecosystems. It can be argued that the Anthropocene era was initiated when the symbolic order that constituted human beings affected, even at a small scale, the metabolism of local ecosystems and life territories and started the co-evolution of Nature and culture on the planet.

E. Leff, *Political Ecology*, https://doi.org/10.1007/978-3-030-63325-7_2

the process of production, destroying the ecological conditions for economic sustainability (Naredo 2006). The imperative to control and reverse environmental degradation has led to the search for a theory and a method to internalize the values and potentials of Nature to regenerate a process of "sustainable development." However, sustainability is not forged through the adjustment and reform of the economic order. It is "constructed" in the conflictive field of political ecology, in the encounter of different ontological regimes, where alternative scientific principles and political strategies confront one another in their valuation of Nature, shaping its meaning and guiding social action in the perspectives of a sustainable future.

Environmental economics—the neoclassical economics of natural resources and pollution—assumes that the economic system can internalize ecological costs and the preferences of future generations by assigning property rights and establishing market prices on natural resources and environmental services. However, the reintegration of Nature into the economic sphere faces the problem of translating the productivity of Nature, as well as its conservation and restoration costs, into market prices as a standard measure of value. This reductionist approach overlooks the fact that valuing natural resources is not only subject to temporal and spatial ecological conditions of resiliency and reproduction, which are not synchronic nor commensurable with economic cycles; it also depends on social interests and the cultural meanings that define and assign values and norms to Nature. In other words, values and norms orient behavior and practices that determine concrete modes of appropriation of Nature through extra-economic processes—comprising both symbolic and power relations—that in turn affect the modes and the rhythms of extraction and transformation of Nature, and that cannot be translated appropriately into, nor reduced to, market prices.

Ecological economics has contributed to unmask the false pretense of environmental economics that equated the value of Nature to capital. However, sustainability cannot be assessed either by directly measuring matter–energy flows in the economic process. The purpose of quantifying the mass and energy that is consumed and degraded in the process of production—the "*throughput*"—(Hinterberger and Seifert 1995) is useful for designing "cleaner technologies," but cannot account for the social processes of signification from which natural resources derive their usefulness, or for their cultural value that grounds or breaks down ecological sustainability. The *throughput* analysis is useful to measure the input–output balance of matter and energy that is degraded and recycled in the economic process; but it cannot assess the economic, technological, ecological and cultural processes that determine the conditions of productivity, conservation, and/or degradation of an ecosystem, nor the equity and sustainability conditions of a productive system.

Ecological economics considers economic and ecological processes as two interdependent systems. The environment is conceptualized in terms of ecological norms that are internalized into the economic system and become distribution rights of ecological potentialities and environmental pollution costs (Passet 1979). The environment is therefore seen as a cost and limit, rather than as a productive potential. However, a critical analysis of the economic process from a thermodynamic perspective addresses its ineluctable tendency toward entropic degradation (Georgescu-Roegen 1971), a condition that is overlooked by the prevailing techno-economic rationality and its assumed capacity to dematerialize production, without assessing the problem of environmental degradation as a consequence of resource scarcity and its relation to the sustainable growth of the economy. Ecological economics provides a critique of the false principles of mainstream economic theory and advances conceptual instruments for the ecological regulation of the economy; but it does not offer a new theory to ground production on a new material basis, on the ecological conditions for a sustainable economic process. Ecological economics does not furnish criteria to adjudicate and settle the socio-environmental conflicts that lie at the roots of ecological distribution and that manifest as the conflict of interests among social groups and nations for the appropriation of Nature.

Internalizing ecological potentialities, conditions and degradation costs for sustainable development calls for a characterization of the ontological regimes and social processes that underlie different ways of valuing of Nature; this implies the need to assess ethical values and cultural meanings ascribed to Nature, instead of merely quantifying its material flows and chrematistic costs. There is no economic, ecological, or technological instrument that can establish the "real value" of Nature in the economy. Warning against attempts to reduce different environmental values to a standard unit of measurement, William Kapp (1983) noted that heterogeneous physical processes are involved in the comparative evaluation of economic, energy and environmental rationalities. Furthermore, beyond the purpose of unifying the heterogeneous material and symbolic processes that underlie the value of Nature, economics has been left without an objective value theory (Leff 1980, 2000; Altvater 1993; see Chaps. 5, 6, and 7). Environmental costs and the valuation of natural resources cannot be determined by a quantitative measure of labor time or energy flows and throughput, but rather depend on qualitative processes—the ecological metabolism of the biosphere, cultural imaginaries, community rights, and social interests—established outside the market that enact the social appropriation of Nature.

On the other hand, socio-environmental claims are revaluing Nature: the power strategies for the appropriation of Nature are generating a political force that is reflected in the economy by the increase in prices of resources and environmental degradation costs. The environmental movement makes ecological costs visible to the economic system by means of social resistance to the capitalization of Nature; social struggles seeking to improve living conditions and quality of life for all people give rise to new democratic values and cultural rights that reflect how different social groups, in different ecological and social contexts resignify and value Nature. However, the environmental movement does not only increase the ecological costs of economic growth, it also confronts the way in which capital appropriates Nature, and offers alternative strategies for the sustainable management of natural resources, including the bio-cultural patrimony of traditional peoples.[2]

The environmental movement is generating a more decentralized development process and mobilizing the deconstruction of economic

[2] The concept of bio-cultural patrimony was proposed by Eckart Boege (2008, 2018).

rationality by shifting the very bases of production away from its economic center (capital, labor, technology) to its ecological conditions, cultural values, and existential meanings. From this perspective, sustainable development goes beyond the task of making conservation and development compatible, and the unrealistic purpose of internalizing ecological conditions to achieve sustained economic growth. Radical environmentalism is constructing and grounding a new concept of *environment as the potential for alternative development*, that is, for building a new production paradigm based on Nature and culture as productive forces. In this perspective, Nature is no longer a raw material to feed the economic process but becomes a negentropic potential, an ecological means of production and a life-support system that sustains sustainable societies. The environment emerges as a complex system, an object of social appropriation crossed over by power relations involved in the construction of a sustainable world (Leff 1995, 2004).

Sustainability of the economic process cannot be attained by internalizing ecological norms and adopting abstract principles of interdependence and coevolution among cultural, ecological, and technological processes that lead to more rational use of natural resources. Lacking the appropriate instruments to account for the specificity of the different ontological regimes that organize the processes that underlie the valuation and mobilize the transformation of Nature, economic reason cannot evaluate nor internalize the conditions for sustainable development.

Ecological and communal production conditions appear as the bases for new productive rationality where natural, technological, social, and cultural processes converge to generate the eco-technological potential for a new paradigm of *negentropic production* that has been hidden by the prevailing economic order (Leff 1986, 1995, 2019). The principles of social equity, cultural diversity, and political difference open up broader perspectives to build sustainable societies. Environmentalism is innovating theories and legitimizing new ethical and political values that question the prevailing economic order, orienting social action toward the construction of alternative productive rationality founded on the potentialities of Nature and cultural creativity.

Ecological Distribution and Capitalization of Nature

The confluence of ecological economics and political ecology has forged the concept of *ecological distribution* in order to analyze the unequal ecological costs and social effects of economic processes, as well as the study of differentiated ecological assets and the environmental potential of a planet that is traversed by power strategies of opposing interests. As stated by Joan Martínez-Alier, ecological distribution refers to

> the social, spatial and temporal asymmetries or inequalities in the use by humans of environmental resources and services, i.e., in the depletion of natural resources (including the loss of biodiversity) and the burdens of pollution. (Martínez-Alier 1995a, p. 520)

Ecological distribution refers to the unequal distribution of ecological costs and its effects in the variety of ecological movements, including movements of resistance to neoliberal policies, compensation for ecological damage and environmental justice. Ecological distribution designates the differentiated rights of humanity to populate the planet, to consume energy, to discharge its polluting waste to the common environment, on a planet where one inhabitant from the North consumes 40 times as much energy and natural resources as the average inhabitant of countries in the South. The concept signals asymmetries and inequalities in ecological degradation costs and endowments and their effects on social-environmental movements. Nevertheless, the concept has remained inside the influence area of dominant economic rationality, reducing the solutions to these conflicts to an equal distribution of the ecological costs of growth and economic compensations for ecological damage. However, the question addressed by ecological distribution goes beyond the possible equalization of costs and benefits in the use of environmental resources within the current anti-ecological economic rationality.

The category of ecological distribution encompasses the environmental externalities of the economic process as well as social movements that emerge from *distributive conflicts*, that is, from inequalities in ecological costs and their effects on the bio-cultural patrimony, traditional identities, and life territories of different social groups. Collectively these groups organize around environmental justice claims, environmental defense and resistance to both the territorial displacement of peoples and capitalization of Nature. The dominant economic rationality generates these environmental conflicts, but as most environmental goods and services are not "traded" within the market, environmental claims are not negotiated nor solved through economic mechanisms, but disputed in the political arena.

Ecological distribution comprises the extra-economic—ecological and cultural—processes that shift ecological economics to the field of political ecology; it extends to ethical principles and existential conditions that turn it into a political ontology. In an analogy with the concept of economic distribution that opens economic rationality to political economy, socio-environmental conflicts are not settled by any ecological rationality but fall under the dominance of contested interests, values, and meanings that mobilize social actors through power strategies. It is under ecological distribution that the ecological conditions for survival and sustainable production and the social conflicts emerging from the dominant processes of destruction of Nature and pollution collide. Ecological distribution refers to social processes by which Nature is valued in ways that cannot be reduced to economic values and ecological measures, mobilizing social actors for material and symbolic interests (for survival, identity, autonomy, sustainability, existence rights, and quality of life), that go beyond economic struggles for employment, income distribution, and social welfare.

While mainstream economics seeks to internalize environmental externalities by assigning property rights and market prices to natural resources and environmental services—a strategy that responds to the "tragedy of the commons" (Hardin 1968)—ecological economics recognizes that economic rationality and distribution of wealth and income are the root causes of environmental degradation and the devaluation of Nature. The category of ecological distribution unveils the economist approach to conceiving the environment as a condition for sustainability, that by addressing the ecological undervaluation of Nature and poverty persistence uncovers the privileged mechanisms that sustain capital accumulation and the global neo-liberal economic order. This perspective from ecological distribution reveals how the economic strategies dominate both ecological and cultural matters.[3] Nevertheless, as a category, ecological distribution cannot escape the circle of economic rationality, if and when the environment is reduced to a cost of economic processes instead of an alternative potential for sustainable development.

From a radical environmental perspective, ecological distribution appears to be a conciliatory term between ecological economics and political ecology (between economic calculation and environmental rationality). It offers an argument that explains and claims the historical unaccounted ecological debt of rich owed by dominant countries to poor countries, by the political conquest and accumulation of wealth over the displacement of their territories, and the uneven development of the global economy. These days, if any country or social group appropriates biomass in excess of what its geographical space can ecologically produce and sustain, or if it produces polluting wastes beyond its capacities to dilute them or confine them within its territory, it would be considered an ecological debt to those who bear the costs of over-exploitation or over-pollution of Nature. However, under the prevailing economic order these unequal ecological exchanges and the overall damage to the environment cannot be solved or compensated for by a more equitable distribution of ecological costs, through green reforms, or environmental justice movements.

Conflicts of ecological distribution emerge fundamentally as a consequence of the *destructive expropriation of Nature* generated by the hegemonic economic theory's negation of ecological processes and collective property rights (Orstrom 1990) as conditions for sustainability. Thus, movements of resistance to capitalization of Nature and culture emerge as a social response to inequity and injustice under the dominant economic, institutional and juridical order. If ecologism in affluent societies is oriented by post-materialist values (Ingelhart 1991), "environmentalism of the poor" (Guha and Martínez-Alier 1997) is characterized by struggles for survival, for alternative social and productive projects based on principles of diversity, identity, and autonomy, beyond claims that can

[3] See Chap. 6.

be solved through transactions, negotiations and compensations established under the rules of market valuation of the dominant economic and political order.

The ecological debt of the rich nations to poor countries and dispossessed peoples throughout 500 years of ecological imperialism (Cosby 1986) has established a *gap* that will not disappear by ecologizing the economy, nor through the negotiation of better and more just terms of commercial exchanges and economic compensations under the "ecological mechanisms" of a new "green economy." Today, indigenous peoples and peasants' organizations are claiming their rights to reappropriate and self-manage their historical patrimony of natural and cultural resources; to preserve and transform them according to their cultural values and social imaginaries. These principles are opening new paths to the building of sustainable societies, ones that contrast with those established by the new economic-ecologic order. The struggles for the appropriation of Nature, self-management of production, cultural diversity, ethnic identities, and direct democracy are defining socio-environmental conflicts in the emergent field of political ecology. Political ecology's reach goes beyond the restricted vision derived from environmental impacts or ecological costs, debt, and distribution that dominates the discourse on economic globalization and the geopolitics of *sustainable development* (Leff 2002a).

Ecological distribution, as a heuristic concept, is not consistent enough to accommodate the environmental conflicts generated by the economy that affect the living conditions of the people and their complex emergent cultural and political demands; it can only grasp them, once they lie dead, extended, and exhausted, in the benchmark of the market.[4] Viewing socio-environmental conflict as issues of ecological distribution, the "environmental" character of citizens' movements for their collective identities is veiled and distorted. Disputes between the private and the collective appropriation of the environment are not solved through technical criteria of environmental impacts or cost–benefit analysis, nor with

[4] As the popular African saying goes, "You can only measure the toad once it is dead." I heard this saying during the Seminar Conservation with Equity. Strategies for Sustainable Development, Ottawa, Canada, May, 1986.

political negotiations and economic compensations. New socio-environmental movements are emerging by means of cultural resistance in defense of livelihoods, and for the *empropriation* of their bio-cultural patrimony of natural resources grounded on existential ontologies that are incommensurable with economic values.

Thus, the idea of ecological distribution that emerges from an analogical transposition from economics to ecological externalities does not attain the status of a theoretical concept capable of grasping and accounting for the incommensurable values involved. Ecological distribution in the context of ecological economics does not assume the position that economic distribution has in Marxist or Sraffian economics because it does not strictly follow the principle of incommensurability. Ecological distribution recognizes extra-economic factors that value the environment, but it does not apprehend the specificity of these processes: the conditions of ecological stability and productivity; the cultural meanings assigned to Nature; the power strategies in the valuation of externalities; the social processes for the appropriation of natural resources and production processes. All these processes convey values that transcend the re-functionalizing of economic rationality in order to internalize ecological externalities. The environmental question calls for an alternative economic paradigm where the environment is no longer an externality of the prevailing economic order, but rather a potential for new productive rationality.

Notwithstanding its operative value to *ecologize* political economy, the category of ecological distribution does not deconstruct the basic structure of economic rationality that causes ecological maldistribution and degradation. Similarly, building a bridge from ecological economics to political ecology does not bridge the gap, nor erase the ecological footprint, nor compensate the historical, ecological debt owed to underdeveloped countries. Indeed, such efforts do not dissolve the inequality gap opened by the disjunction between the Real and the Symbolic, and its metaphysical consequences, through the domination of techno-economic rationality over the degraded conditions of life in the planet. The analogic use of the concept distribution and its application in the field of externalities does not fulfil the purpose of internalizing environmental conflict into economics, nor does it establish a new paradigm for sustainable production. Instead, a productive rationality founded on the negentropic potentials of the biosphere, on the resiliency of its ecosystems and in the cultural creativity of the people, can eliminate the root causes of inequity and unsustainability.

Political Ecology: Ontological Difference, Environmental Incommensurability and Social Conflict

Environmental processes and values are ontologically incommensurable and irreducible to market measurements. Environmental *knowledge/savoirs* (Leff 1998b/2002) emerges in the exteriority of the *logocentrism of sciences* (Derrida 1982); it internalizes diversity and incommensurability as epistemological and political principles that challenge the homogenizing dominant order imposed by scientific reason and economic rationality. Incommensurability in the field of the environment refers not only to difficulties in translating energy and ecological variables into market prices, but also to the impossibility of establishing a standard measure for extra-economic costs and benefits. Confronting economic rationality with environmental rationality radicalizes the principle of incommensurability—as an opposition of irreducible ontological regimes and paradigms (Kuhn 1962)—in which environmental processes and values cannot be translated to market prices nor recodified as capital (O'Connor 1993b; Leff 1994, 2001, 2004).

Incommensurability between economy and ecology implies the impossibility of establishing economic measures for Nature and for cultural values that are independent from property rights, income distribution, and risk considerations to uncertain future contingencies. The profound meaning of incommensurability for sustainable development emerges from understanding the environment as a complex system integrated by ecological, technological, and cultural processes, where the material and the ideal hybridize, and where diverse rationalities convey different values and meanings to Nature that underlie ecological sustainability.[5] By contesting the asymmetries and inequalities, and the entropic degradation enacted by "greening the economy," environmental rationality mobilizes the construction of an alternative paradigm of production. Thus, new concepts must be developed in order to apprehend the processes that constitute a sustainable economy, based on *negentropic productivity*, ecological resiliency, social equity, cultural diversity, and environmental justice. Environmental rationality blends new ethical and material principles in the valuation of Nature, new strategies for the reappropriation of productive processes and new meanings that mobilize the reorganization of society toward a sustainable future.

The principle of incommensurability of the different processes that integrate a socio-environmental system acquires a more ample and concrete meaning in the perspective of a new production paradigm. Thus, the environment becomes a productive system grounded in the functional structures of ecosystems, in their conditions of resiliency, productivity, stability, and sustainability that cannot be reduced and assessed through market values. The ecological, cultural, social, and institutional processes that appear as externalities of the economic process but establish its conditions for sustainability, cannot be translated into present economic costs and benefits as it is not possible to assign discount rates that can actualize complex, uncertain and long-term ecological processes, nor the future preferences of consumers.

A paradigm of sustainable eco-technological productivity results from the integration of ecological, technological, and cultural processes that determine the modes of production and the rhythms and scales of Nature's appropriation and transformation. This productive rationality cannot be constructed from the top down, by means of theory and planning, as a process that could impose the laws of a new global ecological order on nations and communities. The new production paradigm is grounded geographically and socially, but functions by integrating ecological, technological, and cultural principles, as well as the potential that emerges from productive practices of local communities and

[5] Laclau and Mouffe assert: "the logic of equivalence is a logic of simplification, while the logic of difference is a logic of its expansion and increasing complexity" (Laclau and Mouffe 1985, p. 130).

self-management of their livelihoods. The construction of this emergent social order is guided by diverse cultural values and different social interests; its processes are interwoven in power relations for the reappropriation of Nature; for the construction of new modes of material production and alternative ways of existence: new modes of understanding life conditions and new principles to inhabit the planet sustainably.

It is at the community level where the principles of environmental rationality acquire their full meaning and potential to build this new paradigm of sustainable production. The social construction of environmental rationality involves the emancipation of the *peoples of the Earth*, the historical transcendence of their modes of being-in-the-world, the reinvention of their cultural identities, and the role of traditional practices in the construction of new life territories, in a horizon opened to the future by upholding the sustainability of life on the planet. This process takes the principle of incommensurability to its ultimate consequences—by displacing the ontology of the universal and the general, of unity and totality, to make the ontological shift to the principle of difference and the ontology of life, of the multiple and the diverse—and situates the irreducibility and specificity of material processes to enable different forms of cultural signification to define the environmental potentials and conditions for sustainability. No quantitative and homogeneous measure can unify the differentiated processes that determine the sustainable production of use values, or that can measure its effects in the quality of life defined for different social norms and cultural values.

Negentropic production cannot be reduced to a unitary measure of matter and energy or to a quantitative value calculation (a quantum of socially necessary labor time). A sustainable economic process under environmental rationality results from the integration of differentiated levels of ecological, technological and cultural productivity; from the balance of negentropic biomass production through photosynthesis and the dissipation of entropy generated by ecological and technological transformation of matter and energy in the overall ecological-economic productive process. From this perspective, sustainable development is rooted in the conditions of ecological and cultural diversity. Singular, irreducible material and symbolic processes depend on the functional structures of ecosystems, the life-support-systems that sustain the production of biotic resources and environmental services. They also rely on the energy efficiency of technological transformation processes, and on the symbolic structures, ideologies, and social imaginaries that underlie the cultural meaning and value of natural resources, as well as on the power strategies and political conditions that drive the social appropriation of Nature, which in turn determines the destiny of life in the planet.

If the marginal theory is incapable of internalizing these complex environmental externalities through market mechanisms, then the social and environmental movements can contribute, to a certain extent, to the transfer of ecological costs to economic calculation (Leff 1995, Chapter 8). However, social resistance to the capitalist appropriation of Nature expressed through such movements, and the compensatory actions of environmental justice movements, cannot collect and convey the real value of environmental externalities to the economic system. Poor people are dispossessed and forced to sell their labor, their natural resources, and environmental assets cheaply. Sustainability with equity and diversity cannot be achieved by equalizing income and reaching a better ecological distribution under the current one-dimensional, homogeneous unsustainable economic paradigm; it can only emerge from the emancipation and *re-existence* of the people, under an ontology of life and an alternative productive rationality that is built from the ecological and cultural potentials of their life territories, within an ontology of diversity and a politics of difference (Leff 2015).

The power strategies that mobilize socio-environmental movements for the reappropriation of Nature stem from cultural and symbolic values as well as from material processes and social interests, precluding any possibility of solving environmental degradation—which is at the root of distributional conflicts—through market transactions, economic calculations, and political negotiations. Environmental justice establishes values that are beyond economic and ecological rationalities, giving extra-economic meaning to social mobilizations for the defense of cultural rights and the reappropriation of Nature's ecological potentials. The political force and legitimation of these environmental values are drawn from the constitution of new identities and collective rights that mobilize social actors and drive political actions toward the construction of a new social order. However, for these political changes to occur, the environmental movement has to confront, dissolve, and overcome the strategies of simulation, domination, and control exerted by the hegemonic economic and political powers that govern the present unsustainable global world.

In this context, *political empowerment* as a choice to redistribute the overwhelming oppressive power over the people, results in a willful proposal. Power is not a commodity or an object that can be given and distributed at will, but rather a relation of forces that is established from the confrontation of differentiated views and interests in the construction of different *worlds*. According to Derrida, the concept of *differance*—the original point of dissociation of the Symbolic from the Real, of *logos* from *physis*—appears in this conflictive and complex environmental field as "the 'active' moving discord of different forces and of differences of forces, that Nietzsche sets up against the entire system of metaphysical grammar, wherever this system governs culture, philosophy and science" (Derrida 1982, p. 18). This principle of ontological difference extends today to the field of political ecology, to every conflict for the reappropriation of Nature, life, and culture, where individuals and communities confront the globalized world. Environmental rationality entrusts this politics of difference as an antagonistic field of alternative development modes, driven by the "lack in being" (Lacan) that emerges from the unfulfilled human symbolic condition, and disperses to the diverse modes of being-in-the-world and ethnic styles of the existence of the human race.

Power underlies conflicts of "ecological distribution" as they revolve around alternative cultural meanings, social imaginaries, and productive rationalities. The struggle over the appropriation of Nature is the root of socio-environmental conflicts, the motor of historical transcendence driven by the antagonistic dynamism of environmental movements that cannot resolve their claims through economic compensations for ecological damage nor by their participation in decision-making processes, where options and possibilities are limited by economic criteria that dominate the sphere of globalization. Therefore, the geopolitics of sustainable development are exacerbating environmental conflicts, as they deviate radical claims of society for environmental solutions to ecological debts and past damages by implementing economic compensations, and upholding procedures and negotiations set by current juridical rules and practices subject to the dominant techno-economic rationality and political powers.

Beyond the issue of ontological incommensurability and diverse cultural meanings confronted in the social appropriation of Nature, environmental conflicts, and ecological claims take the issue of differentiation of material forces and symbolic processes a step further toward a politics of difference. Incommensurability is the sign of the limits of the *Ontology of Unitary and Universal Being* that recovers from oblivion the *Ontology of Life*, that is, the pursuit of diversity, difference and otherness through power strategies inscribed in the immanence of life. Thus, new social movements are emerging in the political arena that link the defense of Nature with struggles for democracy, autonomy, and self-reliance. Environmental conflicts manifest in a strategic and politically heterogeneous field, where material processes, social interests, and cultural meanings constitute different *socio-environmental* rationalities. In the political context of these emancipatory processes, ecological goals can be subordinated (for historical, cultural, political, strategic, or tactical reasons) to demands for cultural autonomy or political democracy. This has been the case in some of the emergent peasant and indigenous people's movements in Latin America. However, the environmental question underlies all these emergent emancipation movements.

Reappropriating Nature: Cultural Diversity, Social Equity, and Environmental Justice

In the construction of a sustainable world guided by the perspectives opened by environmental rationality, the environment appears as a productive system founded on the conditions of ecosystem stability and productivity, and on the ethnic modes of living-in-the-world of the different cultures that have co-evolved with Nature, building their life-territories. The integration of ecological, technological, and cultural processes determines the modes of appropriation and transformation of Nature in a new paradigm of eco-technological-cultural productivity (Leff 1995). Diverse cultural values guide the construction of this new social rationality that is crossed by conflicting social interests and interwoven in power relationships for reappropriating Nature for the self-management of productive processes.

If the environmental crisis has brought to the political arena the imperative to internalize ecological conditions into the economic process, this cannot be a homogeneous measure that is applied throughout different ecological and cultural contexts in an attempt to build planetary sustainability. Ecological sustainability does not only entail the preservation of Nature, but also the reversal of its degradation and the enhancement of its potentialities, which are all indissolubly linked to social practices and cultural modes of appropriation and transformation of Nature. Environmental degradation generates perverse cycles of poverty that in turn accentuate ecological decay. Thus, social equity cannot be dissociated from environmental sustainability. In other words, beyond the question of a diachronic solidarity, beyond the concern about the rights of future generations to afford enough Nature for their sustainable livelihoods, intragenerational equity or the access of present social groups to the environmental resources of the planet becomes a priority. The conservation and sustainable use of Nature imply the democratic management of its resources. If the incommensurability principle has led to questioning the reduction of ecological distribution to environmental costs through homogeneous measures, the ontology of life questions the possibility of thinking about equity in terms of sameness, beyond that of equal rights to diversity and difference.

The problem posed by the social reappropriation of Nature goes beyond the possibilities of solving the inequity of socio-environmental conflicts through a more just distribution of environmental degradation costs, a better evaluation of the stock of resources in national accounting and equal distribution of income. Similarly, the essential question for constructing sustainability goes beyond the assessment of costs and benefits derived from the distribution of ecological potentialities and damages within the actual forms of exploitation of Nature, assigning market prices or property rights to Nature. The "tragedy of the commons" (Hardin 1968) is not the tragic condition of the diverse modes of "living in common" of the Peoples of the Earth, but that of the forced destruction and displacement of communal modes of inhabiting the planet by capital's epistemological, economic, and juridical means.

The grass-roots level is where the principles of environmentalism take on their full meaning in terms of sustainable productivity, cultural diversity, and social participation, to build this new productive rationality. This process sets forth the specific Nature of biophysical processes in the metabolism of the biosphere through cultural meanings assigned to Nature that define the environmental potential to build sustainable societies. No quantitative and standardized criterion can account for the complex ecological processes that sustain the production of use values or that can measure its effects on the quality of life defined by various cultural codes.

Sustainable production cannot be designed merely through flows of mass and energy—the balance between negentropic biomass formation through photosynthesis and entropy dissipation by technological transformation of matter and energy in productive processes—and by a quantitative calculation of labor-value. From the perspectives of environmental rationality, the roots of sustainable development lie in the limits and potentialities set by the laws of thermodynamics, the conditions of ecological productivity and resilience, and the effects of cultural productive practices on the metabolism of the biosphere. These processes depend on the state of conservation of ecosystems that sustain the production of biotic resources and environmental services; on the energy efficiency of the technological processes; on the social imaginaries and the symbolic processes that underlie the cultural valuation of natural resources; and on the political forces that determine differentiated strategies for the social appropriation of Nature.

Social justice and democracy are necessary conditions for constructing sustainable societies. This statement is being confirmed by the rising demands of indigenous peoples, in their struggles for dignity, autonomy, democracy, participation, and justice. Equity in the diverse processes of cultural reappropriation of Nature is not solved through equalization of the costs and benefits involved in the actual forms of exploitation of Nature and the calculation of economic benefits. Environmental democracy and justice depart from the "politics of equity and equivalence,"[6] to flourish in a new field of *politics of diversity and difference*. This shift in rationality challenges the possibility of achieving environmental justice through the commensurability of costs, the homogenization of needs, and the normalization of demands and rights over Nature that are defined through differentiated cultural meanings and opposing social interests, expressed in struggles and alternative strategies for the reappropriation of Nature.

The conditions for sustainable community existence depend on the legitimation of their property rights to the heritage of natural resources; of their rights to preserve their cultural identity; of autonomy to redefine their lifestyles and reconstruct their production processes. The struggles for legitimation of environmental, communal, and indigenous people's rights, are transforming the norms established by the dominant juridical and legal system in response to emergent social demands for common rights over *their commons*, to construct new life-territories. The claims of indigenous groups to *their commons*, are also their struggles for dignity, autonomy, democracy, participation, and self-management of their bio-cultural patrimony, and go beyond justice defined merely in terms of better distribution of benefits derived from the prevailing mode of production and the dominant juridical system (Leff 2014b).

The reappropriation of Nature in the context of collective rights and cultural diversity is rooted in a principle of *equity in difference and diversity* (Grünberg 1995); it involves the cultural autonomy of communities, the self-determination of needs and self-management of ecological potentials in each region under alternative sustainable lifestyles. These processes define production conditions and the construction of the life-territories of diverse groups of the population for the sustainable management of their environment. Property rights are established through social movements to appropriate Nature and to construct alternative practices for the sustainable use of resources that depend on distinct ecological, cultural, and social conditions. This does not mean that socio-environmental

[6] "Thus, environmental rationality questions the traditional concept of justice," where "Justice is the demand for equity, for 'fair play,' and a share in the benefits of life that are commensurable with one's contribution. In Thomas Jefferson's words, it is 'equal and exact…' based on respect for the principle of equivalence" (Bookchin 1990a, p. 96).

movements act above or against the law, but human rights are legitimized through social changes that transform the norms established by the juridical order as the power device of a society alienated from the conditions of life. The new environmental and territorial rights of indigenous peoples are being legitimated under the prevalent legal framework through social demands to inscribe rights to the commons in the ontology of life. The enforcement of these socio-environmental rights demands the reconstitution of a juridical rationality that recognizes, translates, and codifies into procedural law the community rights of peoples over the commons (Leff 2014b).

A standard pattern of wellbeing cannot define equity. The existential modes of *living well* (*vivir bien*) claimed by the peoples of the Earth do not depend only on a more equal distribution of the stock of natural resources nor of wealth generated by an homogeneous pattern of production and consumption under the dominant economic rationality. Equity can only be attained by subverting and abolishing the dominance of the global economic system over the autonomy of peoples and by creating conditions for appropriating the ecological potential of each region through the cultural values and social interests of each community.

Conversely, ecological distribution and environmental justice are still viewed with modern epistemological lenses where the Real is perceived as differentiated ontological orders but capable of being valued through unified measures. The rational view of the world veils the non-objective character of the meanings attributed by traditional cultures to Nature. From the perspectives of a non-essentialist and post-constructivist political ecology (Escobar 2010), the social appropriation of Nature claims a *politics of difference* that is beyond ontological incommensurability and logical contradiction; it is a confrontation and struggle among alternative paradigms and schemes of cognition and practices, of modes of production of existence, of ways to inhabit the planet and of being-in-the-world, where radical environmentalism addresses the strategies of capital and economic rationality through social conflicts and antagonisms that subvert the dominant social order and confront the hegemony of power with new strategies for the reappropriation of Nature, for living-well under the conditions and the immanence of life.

The strategic question in the construction of this environmental rationality is the possibility of establishing alliances from the constitutive diversity and difference of environmental interests, arising from spaces of marginality and externality. From the otherness of potential alternatives, diverse rationalities and discourses articulate in a dialogue of knowledge/*savoirs* weaving solidarities among different social actors to establish an antagonistic hegemony (Laclau and Mouffe 1985) capable of opposing dominant economic rationality. Thus, beyond essentialism and subjectivism, escaping the determinations of structuralism and holism, the environmental movement is the political expression of new identities in the complex becoming of life. New social subjects and collective interest are being mobilized from their resistance to the capitalization of Nature and culture to an unending construction of meanings that guide the reconstruction of lifestyles, of social identities and modes of being-in-the-world. Thus, the dominance of techno-economic rationality—as the homogeneous ontology in which reality is measured, codified and unified, to be transferred as values onto subjective meanings—is being contested by the ontology of life under which environmental rationality is being constructed.

Environmental Common Rights: Autonomy, Self-management and Democracy

Environmental rationality is defined by the principles of equity and sustainability, autonomy and democracy, diversity, difference, and otherness, beyond the ontological reduction of the world to market prices and the values of conservationism, biocentrism, and global governance of the planet. Indigenous and peasant-farmer community struggles are associating new cultural rights with claims for access to and appropriation of Nature through power strategies that are generating alternative production practices. These emergent cultural and environmental rights are displacing traditional juridical principles founded in individual equal rights, incorporating demands for the self-management of the production conditions and lifestyles of the people. A process of reappropriation of Nature is "taking place" as a basis for the survival of the Peoples of the Earth and a condition for generating an endogenous self-reliant construction of sustainable life-territories (Moguel et al. 1992; Leff 1996).

Environmental justice opens a fundamental question: Who owns Nature? And, consequently, Who grants the rights to inhabit the planet, to exploit the Earth and its natural resources, to pollute the environment? Is it a decision that descends from the heights of power over people as the fatality of a natural law: as the destination of the history of being, as the ineluctable objective law of free market of economic rationality? or Is it possible to reopen human history as a consequence of this environmental crisis, rooted in the imaginaries and practices of the peoples of the Earth, mobilized by social action based in an environmental rationality and other ways of being-in-the-world, through innovative power strategies for the reappropriation of the ecological potentials of Nature, to build new life-territories and sustainable worlds?

The emergent socio-environmental conflicts over the reappropriation of Nature bring class struggle onto the social scene once again, not only over means of production—as the industrialized productive forces of Nature—but over the *endowment* of the ecological conditions for resiliency and sustainable production of new life-territories. In contrast to the appropriation of means of production guided by a one-dimensional view of natural forces unleashed and constrained by technology, environmentalism poses the *endowment* of Nature as *enownment* of the ecological conditions for sustainable livelihoods; as embodiment of a bio-cultural patrimony; as the reinscription of humankind in the immanence of life. It encompasses the construction of a complex process of *negentropic production*, based on the ecological-technological-cultural potential of different life-territories that support alternative strategies for the sustainable management of the forces of Nature and ecological potentials subject to the conditions of life on the planet (Leff 1986, 1994, 2019).

In response to the dispossession and marginalization of majority groups of the population and the ineffectiveness of the State and market logic to provide basic goods and services, emergent social groups are claiming their rights to participate in decision-making on public policies that affect their living conditions and the sustainable management of their productive resources. Environmental movements and social struggles are gaining strength with the legitimization of human and cultural rights for self-determination of their modes of existence in an increasingly democratic world.

In the field of environment, new human rights are incorporating the protection of the commons (the environmental goods and services of humanity), as well as the right of all people to develop their full potential. New cultural rights to ethnic territories, indigenous languages, and cultural practices are incorporating community political and economic demands that include collective control of their resources, self-management of their productive processes, and self-determination of their lifestyles. These new social movements have already led to the constitutional recognition of the *rights of Nature* in countries such as Bolivia and Ecuador, and are redefining property rights to ancestral territories, to forms of land ownership, and of sustainable appropriation and use of natural resources. Collective property rights are starting to be recognized as conditions for sustainability in the economic world (Orstrom 1990).

The conservation of biodiversity is a paradigmatic example of opposing interests in the appropriation of Nature. Strategies employed by transnational biotechnology companies to manage the genetic material of biotic resources defy the rights of indigenous peoples over their historical patrimony of natural resources. This issue cannot be solved through economic compensation, not only because it is impossible to calculate the "real" economic value of biodiversity (the result of centuries and millennia of ethno-ecological co-evolution) in terms of capital and labor-time invested in the conservation and production of genetic material, nor by the current market value of its products or the prospection of their future economic value; but mainly because indigenous peoples reject absorption by the ontological regime of economic rationality. The dilemma posed by global biodiversity is the expropriation of Nature by capital through intellectual property rights versus the rights of indigenous peoples over their heritage of natural resources resulting from biological evolution, the cultural selection and domestication of species and the symbolic meaning of their life-territories (Hobbelink 1992; Martínez-Alier 1994).

In that regard, the peoples of the Amazon forest have developed productive strategies for the self-management of "extractive reserves" (Porto Gonçalves 2001a); in Mexico, the decades-long struggle of the peasant farmers to establish Los Chimalapas Biodiversity Reserve is prompting the communities to fight for the legalization of the ownership of their land and the exercise of effective control over the conservation and sustainable use of their biologically diverse resources (García 2014). The entry of indigenous and peasant farmer communities into the globalization process is leading to struggles of resistance and *re-existence*, reorienting development toward the construction of new life-territories, rooted in the reinvention of their identities and legitimized by new socio-environmental rights founded in the politics of cultural diversity, the *ontology of difference*, and an *ethics of otherness* (Escobar 1997; Leff 2014a). Thus, peoples and communities are giving new significance to the discourse on democracy and sustainability to reshape their ethnic-eco-development styles. This democratization process is giving rise to unprecedented movements for the reappropriation and productive management of biodiversity, the habitat in which native communities have developed and where their future projects for life are being redefined.

The possibility of eradicating poverty and improving the quality of life of indigenous and peasant farmer populations depends on the conditions for access, management, and control of their productive resources. Thus, the principle of participatory sustainable management of natural resources is permeating the struggles for autonomy, yielding new forms of a direct and substantive democracy. Democracy in the productive process points toward the reappropriation of natural resources and toward the collective management of the communities' environmental goods and services. In that regard, some of the new social movements in the rural areas of Latin America are transcending traditional claims in the economic sphere: for more employment, better salaries and better distribution of wealth; in the political field: for greater plurality and participation in decision-making and in the institutionalized party system; and in the cultural sphere: for defense of cultural values and ethnic diversity (Giménez 1994).

The emerging rural movements are building solidarity links not only in their rejection of neoliberal policies and the neo-extractivist enterprises and transgenic agri-businesses that carry out the over-exploitation of Nature, economic disparities, political marginalization, cultural segregation, and environmental degradation. Their struggles are not only for greater equity and participation within the established order but rather for building a new social order: for a reform of the State that will include indigenous peoples as equals, which implies recognizing their political differences, their ethnic identities and cultural rights (González Casanova and Roitman 1996; Leff 1996). These social struggles for democracy mobilize the construction of a new political order and a new productive paradigm. Although the environmentalist seed is not always evident in the discursive and political strategies of the emerging popular movements—focused on struggles for cultural rights and political autonomy of indigenous and peasant farmer communities; for democracy as a requirement for reappropriating their bio-cultural patrimony and their ecological means of production—many of them are reinventing their traditional practices of natural resources use, expressing demands for the self-determination of their life-worlds and livelihoods and the self-management of their productive processes (Instituto Indigenista Interamericano 1990; Díaz Polanco 1991; Moguel et al. 1992; Torres 1997; Gómez 1997).

From this perspective, building sustainable territories rooted in ecology and culture marks the difference in the capitalization of Nature and ecologizing the economic order. The construction of environmental rationality is achieved through the socialization of Nature and community management of resources, founded in principles of ecological and cultural diversity. In that regard, democracy and equity redefine their meaning in terms of a politics of diversity, difference, and otherness that informs the social reappropriation of Nature.

However, if the market is incapable of assigning real economic value to ecological and communal conditions for sustainability because it is impossible to discount sustainable futures and reduce diverse and heterogeneous conditions of sustainability to unitary measures, ecology cannot provide a paradigm to integrate environmental ontological diversity into a holistic view. Ecology, posited as the science of biophysical interrelations, is, in fact, a non-spatial and abstract paradigm that lacks specific geographical reference to territorialize space and place. The French geographer Jean Tricart (1978, 1982) saw the need to integrate ecology with geography in an interdisciplinary eco-geo-graphical paradigm. The emergence of geo-reference comes from the need to root sustainability in local conditions. Space and place have thus been reinvented as material supports, to anchor and specify the conditions of sustainability where the rational construction of sciences goes astray. This is not only the need to give empirical support and reference to global rationalities, a balance between the global and the local, or an equilibrium point between ecological conservation and economic growth. Sustainability is not rooted in objective material processes; its meaning and potentialities are supported by ontological principles, cultural values, and existential conditions. The destruction of Nature and culture is reactivating the interpretation of the history of life on the planet, its creative evolution. Our time is one of a renaissance of Nature and culture from the violence of the history of metaphysics, the ontology of being and of a one-dimensional economic and technological rationality. Identities are reborn from the liberation of cultural rationalities flowing anew in the realm of history from the fire of life (Leff 2018).

Bibliography

Altvater, E. (1993), *The future of the market*, New York: Verso.

Boege, E. (2008), *El patrimonio biocultural de los pueblos indígenas de México. Hacia la conservación in situ de la biodiversidad y agrobiodiversidad en los territorios indígenas*, México: INAH/Comisión Nacional para el Desarrollo de los Pueblos Indígenas.

———. (2018), "Hacia una antropología ambiental para la apropiación social del patrimonio biocultural de los pueblos indígenas", in Toledo, V.M. & Alarcón-Cháires, P. (Eds.), *Tópicos bioculturales: reflexiones sobre el concepto de bioculturalidad y la defensa del patrimonio biocultural de México*, México: UNAM.

Bookchin, M. (1990a), *Remaking society: Pathways to a green future*, Boston: South End Press.

Cosby, A.W. (1986), *Ecological imperialism. The biological expansion of Europe, 900–1900*, New York: Cambridge University Press.

Derrida, J. (1982), *Margins of philosophy*, Chicago: Chicago University Press.

Díaz Polanco, H. (1991), *Autonomía regional. La autodeterminación de los pueblos indios*, México, Siglo XXI/UNAM.

Escobar, A. (1997), "Cultural politics and biological diversity: State, capital and social movements in the Pacific coast of Colombia", in O. Starn & R. Fox

(Eds), *Culture and social protest: Between resistance and revolution*, New Brunswick, Rutgers University Press.

———. (2010), "Post-constructivist political ecologies", in Redclift, M. & Woodgate G. (Eds.), *International handbook of environmental sociology*, 2nd edition, Cheltenham/Northampton: Edward Elgar Publishing Limited, pp. 91–105.

García, M.Á. (2014), "La reserva ecológica campesina de Los Chimalapas", en Betancourt, A. et. al., *Del monólogo a la polifonía: proyectos supranacionales y saberes indígenas en la gestión de la áreas naturales protegidas (1990–2010)*, México: UNAM.

Georgescu-Roegen, N. (1971), *The entropy law and the economic process*, Cambridge, Mass.: Harvard University Press.

Giménez, G. (1994), "Los movimientos sociales. Problemas teórico-metodológicos", *Revista mexicana de sociología*, Vol. LVI, Num. 2, pp. 3–14.

Gómez, M. (Coord.) (1997), *Derecho indígena*, México: INI/AMNU.

González Casanova, P. & M. Roitman (Coords.) (1996), *Democracia y Estado multiétnico en América Latina*, México: La Jornada Ediciones/CIICH-UNAM.

Grünberg, G. (Coord.) (1995), *Articulación de la diversidad: pluralidad étnica, autonomías y democratización en América Latina*, Quito: Ediciones Abya-Yala.

Guha, R. & Martínez-Alier, J. (1997), *Varieties of environmentalism. Essays North and South*, London: Earthscan.

Hardin, G. (1968), "The tragedy of the commons", *Science* No. 162: 1243–1248.

Hinterberger, F. & Seifert E. (1995), "Reducing material throughput: A contribution to the measurement of dematerialization and sustainable human development", in J. van der Straaten y A. Tylecote (eds.), *Environment, technology and economic growth: The challenge to sustainable development*, Aldershot, Edward Elgar.

Hobbelink, H. (1992), "La diversidad biológica y la biotecnología agrícola", *Ecología política*, No. 4, Barcelona, Icaria.

Ingelhart, R. (1991), *El cambio cultural en las sociedades industriales avanzadas*, Madrid: Siglo XXI Editores.

Instituto Indigenista Interamericano (1990), "Política indigenista 1991–1995", *América indígena*, Vol. 50, Num. 1.

Kapp, W. (1983), "Social costs in economic development", in J.E. Ullmann (Ed), *Social costs, economic development and environmental disruption*, Lanham, Mass., University Press of America.

Kuhn, T. (1962), *The structure of scientific revolutions*, Chicago: The University of Chicago Press.

Laclau, E. & Mouffe, C. (1985), *Hegemony and socialist strategy*, London: Verso.

Leff, E. (1980), "La teoría del valor en Marx frente a la revolución científico-tecnológica", en Leff, E. (Editor), *Teoría del valor*, México: UNAM.

———. (1994), *Ecología y capital: racionalidad ambiental, democracia participativa y desarrollo sustentable*, México: Siglo XXI Editores.

———. (1995), *Green production: towards an environmental rationality*, New York: Guilford.

———. (1996), "Los nuevos actores sociales del ambientalismo en el medio rural", en Carton de Grammont H. & H. Tejera, *La sociedad rural frente al nuevo milenio*, Vol. 4 "Los nuevos actores sociales y los procesos políticos en el campo", México: UNAM/INAH/UAM/Plaza y Valdez Editores.

———. (1998b), *Saber ambiental: Sustentabilidad, racionalidad, complejidad, poder*, México: Siglo XXI/UNAM/PNUMA (3rd edition, 2002).

———. (2000), "The scientific-technologic revolution, the forces of nature and Marx's theory of value", in *Capitalism, nature, socialism*, Vol. 11 (4):109–130.

———. (2001), *Epistemología ambiental*, Sao Paulo: Cortez Editora.

———. (2002a), "La geopolítica de la biodiversidad y del desarrollo sustentable: economización del mundo, racionalidad ambiental y reapropiación social de la naturaleza", in Ceceña, A.E. & Sader, E., *La guerra infinita: hegemonía y terror mundial*, Buenos Aires: CLACSO-ASDI, pp. 191–216.

———. (2004), *Racionalidad ambiental: la apropiación social de la naturaleza*. México: Siglo XXI Editores.

———. (2014a), *La apuesta por la vida: imaginarios sociales e imaginación sociológica en los territorios del Sur*, México: Siglo XXI Editores.

———. (2014b), "Los Derechos del Ser Colectivo y la Reapropiación Social de la Naturaleza", en Pereira de Cunha, Belinda; Agustín, Sergio; Araújo da Costa, Nálbia Roberta (Org.), *Saberes Ambientais, Sustentabilidade e Olhar Jurídico: visitando a obra de Enrique Leff*, Joao Pessoa e Caxias do Sul, Brasil.

———. (2015), "Encountering political ecology: epistemology and emancipation", in Bryant, Raymond & Soyeun Kim (Eds.), *The International Handbook of Political Ecology*, Cheltenham, UK: Edward Elgar, pp. 44–56.

———. (2018), *El fuego de la vida: Heidegger ante la cuestión ambiental*, Mexico: Siglo XXI Editores.

———. (2019), "Negentropic production", in Kothari, Ashish, Demaria, F., Acosta, A., Salleh, A. and Escobar, A. Eds., *Pluriverse: a postdevelopment lexicon*. London: Zed Books.

———. (Ed.) (1986), *Los problemas del conocimiento y la perspectiva ambiental del desarrollo*, México: Siglo XXI Editores.

Martínez-Alier, J. (1994), "The Merchandizing of Biodiversity", *Etnoecologica*, 2 (3): 69–86, Mexico.

———. (1995a), "Distributional issues in ecological economics", *Review of social economy*, Vol. LIII (4): 511–528.

Moguel, J., C. Botey & L. Hernández (1992), *Autonomía y nuevos sujetos sociales en el desarrollo rural*, México, Siglo XXI Editores.

Naredo, J.M. (2006), *Raíces económicas del deterioro ecológico y social*, Madrid: Siglo XXI Editores.

O'Connor, M. (1993b), "On the misadventures of capitalist nature", *Capitalism, nature, socialism*, 4 (3): 7–40.

Orstrom, E. (1990), *Governing the commons. The evolution of institutions for collective action*, Cambridge: Cambridge University Press.

Passet, R. (1979), *L'économique et le vivant*, Paris: Payot.

Porto-Gonçalves, C.W. (2001a), *Geo-grafías. movimientos sociales, nuevas territorialidades y sustentabilidad*, México: Siglo XXI Editores.

Torres, R. (1997), *Entre lo propio y lo ajeno. Derechos de los pueblos indígenas y propiedad intelectual*, Quito, COICA.

Tricart, J. (1978), "Vocations des terres, ressources ou contraintes et développement rural", *Hérodote*, 12: 65–75.

———. (1982), "Géographie/Écologie", *Hérodote*, 26.

3

Space, Place and Time: The Local Construction of an Environmental Rationality

1. Environmental Crisis: The Ontological Turn of History

Environmental degradation and ecological decay have become the most eloquent expressions of dominated and over-exploited Nature. Space and place are being resignified as cornerstones of a decentralized and democratic reconstruction of Nature. They emerge as geo-referential categories for embedding strategies of appropriation of Nature in ecological processes; they embody Nature in local identities, giving cultural meaning to sustainable development. Space and place, as geographical referents, are being signified by time as an existential condition to root ecological potentials in cultural imaginaries and practices. From this perspective, time is not only the chronological flow of ontological events, but it is also conceived as phenomenological time, that elusive and intangible "dimension" in the *production of existence* that forges being and becoming through meaningful events. Time rules the genealogy of

---

This chapter is based on an unpublished paper written by invitation of Arturo Escobar, presented at the Conference Neo-Liberalism in the Space, Place and Nation: Reconstructing Americas, Universidad de Amherst, 20–21 November, 1998. A Spanish version was published in Leff, E. (2000), Espacio, Lugar y Tiempo: la reapropiación social de la naturaleza y la construcción local de la racionalidad ambiental, in *Desenvolvimento e Meio Ambiente*, No. 1, Editora UFPR, Curitiba, Brazil, pp. 57–69.

E. Leff, *Political Ecology*, https://doi.org/10.1007/978-3-030-63325-7_3

rationality, the permanence, endurance, and becoming of past beings and events. Time occurs in the realm of utopia, in the possible "to be" of the "not yet" that is embodied in the shaping of new identities, at the crossroads of differentiated histories and in building possible futures. This vision of future emerges from the dialogue of knowledge/ *savoirs*, from the constraints of modernity—of techno-economic rationality that dominates life—and from the entropic decay of the planet and the limits of human existence. Time is the texture of the hermeneutical historical reconstruction of Nature–culture relations that aligns strategic thinking in the transition toward a sustainable world order, grounded on ecological diversity, cultural difference, and political autonomy. Time is the signifier that guides social reappropriation of Nature and local reconstruction of environmental rationality.

In an economized world, time is money. But in the symbolic order of culture and history, time is forged through the means of meaning. The density of destiny, that is the elasticity of symbolic time in human consciousness and actions, guides the transcendence of life in history through significations embodied in cultural identities and instituted by the social imaginaries (Castoriadis 1998).[1] In turn, through productive practices, these cultural identities and social imaginaries are rooted in life territories that mobilize the metabolism of the biosphere, and evolve through bio-cultural genealogies to the generations to come. The enigmatic survival and reactivation of traditional cultures and their hybridization—their submission, resistance or *re-existence*—with the discourse and politics of globalization are expressed in the reinvention of social imaginaries of the peoples of the Earth. The world crisis, triggered by the abstract scientific rationality where globalization is suspended, takes a new turn when envisioned by environmental rationality and mobilized by a dialogue of knowledge/ *savoirs*. This new turn aims at transcending history toward the sustainability of life.

2. Culture and Sustainability

Any given cultural organization is a complex system of values, ideologies, significations, productive practices, and modes of existence. Culture

[1] See note 8, Chap. 1.

co-evolved throughout history in different ecological and geographical contexts. The principles of an "ecological culture" that mobilizes social processes toward a sustainable world are therefore embedded in a cultural rationale that is constituted by the different forms of productive organizing and ethnic modalities of traditional societies, indigenous peoples, and peasant communities. Environmental degradation and destruction of their livelihoods, has led to the dissolution of cultural values, identities, and traditional productive practices. The politics of sustainability based on cultural diversity legitimizes the rights of communities over their ethnic territories, their cultural values, their traditional practices and social institutions, while fostering self-management of their heritage over their natural resources.

The principles of diversity in environmentalism challenge the homogenizing of productive and cultural patterns governed by economic globalization. They defend the values of cultural plurality and the preservation of people's identities. These ontological principles are a condition for local sustainable practices and for attaining global sustainability. The sustainability conditions that are integrated in their production practices are a reflection of creativity and adaptation of the technical instruments employed by every culture, based in long processes of Nature–culture co-evolution, environmental transformation, and cultural assimilation. Productive practices founded in cultural symbolization of the environment, religious beliefs, and social meanings assigned to Nature are inscribed in their social imaginaries, in their ways of perceiving and appropriating Nature, in social rules of access and use, ecosystem management practices, and cultural patterns of use and consumption of resources. Thus, different forms of "traditional agricultural ideologies" (Alcorn 1993) have guided the "Mesoamerican strategies for production" (Boege 1988) based on the multiple uses of "ecosystem-resources" (Morello 1986).

To understand the functioning of these cultural strategies in the sustainable management of natural resources, it is not only necessary to know the rationale of traditional classifications and folk-taxonomies that reflect the local knowledge of different ethnic groups. What is essential is an understanding of the system of beliefs and wisdom, of myths and rituals, of imaginaries and practices that form the holistic models of perception and use of natural resources by traditional cultures (Pitt 1985; Infgold 2000). These forms of signification are intimately related to the economic organization and productive practices of traditional societies. Ethnic identities organize differentiated practices of resource use, which together constitute a patrimony of natural and cultural resources of indigenous peoples and peasant societies.

This cultural organization readjusts to processes of acculturation and technological change that reaffirm and transform their identities. Indigenous cultures of the Americas have preserved and redefined their ethnic identities through processes of miscegenation and racial intermingling (*mestizaje*) since the Spanish conquest. Seemingly, today the survival of indigenous people's identities despite the impacts of modernity and the economic and technological strategies of globalization to r-appropriate Nature and biodiversity is leading to resignify and hybridize traditional ethnic identities (Escobar 1997). Culture is being revalued as a "social resource for sustainable development." Culture is

> The complement of natural resources in peasant and indigenous peoples' productive systems […] orienting resource use, while nature sets certain conditions to the life options of the ethnic group. Conceived in this way, culture is a social resource, capable of being used destructively or rationally, of being lost or developed. (Varèse and Martin 1993, p. 738)

In this perspective, the cultural legacy of indigenous peoples in Latin America appears as an integrated part of their heritage of natural resources, defined through the symbolic and productive relations that have led the co-evolution of Nature and culture over time. Every cultural group's organization of space and time establishes a system of social relations of production that supports the integrated and sustainable management of natural resources. Jorge Morello relates the cultural patrimony to the environmental resources of the region:

> Cultural inheritance is an essential resource for the region. Attaining sustainability in the Andean agricultural strategies, in the tropical rainforests, and the wetlands will require the incorporation of technologies used by the Maya, Inca, and Aztec civilizations, as well as from other ethnic groups. Those peasant ethnicities possess a rich technological heritage, and its dete-

rioration has generated significant ecologic costs in numerous countries, especially in Mexico and Peru. They were able to solve problems where technologies from the North have failed [...] such as integrating agricultural poly-cultures in small areas with extensive use of the rainforest (as descendants from the Maya still do); or mixed multipurpose ranching to survive in highly variable semiarid environments (the Aymaras and the Quechuas in Bolivia); the management of deciduous forest to transform it into a poly-productive ecosystem, even at times of extraordinary drought; or developing germplasm to respond to unpredictable climate and rain seasons (varieties of maize and beans of very short, short and long cycles); the development of varieties adaptable to different altitudes (Incas, Mayas) and different latitudinal gradient (Andean and Maya peoples). (Morello 1990)

Several authors have discussed the environmental rationality embedded in "Pre-Hispanic development styles" (Gligo and Morello 1980), the "peasant mode of production" (Toledo 1980), "Andean utopia" (Burgoa and Flores Galindo 1982) and "vertical eco-symbiotic complementarities" (Condarco and Murra 1987). These pre-Hispanic cultures had internalized ecological foundations and developed the potential of cultural organization for sustainable land and resource use in their production strategies. Practices took into account the complementary elements of diverse ecological and geographical spaces, integrating regions that expanded beyond the territories of any particular ethnic group. This strategy permitted traditional societies to optimize the ecological productivity of their diverse geographies, as well as the seasonal use of different crops and labor, productive spaces, ecological cycles, land fertility, and biological regeneration processes. They integrated local production and territorial space through interregional trade and intercommunal exchanges of economic surpluses.

The ethnic spaces of tropical America were, and continue to be, sites of innovation in strategies for survival and ethnic-eco-evolution, and of adaptation and transformation of the environment. Important technologies and public works have been developed for the sustainable use of hydrological resources and the enhancement of soil fertility through techniques for water conservation and erosion prevention, and with innovative management strategies such as those of *terrazas*, *chinampas*, *andenes* and *camellones* (Murra 1956, 1975; Romanini 1976; Denevan 1980; Masuda et al. 1985; CEPAL/PNUMA 1983; de la Torre and Burgoa 1986; Uribe 1988; Leff and Carabias 1993). Thus, a vast repertoire of technical knowledge and productive practices developed, in turn allowing the evolution of the different cultures that inhabited the American tropical regions.

3. Ecological Complementarities, Cultural Temporalities and Ethnic Identities

The productive strategies developed by the different American indigenous cultures not only implied the development of technologies for resource use by ethnic groups. This ecological culture was also integrated into the social relations and productive forces of traditional rural societies. Cultural and ecological organization of production operated through a system of complementarities of territorial spaces and ecological cycles for the sustainable, productive management of natural resources: seasons of rain and drought; annual distribution of different crops selected by the differentiated growth patterns and climatic conditions of each season and territory; the integral use of specific plants and the integrated management of genetic varieties of different species (maize, potato), depending on the topographic conditions and on the type and quality of soils. Spatial and seasonal ecological practices also included different strategies of final use (for self-consumption or the market) and different technological inputs (machines, fertilizers). These strategies of diversification and complementarity of ecological functions generated practices of mixed and multiple cropping, as well as the integrated use of natural resources (family orchards, *milpas*, and *acahuales*), management of ecosystem's secondary succession and selective regeneration processes and sustainable management of tropical forests.

These "pre-Hispanic styles of sustainable development" were characterized by the productive integration of the different ecosystems and ethnic spaces, as well as the perception of Nature as integrated and synergic processes, rather than as a stock of discrete resources. These strategies optimized the use of labor through diversifying production, and adjusted to the ecological conditions and the environmental potential of each region, integrating agroforestry with hunting and gathering of biological resources, as well as with fisheries. This style of development fostered complementary labor processes and interethnic cooperation practices for the integrated management of natural resources. Therefore, the integration of family and communal economies enabled the exchange and consumption of products from a larger territory. At the same time, social rules were established for collective regulation of production, through long processes of observation of Nature, productive experimentation, technical innovation and exchange of knowledge.

Culture is therefore an integral part of the general conditions of sustainable production, in the sense that ethnic identities and cultural values, as well as traditional practices for collective management of Nature, are the basis for signifying the environmental potential for sustainable development. Today, cultural organization becomes an active principle for the development of productive forces in an alternative paradigm for sustainable production, and ecological productivity and technological innovation are interwoven with cultural processes that define global social productivity. Ethnic identities and cultural imaginaries define a system of natural resources, and culture becomes the basis for sustainable productive practices (Leff 1993, 1995).

Studies of ethnic modalities for the appropriation of the ecological resources in different regions in the world have shown how their in-depth knowledge of soils allowed indigenous people to use their productive base efficiently, obtaining sustainable yields under restrictive environmental and socioeconomic conditions, while preserving their productive potentials and ecological resilience. These indigenous practices are still in use in different places as strategies for cultural survival and sustainable livelihoods (Parra 1993). In traditional cultures, knowledge and practices are interwoven in complex taxonomic systems, symbolic codes, worldviews, and meaning-values that guide the forms of cognition and appropriation of Nature. The multiple functions of Nature in traditional agriculture preserve ecological processes in their protection of water and wind erosion of soil; in the conservation of biological and genetic diversity, and in the selective regeneration of useful species; in the preservation of soil fertility by enhancing its physical, chemical, and biological characteristics, as well as its capacity to retain water; in the integrated management of wild and cultivated plants, and in the innovation of highly productive agro-ecological systems such as Peruvian *camellones* and Mexican *chinampas*.

In rural areas of the Third World, these social and productive practices are intimately connected with cultural values and symbolic processes that organize socioeconomic formations and mediate the social appropriation and transformation of the environment. Culture regulates the use of resources to satisfy the needs of its members. Through symbolic processes, cultural norms are constructed that restrain social access to Nature, harness the use of technological means and regulate the rhythms of extraction and transformation of resources. Kinship relations, social reciprocity, communal property, and territorial rights favor the sustainable use of natural resources. Socially sanctioned access to Nature is defined through cultural traditions that establish forms of land tenure, property, and management of environmental resources, of gender, age, and time division of labor and distribution of productive activities. These social conditions determine specific cultural rationality for the use of natural resources.

Among the best examples are traditional cultures in India that have established specific relations with the environment, developing conservation practices through diversity and complementarity of ecological niches occupied by different groups in one same region. Each family, tribe or social caste has specific rights to exploit their land or certain natural resources under communal regulation (Gagdil and Iyer 1993). These practices include restrictions over the territory that any given social group can occupy and where that group can extract resources, over the techniques, methods and periods authorized for the exploitation of forests and animals, establishing an ecological specialization for every caste in the appropriation of Nature (Gagdil 1985).

These practices of multiple and integrated uses of natural resources, based on cultural norms and traditional knowledge, "decode a great variety of microenvironments, developing productive practices that preserve biodiversity and increase the self-satisfaction of material needs of the community" (Toledo and Argueta 1993). Nature is at once an economic resource and a bio-cultural patrimony. These cultural strategies for productive management of Nature enhance the sustainable productivity of natural resources while preserving the conditions for diversified practices and rationalities blending Nature and culture in the ever more complex and creative evolution of life.

Cultural perception of the environment plays a fundamental role in the cognitive systems of traditional societies. Imaginaries and worldviews of traditional societies integrate religious myths and social rituals to productive practices. Their knowledge of geophysical phenomena (lunar phases and seasons; bio-geo-chemical, ecological and hydrological cycles) is associated to knowledge of different soil types and topographic conditions, allowing a more integrated and complimentary use of ecological space, generating strategies of multiple and integrated uses of resources. Cultural values are interwoven with community knowledge of the conditions for sustainable production, with social organization and ideological formations. These forms of social cohesion and self-sufficiency support the survival of different populations today. The reinvention of self-reliant practices for multiple and sustainable uses of natural resources is forging new identities, establishing new cultural territories and environmental rationalities for the sustainable livelihood of the peoples of Earth.

4. Localizing Globalization and the Symbolic Appropriation of Nature

Today, the inertia of the globalizing process and the reflexivity of modernity is reaching the limits of *scientific logocentrism*, of economic reason and concentration of power. A new centrifugal force is emerging, one that is driven by the ontology of life—by the principles of diversity, difference, and otherness—leading to economic decentralization, participatory democracy in a dialogue among different forms of wisdom and knowledge. A new understanding of living conditions on the planet is reorienting the odyssey of human civilization. Although seduced by economic rationality and assimilated in policies for the capitalization of Nature, sustainability is grounding its principles at the local level through the construction of a new productive rationale. This new paradigm for sustainable production is embedded in diverse cultural values and meanings, the ecological potential of Nature, and a critical appropriation of science and technology. As globalization promotes an economic logic that infiltrates every territory, every ecosystem, permeating every culture, and every individual, the politics of locality are building a sustainable global world from the specificity of ecosystems, cultural identities, and local autonomies.

Environmental rationality is being constructed by unearthing the ecological conditions of sustainability, the resignification of traditional cultures and actualization of identities, unfreezing historical times. The defense of local environments depends not only on the economic valuation of Nature and culture, but also on the reconfiguration of new identities, and a reassessment of the sustainability of life on the planet by means of the negentropic–entropic metabolism of the biosphere. This implies the organization of new social actors who can build a productive rationality based on the ecological potentials of their territories defined through their cultural meanings and values.

Environmental rationality opens up the question of the relation of cultural meanings and local knowledge to the sustainable management of Nature. Human beings apprehend the Real as world visions, through their imaginary intuitions, perceptual senses, linguistic signifiers, conceptual lenses, and paradigmatic armors. Environmental knowledge goes beyond grasping the objectivity of Nature in the established scientific rationality and economic order; the emergent potential of Nature is being reappropriated through the reconstruction of knowledge, guided by cultural imaginaries, wisdom, identities, meanings, and practices (Leff 1998b/2002, 2010). Local knowledge is not merely knowledge constructed through practice nor is it formed by testing abstract knowledge under empirical conditions. It is not the submission of local particularities to the universal, hegemonic, and domineering rationality. Local knowledge is forged by meanings built through symbolic processes that are cast in diverse ethnic modalities (Leroi-Gourhan 1964–1965). The question of time emerges as an inquiry about the forging, sedimentation, coupling, survival, and reinvention of identities of the diverse modes of being-in-the-world. These modes of existence are rooted through cultural imaginaries and meanings in landscapes and life-territories, in different times and places where life blends and flows again through history, envisioning new horizons for the sustainability of life on the planet, revaluing

Nature and culture through the claims for autonomy and the rights of existence of the peoples of Earth.

Prior subdued and dominated knowledge is reconstructed in new local and personal knowledge that hybridizes with new discourses of globalization and sustainability, with modern sciences and techniques, opening unedited paths in history. Sustainability is a question of time and being in the creative evolution of life and not the subordination of Nature to economics. Becoming is not the unwinding of ontological essences and pre-established rationalities, but rather the awaking of the potentialities of life through the reinvention of identities to discover new sustainable worlds, to unveil new horizons in history, and to construct new life territories.

Globalization, guided by one-dimensional technological rationality, the short-sightedness of time, the ontological reduction of being to economic calculation and the profit motive, has resulted in the ecological degradation of a polluted planet and a saturated world. From the limits of economic rationality, history is giving way to new paths to a sustainable future, fertilized by the flow of diverse beings and divergent times. The emergence of complexity, difference, autonomy, and otherness is encountering the ontological and epistemological conceptions of the world founded by the original *logos* of ancient Greek civilization, "enlightened" and technified by capital accumulation, the industrial revolution, and modern rationality. Those seemingly dead, "primitive," and "traditional" societies, the co-evolution of life and ethnicity, the cognitive constructions of Nature through culture, are giving birth to new definitions of self and other, subject and object, culture and Nature through the interplay of the original disjunction of the Real and the Symbolic. From the melting pot of the Material and the Ideal (Godelier 1984), different denominations of Nature emerge, and other worlds are being constructed through different meanings and the positioning of cultural beings in diverse environmental conditions.

Traditional societies and local economies do not produce only use values and exchange values. They create "use meanings" which reflect the complex relation of the material and ideal, the hybridization of the symbolic–cultural order with the thermodynamic–ecological conditions in economic–political relations of production. Nature is not dominated by economic rationality but is set in the creative evolution of life. Human inventions that guide the metabolism of life are recorded in the memory and decoded by the cultural imaginaries of the peoples. Past "cultural beings" are not only remembered. The recovery of what was once deposited in cultural memory is signified by the Spanish word *recordar*, which means remembering and rewinding; it is the unwinding and a new binding of the past with the present, projecting life toward a sustainable future.

From the perspective of a politics of space, place and time, peoples' self-reliance cannot be conceived as the empowerment of the oppressed from above. Cultural autonomy cannot be graciously granted, even in this era of democracy, citizenry, and human rights. The claims for emancipation emerge from oppressed modes and rights of being-in-the-world. These social demands arise from the actualization of identities, from the hybridization in time and space of different forms of being: of local cultures, local knowledge, and local people. Resistance to the capitalization of Nature is the expression of *cultural resiliency* as the capacity to preserve its identity from external perturbations. It is through these processes of *re-existence*—of the reconstitution of their "rights of being"—that autonomies can lead to the self-management of people's living conditions and the self-production of their modes of existence.

5. The Ontology of Being and Time and the Politics of Place and Space

The politics of space and place that emerge from social claims of local people are being legitimized by more democratic rules of citizenship, cultural and environmental rights, and the rights to existence of indigenous peoples. Socio-environmental claims are struggles over the meaning of "the Real" over "Truth." Economic rationality appears today as the highest degree of rationality attained by human Nature. This mechanistic rationality that ignores life and Nature, has produced the entropic decay of an unsustainable world. It has been led by the economic rationality that governs the destiny of life and human existence. The place is the territory where sustainability is rooted in ecological grounds and cultural identities. It is the social space where local social actors exert power to control environmental degradation and to mobilize environmental potentials in self-reliant projects geared to satisfy the needs, aspirations, and wishes of the people that globalization cannot comply with.

Place is the locus of people's demands and claims arising from the degradation of their environment, while harnessing their capacity to reconstruct their life-worlds. The local level is where cultural identities are forged and expressed as the social valuation of economic resources and as strategies for reappropriation of Nature. If globalization is the space where negative synergies manifest the limits of growth, the local is the place where positive synergies of environmental rationality are rooted and may flourish. The local space is the point of encounter of synchronic and diachronic processes, where differentiated times and spaces collide, precipitate and are projected to new horizons of existence. The local is the place where the integration of specific cultural identities and ecological potentials occur. It is also the convergence of times of sustainability: processes of ecological restoration and productivity; of technological innovation and assimilation; of reconstitution of cultural identities and *re-existence* of the peoples of the Earth.

Time is structured around meaningful social and economic events (Evans Pritchard 1962), and each culture defines different temporal systems (Pitt 1985). Time is territorialized as the integration of different temporalities that coexist in the geographical space (Santos 1996). Time is not only the measure of external events (natural phenomena, regeneration processes, ecological cycles, and life experience), but the realm of internal meaningful events, the permanence of different "presences" through history, the updating of ethnic identities and "cultural beings," of existential time, a time that forges the production of existence. Time reopens the process from de-signed Nature that fixes Nature to predetermined visions to a signified Nature that open up new ways of reconstructing Nature and culture, time and being.

A new politics of space and place redefined by meaningful time is under construction in present struggles for identity, autonomy, and territory. Environmental politics is a *politics of being*, but overall a *politics of life and existence*, a claim for the recognition of people's rights for survival, cultural difference, ethnic styles, and autonomous modes of being-in-the-world. It is also a *politics of becoming* that values the meaning of utopia and the potentialities of life as the rights of Nature to be and of all communities to forge their future. As a result, cultural territories are being fertilized by meaningful time. It is not only the claims for cultural rights, including the preservation of language and traditional practices, but also that of cultural politics for reconstruction of identities and existential meanings, for projecting collective beings to unprecedented futures, as resistance to homogenization, and for the *re-existence* of life as diverse forms of being. It is the unwinding of creative diversification of world-lives built from cultural and ecological heterogenesis.

Identities are never identical. On the contrary, the claim for equality and sustainability is a struggle for diversity that emerges from the deconstruction of the ontology of Being—of the One and the Universal—to a new political culture of difference and plurality. It is the claim for the right to be different and to forge difference against the forced homogeneity of the one-world order. For these politics of time, being and becoming is emerging in the recasting of identities, and in the innovation of cultural projects during the transition to democracy and sustainability. The construction of environmental rationality is being forged in emergent indigenous people's movements. It is a politics that "emphasizes becoming rather than being; that involves positioning rather than essence, discontinuity as well as continuity" (Escobar 1997, p. 18).

Cultural politics of emergent indigenous people's movements is apparent in the recent Black Movement of Colombia's Pacific Coast:

> The movement's main principles for political organization [are, first] the right to an identity, that is, the right to being black according to cultural logic and world view rooted in black experience [...] for the reconstituting of black consciousness itself and the rejection of the dominant discourse of "equality" with its concomitant obliteration of difference. Second is the right to the territory as the essential space for being and for the development of culture. The third is the right to political autonomy as a prerequisite for the practice of being, with the possibility of fostering social and economic autonomy. Fourth is the right to construct their vision of the future, development, and social practice based on the common forms of production and social organization. Fifth, the principle of solidarity with the struggles of black peoples for alternative visions throughout the world [...] Activists conceptualise the territory as a space for the creating of futures, for hope and the continuation of existence. (Escobar 1997, pp. 14–16)

Indigenous peoples are in the middle of a struggle for new cultural rights to regain control over their life territories as ecological, productive, and cultural spaces, and to reappropriate their cultural heritage and patrimony of natural resources. New social actors are constructing environmental rationality through their political demands that guide new principles to revalue the environment and to reappropriate Nature under the conditions of life on the planet.

6. Environmental Rationality and Life Times

The environmental crisis is a turning point in history, a crossroads of the creative evolution of life and the techno-economic regime that has repressed its potentialities; it is the "melting point" of frozen immemorial times at the limits of the social processes that have degraded the complex web of life of the planet formed through four eons of evolution. The irruption of the global ecological crises in 1972 exposed the limits of growth and the re-thinking of history toward the ontology of life. The year 1992 was a time of configuration of sustainability in global economic rationality and the rise of a new consciousness of the indigenous peoples in the Americas after 500 years of conquest and exploitation. Through these recent events, internal forces are loosening the subjugation bonds established by the dominant structures now destabilized at this breaking point of history in the transition toward a sustainable world.

The present is the crossroads of heterogeneous times, from the explosion and activation of frozen traditions once blocked by domination and subjugation to the interweaving of different rationalities that have forged diverse forms of relations between Nature and culture. It is the survival and transposition of past time to present times, the configuration of new identities and meanings, and the construction of new utopias that drive the world toward alternative futures. Today, history is forged by the re-emergence of a meaningful life. Quiet histories that seemed to have lost their memory and their traditions are being reborn, leading to the politics of time and being, to the reconstruction of past historical events through the inquiry of oppressed and repressed meanings. Heidegger underlines the ways of quietness in history:

> This quietness of occurrence is not an absence from history, but the primary form of its makings. What we know today as the past and what we represent as such, almost always is only the "actuality" of a past moment [...] what always pertains to history but is not properly history. Mere past does not exhaust what has been. This is still present, and its form of being is a peculiar quietness of occurrence, which form is determined from that which occurs. Quietness is only a movement arrested in itself, and often it is more restless than any event. (Heidegger 1975, pp. 44–45)

This relentless quietness is acquiring new voices, forging new identities, mobilizing new actors, fertilizing new beings and opening new horizons in history. It is more than just the encountering of diverse histories, of the hybridization of differentiated ontological regimes, or of the confluence of synergic processes to construct a sustainable future. Beyond the transgenic modifications of life and the economization of Nature, new identities are forging a historical mutation and the emergence of new times. These new identities are not only updating ancient beings in present times, spaces, and places; they are constituted in emancipation processes in the context of the current reality, in resistance and confrontation with the dominating global hegemony (Laclau 1996). Hybrid identities are being forged through political struggles and antagonistic encounters for the reappropriation of Nature and the reconstruction of life-worlds. These struggles go beyond the right for multiculturalism, for the expression of difference in a finalized global order and a non-ambiguous political system. They transcend the circular logic of representative democracy giving way to the expression of differences and particularities over the closure of a universal regime of being in a globalized world.

By its will to know and its need to appropriate Nature, humankind has transformed the world, changed its laws, deviated life from its natural courses, generated new meanings, and forged new beings. *Environmental complexity* is emerging at the turn of the third millennium.[2] This emer-

[2] I have defined environmental complexity as an emergent ontological order, arising from the intervention of techno-economic rationality in the ontology of life (Leff 2000a, 2001).

gence goes beyond the transition from unachieved modernity to undefined postmodernity (García Canclini 2005), to build a new environmental rationality. It is by envisioning this alternative rationality that history can be reopened. Diversity, difference, and otherness become principles that transcend dialectical negativity as oppositions and antagonisms within an established power system. Environmental rationality opens the logic of identity to the ontology of difference (Heidegger 1957/1969; Derrida 1978, 1982), to confront the *other* as an absolute Other (Levinas 1977/1997), as a radical otherness that constructs new cultural identities, new knowledge/ *savoirs*, new life territories, new modes of being-in-the-world (Leff 2000a, 2004, 2020).

The time that forges a sustainable future is not only that dimension of time that is inherent to events and processes in the evolution of life that stems from the emergent potentiality of *physis*. Sustainability has been encountered by the understanding of *physis* by the *human logos*, by the objectification of the world and the *logocentrism of science*, and by the intervention of technology in the courses of life. In the emancipation of life from the domination of scientific and economic reason, phenomenological and existential time finds its place in the web of life, rethinking history to envision new horizons and reorienting the destiny of life on the planet. Time shapes different life-world and is embodied in identities that engage human beings in the creative evolution of life. Time winds and unwinds the world with various meanings that enact the complex negentropic/entropic material and energy flows in the metabolism of the biosphere. Time constructs new modes of being-in-the-world, new ways of inhabiting the planet under the conditions of life on Earth (Leff 2018).

The civilizing project that pretended to unify the world around one God, the absolute idea, the unity of science, and globalization of the market is dead. The ontology of life as the realm of diversity–difference–otherness transcends historical materialism driven by natural and social dialectics. History is reborn to a diversity of life-worlds and productive rationalities driven by the potentialities of Nature and by the meaningfulness of human existence.

Bibliography

Alcorn, J. (1993), "Los procesos como recursos. La ideología agrícola tradicional del manejo de los recursos entre los boras y huastecos y sus implicaciones para la investigación", in Leff & Carabias (Eds.), *Cultura y manejo de los recursos naturales*, México: UNAM, Miguel Ángel Porrúa, Vol. 2, pp. 329–365.

Boege, E. (1988), *Los mazatecos ante la nación*, México: Siglo XXI Editores.

Burgoa, M. & Flores Galindo, A. (1982), "La utopía andina", *Allpanchis*, Vol. 20.

Castoriadis, C. (1998), *The imaginary institution of society*, Cambridge: MIT Press.

CEPAL/PNUMA (1983), *Sobrevivencia campesina en ecosistemas de altura*, 2 Vols., Santiago de Chile.

Condarco, R. & Murra, J. (1987), *La teoría de la complementariedad vertical eco-simbiótica*, La Paz: Hisbol.

De la Torre, C. & M. Burgoa (Comps.) (1986), *Andenes y camellones en el Perú andino*, Lima: Concytec.

Denevan, W.M. (1980), "Tipología de configuraciones agrícolas prehispánicas", in *América Indígena*, No. 40, pp. 610–652.

Derrida, J. (1978), *Writing and difference*, Chicago: The University of Chicago Press.

———. (1982), *Margins of philosophy*, Chicago: Chicago University Press.

Escobar, A. (1997), "Cultural politics and biological diversity: State, capital and social movements in the Pacific coast of Colombia", in O. Starn & R. Fox (Eds), *Culture and social protest: Between resistance and revolution*, New Brunswick, Rutgers University Press.

Evans Pritchard, E. E. (1962), *Social anthropology and other essays*, New York: The Free Press.

Gagdil, M. (1985), "Social restraints on resource utilization: the Indian experience", in McNeely & Pitt, op. cit.

Gagdil, M. & P. Iyer (1993), "La diversificación en el uso de los recursos de propiedad común en la sociedad india", in Leff & Carabias, Vol 2, op. cit., pp. 551–573.

García Canclini, N. (2005), *Hybrid cultures: strategies for entering and leaving modernity*, Minnesota: University of Minnesota Press.

Gligo, N. & Morello, J. (1980), "Notas sobre la historia ecológica de América Latina", In Súnkel, O. & N. Gligo, *Estilos de desarrollo y medio ambiente en la América Latina*, México: Fondo de Cultura Económica.

Godelier, M.M. (1984), *L'Idéel et le Matériel*, Paris, Fayard.

Heidegger, M. (1957/1969), *Identity and difference*, Chicago: Chicago University Press.

———. (1975), *La pregunta por la cosa*, Buenos Aires: Alfa Ed.

Infgold, T. (2000), *The perception of the environment*, Londres, Routledge.

Laclau, E. (1996), *Emancipations*, London & New York: Verso.

Leff, E. (1993), "La dimensión cultural del manejo integrado sustentable y sostenido de los recursos naturales", en Leff E. & J. Carabias (Eds), *Cultura y manejo sustentable de los recursos naturales*, México: CIIH-UNAM/Miguel Angel Porrua.

———. (1995), *Green production: towards an environmental rationality*, New York: Guilford.

———. (1998b), *Saber ambiental: Sustentabilidad, racionalidad, complejidad, poder*, México: Siglo XXI/UNAM/PNUMA (3rd revised and extended edition, 2002).

———. (2000a), "Pensar la complejidad ambiental", en Leff, E. (Coord.), *La complejidad ambiental*, Colección "Aprender a aprender", México: Siglo XXI Editores (2nd revised edition, 2003).

———. (2001), *Epistemología ambiental*, Sao Paulo: Cortez Editora.

———. (2004), *Racionalidad ambiental: la apropiación social de la naturaleza*. México: Siglo XXI Editores.

———. (2010), "Imaginarios sociales y sustentabilidad", in *Cultura y representaciones sociales*. Num. 9, México, pp. 42–121.

———. (2018), *El fuego de la vida: Heidegger ante la cuestión ambiental*, Mexico: Siglo XXI Editores.

———. (2020), "Heráclito: A Φύσις e o Desocultamento da Vida", *Desenvolvimento e meio ambiente*, 54.

Leff, E. & J. Carabias (Eds.), (1993), *Cultura y manejo de los recursos naturales*, México: UNAM, Miguel Ángel Porrúa, 2 Vols.

Leroi-Gourhan, A. (1964–1965), *Le geste et la parole*, 2 Vols., Paris: Albin Michel.

Levinas, E. (1977/1997), *Totalidad e infinito: ensayo sobre la exterioridad*, Salamanca: Sígueme, 4th edition [*Totality and infinity: an essay on exteriority*, Pittsburgh: Duquesne University Press, 1969].

Masuda, S, Shimada, I & Morris, C. (1985), *Andean ecology and civilization: an interdisciplinary perspective on Andean ecological complementarity*, Tokyo; University of Tokyo Press.

Morello, J. (1986), "Conceptos para un manejo integrado de los recursos naturales", en Leff, E. (Ed.), *Los problemas del conocimiento y la perspectiva ambiental del desarrollo*, México, Siglo XXI Editores.

———. (1990), "Insumos para la agenda ambiental latino-americana", in *Nuestra propia agenda*, BID/PNUD.
Murra, J. (1956), *The economic organization of the Inca state*, Chicago: University of Chicago.
———. (1975), *Formaciones económicas y políticas del mundo andino*, Lima: IEP
Parra, M. (1993) "La producción silvo-agropecuaria de los indígenas de los Altos de Chiapas", en Leff & Carabias, Vol. 2, op. cit., pp. 445–487.
Pitt, D. (1985), "Towards ethnoconservation", in McNeely J. & D. Pitt, *Culture and Conservation: the human dimension in environmental planning*, London: IUCN/Croom Helm.
Romanini, C. (1976), *Ecotécnicas para el trópico húmedo*, México: Cecodes-Conacyt/PNUMA
Santos, M. (1996), *A natureza do espaço: técnica e tempo; razão e emoção*, São Paulo: Hucitec.
Toledo, V.M. (1980), "Ecología del modo campesino de producción", *Antropología y marxismo* No. 3, pp. 35–55.
Toledo, V.M. & A. Argueta (1993), "Naturaleza, producción y cultura en una región indígena de México: las lecciones de Pátzcuaro", in Leff & Carabias, Vol. 2, op. cit., pp. 413–443.
Uribe, R. (org.) (1988), *Medio ambiente y comunidades indígenas del sureste: prácticas tradicionales de producción, rituales y manejo de recursos*, Villahermosa, Tabasco, México: UNESCO/Gob. de Tabasco.
Varèse, S. & Martin, G. (1993), "Ecología y producción en dos áreas indígenas de México y Perú. Experiencias y propuestas para un desarrollo culturalmente sustentable", in Leff & Carabias, op. cit., pp. 717–740.

4

Environmental Rationality and the End of Natural Dialectic

Introduction

The environmental crisis has brought about the questioning of its historical causes, the structure of the dominant modes of thinking and understanding the ontology of Being and beings—the Real, history, world, life, things—that ended up alienating *res cogitans* from *res extensa*, culture from Nature. Thinking the environmental question has triggered the deconstruction of the history of metaphysics to unearth the understanding of the conditions of the sustainability of life on the planet. The becoming of life through socio-environmental conflicts and dialogue of knowledge/*savoirs* questions the understanding of history driven by Kant and Hegel's transcendental logic and by Marx and Engels' dialectics of history and Nature. However, the nascent political ecology, emerging from Marxism, did not avoid the temptation of founding its political thinking and praxis on the dialectical ground.

Published in *Capitalism, Nature, Socialism*, 9(4), No. 36, December 1998, pp. 67–93. The original Spanish version was published in Leff, E. (1999), La Racionalidad Ambiental y el Fin del Naturalismo Dialéctico, in *Persona y Sociedad*, XIII(1), ILADES/Universidad Alberto Hurtado, Santiago de Chile, pp. 79–99. A revised version was included as chapter 2 in my book *Racionalidad Ambiental: la reapropiación social de la naturaleza*, Siglo XXI Editores, 2004.

E. Leff, *Political Ecology*, https://doi.org/10.1007/978-3-030-63325-7_4

One precursor of this neo-Marxist political ecology is Murray Bookchin, who earned well-deserved recognition as a pioneer ideologue of the ecological movement, in particular of its eco-anarchist branch. His writings have influenced environmental organizations in the North and inspired eco-community groups in the South. Bookchin's social ecology orients new lifestyles and forms of social organization based on a theory of eco-social evolution that seeks to legitimate ecological values as "objective liberating truth." However, his project to root an ecological society and the political philosophy of environmentalism in natural dialectics and an "ethical ontology" faces problems of theory and praxis.

Here I do not intend to address the theoretical foundations of Bookchin's social ecology comprehensively. I will restrain my argument to his claim to ground his theory in "natural philosophy," specifically in "ecological dialectics." Thus I will focus on the following questions: (a) environmental ethics and the naturalization of society; (b) epistemology, ontological monism, and ecological reductionism; (c) dialectics and totality; (d) dialectics and the politics of environmentalism. I will explore and question Bookchin's positions on these issues, highlighting some of their implications for political ecology and the construction of an ecological society.

Environmental Ethics and the Naturalization of Society

In modern times, social theory has taken its stands in two divided fields: critical theory and an empirical-analytical-positivist approach to reality. Dialectics became the hallmark of critical rationalism and has somehow survived the fall of really existing socialism. Even if it has lost comprehensiveness and coherency as an ontological ground or epistemological theory, dialectic rhetoric still colors today's theoretical and political discourse, particularly that of political ecology.

Bookchin seeks to construct an ecological society by inscribing dialectical reason in the material movement of Nature, instead of grasping dialectics in power strategies and discursive formations that emerge from conflicting social interests for the appropriation of Nature. He realizes that ecological degradation has its origins in the domination of man by man, rather than in human domination of Nature, and traces the historical moments in which hierarchies were established; but he does not recognize the ontological grounds of the will to power that found domination of Nature and among human beings. Bookchin's philosophy of Nature is configured within an evolutionary theory that in its becoming will give birth to an eco-communitarian society. Natural dialectics conflates the divide of Nature and culture, the original difference between *physis* and *logos* in a "graded evolutionary continuum" that should bring about a "fully self-conscious Nature." The sources of human rationality, of dialectical reason and social change, are grounded in the oneness of Nature, not in critical reason and strategic action founded in the deconstruction of the naturalization of the symbolic, cultural and social order.

Thus, Bookchin states that:

> What unites society with nature in a graded evolutionary continuum is the remarkable extent to which human beings, living in a rational, ecologically oriented society, could embody the creativity of nature [...] Life forms that create and consciously alter the environment, hopefully in ways that make it more rational and ecological, represent a vast and indefinite extension of nature into [...] the evolution of a fully *self-conscious* nature [...] Given this conception of nature as the cumulative history of more differentiated levels of material organization (especially of life forms) and of increasing subjectivity, social ecology establishes a basis for a meaningful understanding of humanity and society's place in natural evolution [...] Humanity, in effect, becomes the potential voice of a nature rendered self-conscious and self-formative. (Bookchin 1990a, pp. 35–37, 201)

Bookchin suggests that human behavior should not be viewed merely as an adaptive response, of natural selection, to the environment. Instead, consciousness can orient individual actions and social evolution toward freedom, based on the potentialities and creativity of Nature:

> By their very own biologically rooted mental power, they [human beings] are literally constituted by evolution to intervene into the biosphere [...] their presence in the world of life marks a crucial change in evolution's

direction from one that is mostly adaptive, to one that is, at least, potentially creative and moral. (ibid., p. 72)

Bookchin roots his ecological dialectics in a naturalist view of society; rationality is defined by what is ecological, as if Nature would have evolved "naturally" toward the ecologically rational. Human subjectivity and knowledge are extensions of natural evolution to its ultimate state of self-consciousness. This discourse cannot give theoretical support to a political ecology capable of guiding the reconstruction of economic rationality dominating our present society, nor envisioning the power strategies and social interests that guide the construction of alternative environmental rationality (Leff 1995, 2004). Bookchin claims the need to "re-enter the continuum of natural evolution and play a creative role" (ibid., p. 73). However, he offers no strategic thinking to guide this social reconstruction, other than by the spontaneous configuration of ecological ethics emerging from natural evolution.

For Bookchin, ecological consciousness will generate a world libertarian movement supported by a unified human interest guided by eco-anarchist values evolving from the generalized post-scarcity condition of present society.[1] However, the idea of a global process leading to a generalized state of post-scarcity is untenable, when the destruction of Nature geared by the progression of capital growth that induces the entropic degradation of the environment is accelerating the production of scarcity of natural goods and services, and poverty of a growing population which is far from satisfying its basic needs. Bookchin does not seem to perceive the roots of inequity in the present ecological–economic order, nor the power strategies that block the diffusion of his "liberating technology." The hypothesis that the affluent society will lead to the transition to an ecological society is less convincing; there is no clear evidence of a generalized rejection of the trends of capital accumulation toward the

[1] "Our foremost need is to create a general human interest that can unify humanity as a whole [...] There is not the remotest chance that it [a free ecological society] can be achieved today unless humanity is free to reject bourgeois notions of abundance precisely because abundance is available to all" (ibid., pp. 170–171). Bookchin's search of general interest to unify humanity through the emergence of ecological consciousness is as ill-founded as the strategies to build a consensus for "sustainable development" guided by the neoliberal strategies to capitalize Nature through free market mechanisms.

over-exploitation of Nature and dispossession of peoples (Harvey 2003), nor of over-consumption and a shift in people's preferences favoring the new values of anarchism, frugality, and solidarity, and for the construction of an alternative, sustainable, productive rationality.[2] There is no way that "abundance" generated through capital's mechanistic-entropic mode of production will automatically trigger the ontological turn that can cause an ecological society to emerge without grounding it on production rationality founded on the environmental conditions that support a negentropic sustainable production process.

Bookchin relies on philosophical and moral principles that are far from being confirmed by history, to support this transition to a new "ecological society." For Bookchin, these natural morals would unify the human race in a new (r)evolutionary enterprise. He ignores the power strategies and the mechanisms of ideological simulation and economic manipulation that impede the internalization of ecological conditions and principles of social equity and political difference in new forms of productive and social organization. Bookchin argues at length to distinguish his position from the bioethics of deep ecology, grounding his moral principles on an evolutionary consciousness that should guide the liberation process of humanity in its ecological era:

> The ecological principles that shaped organic societies re-emerge in the form of social principles to shape utopia [...] the re-emergence of "the People" [...] must deal with problems that are best defined as ethical, not merely economic. Only by a supreme act of consciousness and ethical probity can this society be changed fundamentally. (Bookchin 1971/1990, pp. 23, 41)

In a similar vein to Ivan Illich (1973), Bookchin envisions social ecology as a moral force capable of harnessing and adjusting technology to ecological conditions of conviviality and sustainability. Bookchin finds in Nature an ethical conception of society, rather than the fecundity and potentials of the ecological organization of Nature that gives support to

[2] Bookchin has reviewed his position on the general affluence of a post-scarcity society in the introduction to the second edition (1990) of his *Post-scarcity Society*. However, this does not change the underlying ethical and theoretical assumptions that guide his social ecology.

new *environmental rationality* and a new production paradigm founded on the negentropic potentialities of the biosphere and the ecological conditions of the planet.[3]

Bookchin idealizes utopia as a "total and complete revolution." He presupposes that the tension between potentiality and actuality "will continue to 'rise' until utopia itself is achieved [...] this lifestyle and the processes leading to it are indispensable in remaking the revolutionary, in awakening its sensibilities to how much must be changed if the revolution is to be complete" (ibid., pp. 17, 18). This view of utopia contrasts with the Trotskyist permanent revolutionary process and with the "structural" incompleteness of human desire (Lacan). Utopia as a purposeful political construction is rooted in the potentialities of Nature, technology, and society, but is accomplished through political theory and strategic action, rather than by simple actualization of the Real, particularly when the evolution of Nature has been intervened by the actual techno-scientific-economic rationality that governs the world. Instead, Bookchin stirs the sediments of dialectical thinking to articulate his messianic liberation rhetoric with a non-critical view of social change. His narrative of a classless, non-propertied and non-alienated society ignores the drives of human desire, the dialectics of reason and the will to power that constitute "human nature," as well as strategic theory and practice necessary to deconstruct the prevalent dominant world-system and to construct an alternative social order founded on the conditions of life—and of human life—in this living planet.

Bookchin views anarchism as a "libidinal upsurge of the people, a stirring of the social unconscious that reaches back [...] to the earliest struggles of humanity against domination and authority." Anarchism "links the remaking of society to the remaking of the psyche" (ibid., p. 21). By linking the traits of spontaneity in human nature with biological evolution, Bookchin blurs the specificity of human nature that drives the entropic degradation of the natural order in the biosphere and erases the obstacles that hinder the transition to an equitable and sustainable world order through symbolic and political strategies:

---

[3] I have developed the concept of *environmental rationality* in previous writings (Leff 1994, 1995, 2004, 2014, 2018).

> Belief in spontaneous action is part of a still larger belief—the belief in spontaneous development. Every development must be free to find its own equilibrium [… it] involves releasing the inner forces of development to see their authentic order and stability. (ibid., p. 23)

This "natural philosophy" is nothing more than wishful thinking. Its ecological ideals are nor founded on material grounds; its political intentions can only result in passivity, awaiting the forces of Nature to be actualized in society by natural self-consciousness and spontaneous evolution. This "dialectical naturalism" cannot explain the environmental crisis or develop an effective strategy for the reconstruction of a new "ecological" society.

Negating Natural Dialectics: Ecological Reductionism and Ontological Monism

To establish a unified theoretical field to view his evolutionary continuum and expand the workings of Nature (its dialectics?) to the realm of society and thought, Bookchin resorts to ontological monism. The categories of Nature and being; the distinction between the natural, the cultural, and the social lose their ontological and epistemological specificity. The "nature" of Nature, the "nature" of things and the "nature" of being, are all conflated, together with the productions of human beings (thinking, culture, and history) as "second nature." *Differance* (Derrida), the original disjunction of the realm of the Symbolic order from the Real, is just ignored in Bookchin's thinking.

Bookchin searches in dialectics the grounds from which to "derive second nature from first nature organically […] by using a mode of thought that distinguishes the phases of the evolutionary continuum from which second nature emerges and yet preserves first nature as part of the process" (Bookchin 1990b, p. 164). He finds in "natural dialectics" what other authors have thought through "complex thinking" as the development from the "self-organization of *physis*" to a "generalized ecology" (Morin 1977, 1980). Second nature thus appears as a "natural" evolution of Nature. This approach cannot explain the emergence of power and domination, of "hierarchy, patriarchy, classes and the State that unfolded with this development." It cannot grasp the "original difference" between the Real and the Symbolic, the "point of inflection and deflection" where Nature transforms into a cultural order, where the emergence of the symbolic order drives the potentialities of life into the *will to power* that dominates Nature.

Bookchin intends to recast the dialectics of Nature in the mold of ecology, to develop new "forms of reason that are organic and yet retain their critical qualities" (Bookchin 1990b, p. 11). He praises dialectics and tries to establish its superiority over analytic reason and systems theory.[4] However, his ecological rationality does not recognize that empirical social reality, as perceived by the subject, is the objectification of a determined world vision, a way of understanding beings that is all-ready a symbolic and social construction of reality. With its obsession for the objectivity of data, variables, and things, it loses sight of the fertility of the Real, its potentiality to become, to develop possibilities and options. It negates utopia as a prospective human project that generates social change from the potentialities of Nature.

Dialectical thought offers general principles to perceive the transformation (development) of the Real. However, for this logic to apprehend reality as concrete knowledge, there must be a correspondence between thought and the movement of material processes. Marx founded historical dialectics in social contradiction as the structural opposition of class interests. He viewed concreteness of the concept as the articulation of multiple determinations that renders reality intelligible to the mind. Thus, he reversed Hegel's dialectical idealism and established dialectical materialism. Here, dialectic is no longer a form of logic emerging from the mind and imposed on reality. Dialectic reason finds its referent in reality generated by social conflict and the contradictions of capitalism as a specific mode of production (Marx 1965).

Engels tried to extend the dialectics of historical materialism to dialectical materialism by rooting dialectics in the ontology and workings of

[4] "Conventional reason rests on identity, not change [...] What dialectical thinkers from Heraclitus onward had in common [...] is a view of reality as developmental—of Being as an ever-unfolding Becoming" (Bookchin 1990b, pp. 12, 13).

Nature. Beyond the ontological prevalence of being over-thinking, of praxis over theory, stated by Marx, Engels went further to root dialectical thought in the materiality of natural processes, adjusting the laws of Nature to the general principles of dialectics (Engels 1968). In this adventure, Engels surpassed the resources of metaphor and fell into rhetoric simulation. The general principles of dialectics—totality, negation, and contradiction; change of quantity into quality—can be mirrored by reality. However, these principles only represent a meta-theoretical approach; particular concepts and methods are still needed to apprehend the concrete causality, dynamics, and transformations of any specific material process (Althusser 1969). This is what the development of science intended to produce in the nineteenth century, from evolutionary biology to structural linguistics and historical materialism; from thermodynamics to psychoanalysis (Foucault 1966).

Bookchin rightly points out the misuse of dialectics in Hegelian idealism and Engels' materialism:

> Hegel rarefied it [dialectics] into a cosmological system that verged on theology by trying to reconcile it with idealism, absolute knowledge, and mystical unfolding logos [...] By the same token, dialectic was also intermingled with a crude materialism when Friedrich Engels dressed it up in the "laws" of dialectical materialism [...] Indeed, so enamored was Engels of matter and motion as the irreducible "attributes" of Being that in his work, kinetics based on mere motion tended to invade his dialectic of organic development. (Bookchin 1990b, p. 15)

Bookchin is right in his critique of Engels. However, the same can be said of himself; as so enamored he is with evolution and ecology as a model for the dialectics of Nature, that he transferred it to the social order.[5] Bookchin seeks to rescue dialectic thinking through its analogical parallelism with biological evolution and to establish a philosophy of Nature that can guide social action through rational and objective laws. The result is an organismic ontology and a generalized ecology that

[5] Thus, Bookchin admires "the extraordinary coherence dialectical reason offers and its extraordinary applicability to ecology—particularly to an ecology rooted in evolutionary development" (ibid., pp. 16).

contributes nothing to biological science or reconstruction of dialectics, giving false philosophical bases to the theory and praxis of environmentalism. Bookchin states:

> Dialectical naturalism can answer such questions as what is nature, what is humanity's place in nature, what is the thrust of natural evolution and what is society's relationship with the natural world [...dialectical naturalism] can bring coherence [...and] add an evolutionary perspective to ecological thinking—despite Hegel's rejection of natural evolution and Engel's recourse to the mechanistic evolutionary theories that were in vogue a century ago. (ibid., p. 16)

However, how could the general principles of dialectical reason give coherence and offer an evolutionary perspective to ecology? The evolutionary character of ecology stems from the ontology of life, from evolutionary biology and complex open systems, not from metaphysics. Undoubtedly, ecology can inform social organization to internalize the ecological conditions of sustainable production. However, this does not imply the necessity to "ecologize" human thinking and extend it as a general method to guide scientific research, social consciousness, and political action. Ecological approaches contribute to the analysis of complex emergent systems (García 1986, 1994; Funtowicz and Ravetz 1994). Nevertheless, the critique of prevailing mechanistic social theories does not authorize reducing the social order to an ecological system—the totalitarian project of a generalized ecology—and to found an "ecological society" on principles of "dialectical naturalism."

Dialectical naturalism is challenged today when the concept of Nature is being conceived as socially constructed and culturally mediated. Natural resources are defined by cultural meanings, by social interests and economic powers. Bookchin does not see the threads of power interwoven in the discourses of science, as well as in his narratives of ecological anarchism and dialectical naturalism. These are inscribed in the power strategies of knowledge that determine the theoretical and political fields of the environmental question and social appropriation of Nature.

Bookchin is in search of a philosophy of Nature to support the morals of "isonomy." In his view, this ethical approach has been corrupted by the predominance of epistemology (the condition for cognitive apprehension of the real) over a metaphysical reflection on the ontological conditions of being. However, the original and fundamental fault is that of the hegemony of the ontology of being in the history of metaphysics and its oblivion of the ontology of life: of diversity, difference, and otherness (Leff 2018). After Descartes, Kant, and Hegel, epistemology was founded as the inquiry on the conditions of knowledge. This inquiry progressed from the metaphysics of being and transcendental reason to a reflection on the social construction of reality. However, the renewal of metaphysics of Nature as the evolution from the self-organization of life to human consciousness results in an intranscendental idealism. It constitutes thought as the self-reflection of biological evolution in the self-consciousness of the knowing and acting subject. For Bookchin, ecological essentialism and natural morality are his tools to disarm the world dominated by scientific and technological reason.

Notwithstanding the effects of Cartesian dualisms on the alienation of culture from Nature and its impact on the environmental crisis, ontological monism founded on the "nature" of *physis* has been displaced by the unfolding of the *logos*, as history is perceived and constructed through language and culture, as life is being intervened by technology, and social reality depends on power strategies built from and upon the symbolic order that rules the world. Moreover, the subject of environmental consciousness has been decentered by psychoanalyses, finding the unconscious roots of desire and power in the "lack of being" of the subject. Nothing is less natural than the subject and its consciousness. The meaning of dialectics in psychoanalysis is the "negation" of organic life by the emergence of the symbolic unconscious order. Jacques Lacan used a controversial dialectical figure to state the place of truth and wisdom: "Truth is what which wisdom can find out about its knowledge only by acting on its ignorance" (Lacan 1976, p. 309). The subject cannot attain the absolute idea and total knowledge; he will not achieve its completion by the reflection of Nature on knowledge as the self-consciousness of Nature.

Dialectic and Totality. Ecology and Systems

Engels' dialectical materialism—in which he intended to unify thought and matter—did not survive the proof of history and critical reason. However, dialectical thought found fertile soil in systems theories and ecological thinking. The category of *totality* renewed the basis of the dialectical method after Marxist authors like Lukacs, Goldmann, and Kosik, who privileged its "revolutionary character" over principles of negation and contradiction:

> It is not the predominance of the economic motifs in the explanation of history that distinguish in a decisive manner Marxism from bourgeois science, but the standpoint of totality. The category of totality—the determinant and universal predominance of the whole over the parts—constitutes the essence of the method that Marx has taken from Hegel and transformed it to construct the original basis of an entirely new science [...] *The predominance of the category of totality is the support of the revolutionary principle in society*. (Lukács 1923, p. 39)

Totality became the Trojan horse through which the Absolute Idea was implanted in the land of dialectical materialism. With legitimization of systems theory as trans-disciplinary science in recent times, the category of totality ceased to be a novelty and lost its revolutionary flavor. Bookchin has rightly criticized general systems theory for its positivist approach and lack of ontological basis; instead, he posits ecology as the material ground of the self-organized process of life evolving dialectically toward self-completion and totality.

Undoubtedly, there are common traits among the evolution of natural ecosystems, complex systems, and the principle of totality in dialectical thought (i.e., the emergence of novelty). However, by subsuming dialectics in ecology, its method of thinking and argumentation and its fertility at the discursive level (the opposition of contraries, negation), Bookchin dissolves critical reason in the naturalness of biological evolution. Bookchin establishes a parallel between organic thought and dialectics that opposes analytic thinking and formal abstraction. Thus, organismic-dialectical thought would be superior to systems theory precisely because this approach can account for the differentiated material processes that are reduced by systems theory to their common analogical structures:

> The power of the West's organismic—more precisely dialectical—tradition (even at Hegel's highly conceptual level) lies *in building up* the difference of natural and social phenomena from what is implicit in their abstract level—not in [...] *reducing* their richly articulated concreteness to abstract, logically manipulative "data" [...] Dialectic [...] tries to *elicit* the development of phenomena from their level of abstract "homogeneity", latent with the rich differentiation that will mark their maturity, while systems theory tries to *reduce* phenomena from their highly articulated particularity to the level of homogeneous abstraction so necessary for mathematical symbolization. Dialectic [...] is a logic of evolution from abstraction toward differentiation; systems theory is a logic of devolution from differentiation toward abstraction. (Bookchin 1990b, p. 153)

I can agree with Bookchin's critique of the a-ontological, mechanistic and reductionist character of systems' theory. However, his praise for differentiation of material processes is not coherent with his ontological dialectical monism. Bookchin relies too heavily on natural dialectics to apprehend the specificity of the different processes emerging with the evolution of Nature, from physical matter to living organisms developing to the symbolic order. He overlooks that the differentiation of material processes derived from *physis* has been intervened by the human *logos*, generating the ontological regimes that drive the creative evolution of life toward environmental complexity and degradation (Leff 2000a, 2018). Moreover, grasping the core of these differentiated ontological regimes implies the necessity to derive their specific laws and regularities. These can only be apprehended by specific theoretical concepts, not through ecology and evolution biology as general meta-theoretical categories.

This problem leads to the question of integrating the different levels of materiality that constitute the real complex structure of the environment, and the articulation of sciences and other forms of symbolic appropriation of these processes.[6] The differentiation of matter and concept (the

[6] Even Marxist dialecticians like Kosik (1970) have yielded to systems theory—searching for the unity of science in the structural analogies of different material orders—and to genetic structuralism

only way of grasping the concreteness of the articulated processes that constitute the environment) cannot be reduced to the emergence of new traits, characters, and forms in the self-organization of living matter. The "evolution" of life entails the development of different forms of organization of matter that cannot be reduced to an ontological monism based on the generalization of ecological principles and the identity of matter and idea. The production of scientific concepts is necessary to apprehend the specificity of differentiated material orders. Such concepts cannot be reduced to general categories of dialectics nor subsumed in evolutionary biology as an organic totalitarian trans-disciplinary theory to unify the realm of Nature and thought. These ontological and epistemological principles are conditions to apprehend the articulation of the different orders of the Real and material processes: physical, biological, historical, and unconscious.[7]

(Goldmann 1959)—viewing the evolution of thought and scientific concepts as parallel to the emergence of complex matter. Thus, Kosik sees "the reunification of the real through the structural analogies between the most diverse domains of reality," to the extent to which "all the spheres of objective reality are systems or aggregates of elements that exert, ones on others, a reciprocal influence." He takes a critical epistemological stand against monist reductionism, by asserting that "only a dialectical conception of the ontological and epistemological aspects of a structure or a system can yield a fertile solution and avoid the extremes of mathematical formalism and metaphysical ontology [...] the structural analogies between the different forms of human relations (language, economics, kinship, etc.) can lead to deeper understanding and explanation of social reality, only if one respects both the structural analogies and the specificity of the phenomena under consideration" (Kosik 1970, p. 31). However, Kosik thinks of the ontological differentiation of matter and being as a hierarchy in the degree of complexity of different structures in the evolving transformation of concrete totality, in an approach that would integrate genetic structuralism and dialectical thought: "In dialectical thought, reality is conceived and represented by the totality, that is not only an ensemble of relations, of facts and processes but also their creation, their structure, and their genesis. The dialectic totality encompasses the creation of the ensemble and unity, the unity of contradictions and its genesis" (ibid., p. 34). Moreover, he asserts that the emergent complexity of matter is reflected in an evolutionary process of theoretical production. He alleges that to apprehend material processes of increasing complexity (physical matter, living systems, symbolic order), the categories that apply to the first levels (those of mechanics), serve as a first approach that can then be enriched by more elaborated logical categories.

[7] Contrary to this systemic and ecological view of society, the concept of environment is constructed as a complex category that articulates different material processes in natural and social orders with different symbolic—imaginary and conceptual—means of understanding. This category of environment opposes the tendencies to derive a general law to unite the distinct ontological orders of reality in a totalizing concept; it questions the possibility of finding a unifying principle in the organization of Nature and thinking that would be extended to culture and society. For my critique of general systems and ecological theories see my essays: On the articulation of sciences in Nature-society relations, and Interdisciplinarity and environment in Leff 1994, Chapters 1 and 2).

"Total concreteness" is an epistemological category that can be applied as a general methodological principle to different ontological orders. Concrete totality appears in thought as the concept that apprehends the synthesis of multiple determinations (the essential causes) of a material process. For the concept to be *concrete*, there must be an objective correspondence with reality. This concreteness does not emerge from facts; neither is it a reflection of Nature in subjective consciousness. Nature, matter, and being are organized as distinct ontological orders that do not have any objective self-conscious correspondence at the psychological level, as the phenomenological correspondence of *noesis* of the mind to a *noema* as the substance of the world out there. The subject has no "organic" self-consciousness of his internal "unconscious" processes. The concrete totality of these material orders appears in conceptual thought by constructing its objects of knowledge. This epistemological process gives meaning and value to the Real. It thus apprehends reality through the analogical and metaphorical correspondence between the concept and material processes, in specific ontological orders of the Real (Althusser 1969). This social production of knowledge cannot be reduced by ontological monism where the mind would discover its true, logical, essential nature in the reflection of an evolutionary biological process.

Bookchin criticizes the instrumental character of modern science, having lost its revolutionary contribution to social change through rational critique; nevertheless, he believes that ecology "may yet restore and even transcend the liberationist state of the traditional sciences and philosophies" (Bookchin 1971/1990, p. 80). He thinks that ecology "basically deals with the harmonization of nature and man." However, ecology emerges as a scientific discipline concerned with the relations between biological populations and the physical environment. Human ecology has extended the laws of natural evolution to the human order and tried to integrate ecological conditions for the reproduction and development of the social order. Ecology and general systems theory are not revolutionary because of their integrative approach and their claim to totality. Actually, a generalized ecology has been extended over the domains of history—of the symbolic and social order—without integrating the specific traits of human nature—power relations, social interests, human desires, cultural organization and economic rationality—which cannot be reduced to a generalized paradigm of complexity of the ecological order of the Real.

The will for identity and totality in ontological monism is being challenged today by critical structuralism and post-structural theories. As we view Nature and society from the perspective of culture and the symbolic order—of meaning and values; of the unconscious and desire—no totality can be attained. The *environment* can be conceptualized as a holistic socio-ecological structure internalizing the ecological bases of sustainability and the social conditions of equity and democracy. However, the principles and values that guide the reorganization of society as a new utopia of wholeness and totality appear as the "unfulfillable hole" that no holistic knowledge can satiate. This "lack in being" and "lack of knowledge" cannot be replenished by the progress of science, the power of technology, or the completion of organismic Nature in the consciousness of the people. This questions the search of truth as total reality or as the Hegelian Absolute Idea. The will for unity in Nature and totality in thinking—of "wholeness" (holiness) in knowledge—has enchanted and enchained human beings to an instrumental and homogeneous world, repressing the vitality of difference and diversity in the world order, in knowledge, and culture.

The project to found dialectics in an abstract (idealist) concept of totality, and the will to expand its domain of application to an all-encompassing order of Nature, matter and being, leads to metaphysical rhetoric instead of contributing to critical theory. Thus, dialectics must be reassessed for its contribution to critical knowledge in the construction of new environmental rationality and eco-communitarian society. Dialectics must be contrasted with the new methods of complexity to see what can its general principles offer today to science, critical thought, and social action. While the methods of complexity have opened new approaches to understand the articulation of material processes beyond the limits of specific scientific paradigms and instrumental reason, it preempts critical knowledge—the analysis of opposing social interests, ideological formations that orient the construction of an "ecological society." The fertility of discursive contradiction and conflicting interests that give substance to dialectical thinking, and the methods of complexity emerging from ecology and cybernetics, that define reality as the result of interrelations, interdependencies, interactions, and feedbacks, must include other imaginaries and knowledge/*savoirs* that configure the worldviews, habitus and practices of the people and guide the construction of their life-worlds.

If society is to be reorganized as a system of decentralized eco-communities, the transition to a new social order must be critically defined. While the politics of consensus is trying to adjust the interests of different social actors to a "common future" (WCED 1987) in the "new" unsustainable economic order, dialectic thought can reveal the opposing forces and interests in the appropriation of Nature. Complex systems provide a heuristic scheme to analyze interrelated processes that determine eco-social transformations, while dialectics as utopian thinking illuminates the endless path toward social emancipation—a permanent revolution in thought and social change—that mobilizes society for the construction of a sustainable and meaningful world and a new socio-ecological order.

The Monist/Dualist Dilemma: Sartre's Critique of Dialectical Naturalism

In *Critique of Dialectical Reason*, Sartre developed one of the more critical modern inquiries on dialectical thinking and established his position against dialectic naturalism. Sartre affirms the critical and historical character of dialectical reason. For him, dialectic is the form of thought that makes human praxis in history intelligible. However, he also places dialectics in its historical context, inquiring about the conditions that make this form of reason possible to apprehend the "historical real," reflecting critical reason to the historicity of dialectics. Sartre inquires about the foundations of dialectical reason, a question that was never solved, neither by historical materialism nor by dialectical materialism:

> Historical materialism has this paradoxical trait of being at the same time the only truth of History and a total *indetermination* of Truth. This totalitarian truth has founded everything but its own existence [...] Thus, Marxism appears to us, ideologues, as the unveiling of being and at the

same time as an interrogation that has remained at the stage of an unfulfilled demand. (Sartre 1960, p. 118)

This question is particularly relevant today when both the theoretical status of Marxism and its praxeological meaning to guide social-historical transformations is being challenged. If historical materialism is incapable of accounting for its truth, dialectical materialism has to reassess its ability to conduct both political thought and action. The inquiry on dialectical reason as a specific form of rationality and its methodological and cognitive relations with other critical discourses has become an unavoidable ontological, epistemological, and political question, one that is evaded by Bookchin's ontological monism posited as Nature's self-consciousness. Contrary to critical rationalists who demonstrated how modern physics founded new rationalism (Bachelard 1938, 1972), Bookchin does not reconstruct dialectic thought from the contributions of modern evolutionary biology. He replaces critical reason with biological evolutionism and offers ecological content to dialectics.[8]

Sartre addresses the ontological and epistemological questions involved in dialectical reason:

> Dialectics is a method *and* a movement in the object; it is founded [...] on the fundamental assertion that concerns both the structure of the real and our *praxis* [...] the process of knowledge is of dialectical order, and the movement of the object are *itself* dialectical, and these two dialectics make one [...] they define a world's rationality [...] Dialectical reason goes beyond the realm of methodology. It expresses [...] what the whole universe is: it is not limited to orient research, not even to prejudge the form of appearance of objects [...] *it defines the world as it should be for dialectic knowledge to be possible,* it shows at the same time [...] the movement of the real and that of our thinking. (Sartre 1960, p. 119, my emphasis)

[8] The "evolution" of theoretical concepts is not a progression in their adequacy to reality. As shown by critical materialist epistemology, mechanistic and organismic concepts function as epistemological obstacles (Bachelard 1938/1972) in the construction of concepts that correspond to the concrete organization of each material process and ontological order. Thus the mechanistic view of biological systems veils the substance of life (Canguilhem 1971, 1977); the extension of biological principles to human society veils the specificity of the historical and symbolic orders, of power, desire, culture, and knowledge (Lacan 1971; Foucault 1969).

Here Sartre exposes his Kantian lineage. The dialectic reason appears as *a priori* condition to know the world's dialectic essence. However, this "unity" of dialectics in history and knowledge does not imply a homogeneous movement in matter and mind, nor the reduction of both spheres to an ontological monism:

> The only possible unity of dialectics as a law of historical development and dialectics as the knowledge in movement of this development must be the unity of a dialectical movement. Being is the negation of knowing and knowing comes into being from the negation of being. (ibid., p. 131)

The "dialectical negation" between knowledge and being in the realm of history—the relationship between the method of knowledge and the movement of real social processes—is the sign of incompleteness of knowledge of the Real, of that fundamental original difference that drives history from the incomplete representation of the Real by Knowledge, but also by the way knowledge enacts the Real to create reality. In addition it implies the temporality of knowledge, the fact that thinking always runs behind being, that reality always supersedes the concept that intends to apprehend its essential truth. This condition of knowledge can be exemplified by the "fading out" of value theory in historical materialism. Labor-value theory is negated by the development of productive forces (technological change) generated by the reproduction of capitalist relations of production; it is challenged by the historical movement that eliminates the material basis over which the concept of value—socially necessary labor time—was founded as the source of surplus value and capital accumulation. This transformation of the conditions of production poses the necessity of new conceptual developments within historical materialism to grasp the concrete processes that determine capitalist production in its post-industrial and ecological phase (Baudrillard 1974; Leff 1980, 2000a).[9]

This dialectical process between theoretical concepts and historical change must be applied to all philosophical categories and the relation between ontological reality, critical reason and social praxis. Thus, "We

[9] See Chap. 5.

must ask what are the limits, the validity and the extension of dialectical reason" (Sartre 1960, p. 120). Sartre states that "praxis surpasses knowledge in its real efficacy" (ibid., p. 122); thus dialectical reason—either embedded in the evolution of matter and life or thought—cannot supersede the emergence of novelty from the emergence of life itself, from cultural innovation, from the transformative effect of knowledge and social praxis. The concept cannot predetermine the potentiality (indeterminate creativity) of life and praxis. That is the meaning of dialectics between environmental theory and the environmental movement that is inscribed in its essential difference in environmental rationality.

Bookchin has renewed the theoretical dispute between dualism and monism. However, with the reflection on the conditions of knowledge that developed with modern philosophy and science (after Kant and Hegel), we can no longer eradicate epistemology from the field of philosophical inquiry, as an unavoidable relationship between being and knowing, dissolving it in an ontological biological monism. The history of dialectical reason has been no other than the unsolved question of reconciling dialectics as a law of being and as a form of reason: of their meaning in the realm of Nature, thought and knowledge. Hegel answered this question with the Absolute Idea, where Being and Knowledge are unified, where thinking is at the same time constitutive and constituted. In Sartre's mind, this idealist position raises an epistemological contradiction:

> Is there not an insurmountable contradiction between the knowledge of being and being of knowledge? The mistake has been in trying to reconcile both, presenting thinking *as being* carried by the same movement of history as a whole [...] if knowledge is not the whole, then it would follow its development as an empirical succession of moments, and this experience will render what is experienced as contingency and not a necessity [...] nothing can authorize [knowledge] to decide that the movement of its object follows its movement [that of thought], nor that [thought] rules its movement by that of its object. If material beings, *praxis*, and knowledge are irreducible realities then wouldn't we have to appeal to a pre-established harmony to reconcile their developments? [...] If the research of truth must be dialectical in its approach, how could we give proof, without

idealism, that it rejoins the movement of Being; if contrary to this, knowledge must let Being develop by its laws, how to avoid that the processes—whichever they are—do not appear other than empirical? (ibid., p. 122)

Thus Sartre contests the original sin of metaphysics in its will to unify Thinking and Being, in the oblivion of the essential difference between the Real and the Symbolic. Marxist structuralism offered an answer to this question with the construction of objects of knowledge and the production of scientific concepts for differentiated spheres of being (Althusser 1969). The specific laws of these diverse material processes cannot be reduced to dialectical principles that would correspond to the general laws of Being. These theoretical concepts, in experimentation with reality, prove to "correspond" to the laws and regularities of specific material processes. Instead, Bookchin ignores the problem of knowledge posed by critical dialectical reason and turns to ontological monism. In his dialectical naturalism, Nature becomes self-conscious in an evolutionary process; it is extended to the realm of thought as epiphenomena of the biological organism, without critical reflection on the conditions of being, thought, and knowledge in different ontological regimes and the codes of significance and meaning of different existential ontologies.

Anticipating Bookchin's suppression of the difference between the Real and the Symbolic, Sartre stated that:

> Materialist monism has very happily suppressed dualism from the materiality of thought and total being. However, by so doing, it re-establishes as antinomical [...] dualism between Being and Truth [...] Strictly speaking there is no knowledge. Being is no longer manifest in any way: it evolves by its own laws [...] until it reaches its own self-consciousness without the critical reflection of thought, that until today has given its meaning to dialectics [...] When dialectical materialism seeks to establish a dialectic of nature, the result is not [...] a general synthesis of human knowledge but a simple organization of facts. (Sartre 1960, pp. 123–124)[10]

[10] Heidegger has fallen in a similar ontological monism trap when subsuming thinking to the truth of Being in his understanding of the *Ereignis das Seyn* (Heidegger 1999).

Bookchin's ontological monism and dialectical naturalism result in the self-consciousness of Being, without critical reflection on thought. By ecologizing dialectics, Bookchin falls in the same mistake as Engels. He thinks that the object of dialectical materialism is Nature (or society) as such: "it is in this very *human* rationality that nature ultimately actualized its evolution of subjectivity over long eons of neural and sensory development" (Bookchin 1990b, p. 161). The thought is reduced to a simple signifier "to grasp the idea as a signified thing by things and not as a meaningful act." The concept appears as the reflection of reality in consciousness, not as the production of meaning to apprehend and intervene the Real. Thus,

> The object of thought is Nature such as it is; the study of history is a specification: we must follow the movement that engenders life from matter, the man from the elementary forms of life, social history from the first human communities. This conception has the advantage of dissolving the problem: it presents dialectics as exterior: human nature would then reside outside him in a rule *a priori*, in an extra-human nature, in a history that starts with nebulae [...] everything draws back always to the totality of natural history where history is a specification [...] However, the absolute principle that "nature is dialectical" is not susceptible of being verified. (Sartre 1960, pp. 124–125)

By grounding dialectics in ecology, Bookchin negates dialectic's reason, to introduce it as moments of Being in biological evolution. He ignores the difference between the Real and the Symbolic, and thus the problem of knowledge, while trying to give theoretical foundations to his eco-anarchist discourse. He loses that point of reflexivity without which his narrative revolts around itself without connection between critical theory and social praxis to transform reality. Thus, the reduction of dialectic reason to ecological dialectics leads to passivity, awaiting evolution to actualize the potentialities of Nature in the ecological consciousness of people to dissolve the corruption of reality through history.[11]

---

[11] The man remains "Facing its history as uncertain as in front of nature: the law does not generate by itself knowledge of the law. In the contrary; if it is accepted passively, it transforms its object in

Sartre critiques a monism that imposes dialectical laws on Nature and founds its particular meaning in the movement of society:

> Engels reproaches Hegel imposing to matter the laws of thought. However, that is just what he does when he forces sciences to verify dialectical reason that he has discovered in the social world. Only in the historical and social world [...] do we *genuinely* deal with dialectical reason [...] if anything like dialectical reason exists, this is discovered and founded in and by human praxis to men situated in a specific society, in a particular moment of its development. From this discovery, we must establish the limits and validity of dialectical evidence: dialectics will be effective as a method insofar as it remains a necessary law for the intelligibility, and as the rational structure of being [...] it is at the interior of a society that has its tools and its institutions that we will discover the material facts—poverty or richness of the underground, climatic factors—that condition it [...] while dialectics of Nature cannot be the object but of a metaphysical hypothesis. The steps followed by the spirit that consists in discovering in praxis the rationality of dialectics, to project it as an unconditional law in the inorganic world and to come back from there to the societies claiming that the law of Nature, in its irrational opacity, conditions it, is an aberrant procedure [...] that replaces in the name of monism, the practical rationality of man making history. (ibid., pp. 128–129)

The transposition of social dialectics to the order of life was understandable at a time when science was still constructing its theoretical concepts to apprehend the emergence of life, genetic mutation, and biological evolution. Today Bookchin intends to give ecological content to dialectical thought. Dialectics would have found its lost ontological object. Instead of providing new foundations to historical materialism in a new environmental perspective, Bookchin drowns dialectical thinking in ecological evolutionism.

The paradox in Bookchin's monism is that theory appears as the reflection of natural evolution in thought, for the mind not to be separated from Nature so that ontology can encompass matter and mind. Then, the

passivity, it takes away the possibility of picking up its dust of experiences in a synthetic unity [...] if dialectical reason [...] must be, its *rationality* must offer the Reason of its reasons" (Sartre 1960, p. 127).

Real would become transparent in its expression in Thought. This biological idealism opposes any materialist principles of knowledge, where the dialectic of the concrete must be constructed through theoretical strategies to apprehend the Real, which is not manifest through the senses. That is the inescapable condition of human knowledge, where thought can liberate or corrupt reality through power strategies in knowledge. This delusive character of the concept is what explains even the failed theoretical project of Bookchin because of his epistemological *naïveté*. This domain of discourse and ideology in which any theoretical discourse is embedded, comprises the potential corruption of theory and reality, by the fatal strategies (Baudrillard 1983) that have emerged from the symbolic condition of human nature: the order in the unconscious (Freud, Lacan), the will to power (Nietzsche 1968), and power in knowledge (Foucault 1980).

There is nothing in the natural order that contains the germ of this perversion in the realm of thought; nothing in the realm of Nature can unravel the enigma of the production of meaning that mobilizes social change and the possibility of building a new social order that internalizes ecological principles in human morals, social organization, and sustainable production.

Dialectical Thinking, Political Ecology, and Environmental Rationality

Dialectical thinking as critical historical reason reached its highest point with Marx's historical materialism and its last expressions with the Frankfurt School theorists (Adorno, Horkheimer, Marcuse). The critical inquiry has been overtaken by the sciences, from psychoanalysis to bio-thermodynamics and complex systems. Today, the meaning of dialectical principles—totality, contradiction, and negation—is being challenged by the development of critical reason in these scientific theories. Dialectical thought, once the "method" of critical theory, must be applied to itself to reassess its value and applicability in the construction of critical environmental knowledge (Leff 1998a/2002, 2001). This is a necessary theoretical practice, as dialectical argumentation is being used more as a rhetorical discursive figure than as a concrete theoretical category.[12]

Bookchin relies on technology and the spontaneous development of natural dialectics to dissolve the tension between what exists and what is possible, and to guide the transition to a new social order. In his view, liberty results from the spontaneity of the potentials of technology: "the problems of 'transition' which occupied Marxists for nearly a century, are eliminated not only by the advance of technology but by the social dialectic itself. The problems of social reconstruction have been reduced to practical tasks that can be solved spontaneously by self-liberation acts of society" (Bookchin 1971/1990, p. 62). Thus Bookchin's anarchism yields to methodological individualism, unleashing the potentialities of technology and the powers of the rational (Janicaud 1994), and liberating spontaneity and pragmatism of social actors.

The revolutionary character of dialectical thinking is not contained in the potential for novelty and change inscribed in biological evolution; it is rooted in creativity derived from the negation of reality, the antithesis that generates the emergence of the new, the motor of a permanent revolution in human life.[13] Thus, dialectic appears as the metaphysics of history and of the historicity of the Real that resists the will to freeze history

[12] Bookchin's discourse is profuse with dialectical rhetoric rather than any rethinking of the ontological bases for a dialectical reason. For example, he states: "Just as abundance invades the unconscious to manipulate it, so the unconscious invades abundance to liberate it" (Bookchin 1971/1990, pp. 14, 62). However, the unconscious knows no abundance; it is organized from "lack in being" (Lacan), not from abundance of any sort. Moreover, further: "As we stand on the threshold of a post-scarcity society, the social dialectic begins to mature [...] What we must create to replace bourgeois society is not only the classless society envisioned by socialism, but the non-repressive utopia envisioned by anarchism" (ibid., pp. 15–16). However, liberation from repression depends more on a sound psychoanalytic and social theory, and an effective political strategy for the deconstruction of the domineering hegemonic world order and the construction of alternative environmental rationality, than on emancipatory rhetoric.

[13] In this sense, George Steiner has stated: "Creatively we endure thanks to our imperative capacity to say 'no' to reality, to construct fictions of alterity, of a dreamed 'otherness,' wished and longed for with the purpose that our consciousness might inhabit it" (Steiner 1975/2001, p. 15). Michel Foucault added: "It seems that there exists the possibility to make function fiction in truth; of inducing effects of truth with a discourse of fiction, in a way that the discourse creates, 'produces' something that doesn't yet exist [...] History is 'fictioned' from a political reality that makes it true, you 'fiction' a politic that doesn't yet exist from a historical reality" (Foucault, Power relations penetrate in the bodies, in *Microfísica del Poder*, 1979, p. 162). Thus Cornelius Castoriadis (1998) claimed the creative function of the imaginary institution as the creative force of human life.

and mold future developments in the rigid perspectives of science; that is, to proclaim the end of history to maintain dominating rationality. However, dialectics has also had controversial effects in historical materialism when viewing technology as the means to transcend alienation and necessity. Thus, for Bookchin the reign of liberty would be attained after undergoing the necessary phases of exploitation and domination arising from a material scarcity that would be dissolved by the development of social forces of production and the spontaneous self-realization of Nature's potentialities.

Dialectical naturalism drawing on the fulfilment of potentiality presupposes an intrinsic process in Nature toward totality and completeness, as the realization of truth in the self-consciousness of the human species. This idealist discourse founded in the principles of evolution dissolves dialectics as a critical reason. History is mobilized primarily not by the potentiality of the gene, the mutation of species, and the emergence of biological novelties, but by the conflict of social interests that are not reduced to class struggles in capitalism. Today, these antagonistic interests are expressed in the field of political ecology by socio-environmental conflicts for the territorialization of diverse modes of being-in-the-world and by opposing discursive strategies of sustainability, reflecting the social struggle for the appropriation of Nature (Leff 2014).

Facing the dominating tendencies of genetic structuralism and systems theories, Bookchin attempts to give ecological ground to dialectical thought and to guide eco-community practices inspired by Hegel's concept of *actualization*. Dialectical naturalism appears as an expression that synthesizes the dialectics between the potentiality of the Idea and the transformation of the Real. For Bookchin, the most important trait of dialectics is its eductive quality, which allows us to apprehend the potentiality of being. He stresses the property of "self-development [as] full *actualization* of potentiality in its rich, self-incorporative 'stages' of growth, differentiation, maturation, and wholeness [...which] are never so complete that they cease to be the potentiality for a still further development" (Bookchin 1990b, p. 167).

Bookchin's project to recast this dialectical principle in the mold of ecology leads him to ask:

> What are we obliged to modify in the dialectical philosophy […] to render it an ecological mode of thought? […] An ecological dialectic would have to deal with the fact that Aristotle and Hegel did not work with an evolutionary theory of nature […] as to replace the notion of *scala naturae* with the notion of a richly mediated continuum […] "Actuality", to use Hegel's special term, is almost the momentary culmination of maturity, so that the objectivity of the potential, which I hold to be crucial for developing a truly objective ethics, is subordinated to its actualization. By giving an emphatically historical priority to nature as the ground to intellectualisation, an ecological dialectic obliges us to reformulate Hegelian terms like the "real" and the "actual." (ibid., pp. 167–169)

Bookchin defines the real to mean "what is," not "what is necessarily latent in the potential." "Actuality" remains very much what Hegel meant it to be: the rational realization of the potential as distinguished from the "real as the existential" (ibid., p. 170). The Real is reduced to factual reality, while actuality replaces the concept of the Real as *physis*, as the actual material process in which the potential of becoming is embedded, the Real that is apprehended by its theoretical concept but is never fully grasped and controlled by any theory. Bookchin confounds the real/actuality with Real as becoming/reality as presence, thinking actuality as stemming from the evolutionary process of an ecological world.

Conversely, environmental rationality as thinking embedded in the ontology of life, in the potentialities of the Real, can be actualized—enacted—by social action, not by mere evolution. This social construction of the Real emerges from the realm of possibilities that are contained in the structure of different material processes and ontological regimes: ecological life-support systems, cultural meanings, social imaginaries, technological potentials, and political strategies. These processes define complex socio-environmental systems that are rendered intelligible by their specific scientific objects. The concept—synthesis of multiple determinations—expresses the Real as potentialities that can be enacted, potentially actualized and realized. Instead, Bookchin's dialectical naturalism, purified from contradiction and social conflict, appears as mere natural differentiation; actualization as self-consciousness of Nature:

> An ecological view of the dialectic would tilt dialectic philosophy toward differentiation rather than conflict and redefine progress to emphasize the role of social elaboration rather than social competition. (ibid., p. 170)

For Bookchin, the reign of necessity is transcended by a "growing self-consciousness and mutuality" (ibid.), rather than by a historical movement emerging from different and opposing interests. This projection of his natural ethics founded in ecology is covered by dialectics to legitimate his ideology as scientific and philosophical. Ecologism sweeps away history and culture, the social and symbolic order. Dialectic is recast in ecology following Engels' dream to view dialectic as the reflection of Nature in a more organic way.[14] Thus, Bookchin roots his eco-anarchism in natural philosophy and ecology.

Marcuse (1937/1967) stated: "When reason has been realized as the rational organization of mankind, philosophy is left without an object." Thus, dialectical materialism absorbed the function of critical reason to construct a more rational society. Bookchin criticizes the economic core of historical materialism and diverts toward an ecologically based society. However, if ecology explains the potential of rational social order, if consciousness is the self-reflection in the thought of ecological rationality, then philosophy and critical reason are left without an object.[15]

Human civilization is far from having reached the end of history or a post-scarcity stage. Modern society is evolving toward a homogeneous culture and global order ruled by a consensual democracy—the submission of diversity, difference, and otherness to the rationalization of techno-economic rationality—rather than by contradiction and class

[14] Ecological dialectic "is *development*, not mere 'change'; it is *derivative*, not mere 'motion'; it is *mediation*, not mere 'process'; and it is *cumulative*, not a mere continuum" (ibid., p. 170).

[15] Lucien Goldmann viewed the function of dialectical thought in the relationship between utopian thinking and the transformation of social reality through the concept of meaningful structures: "when we deal with human sciences and above all with history and culture, the principal concept of intelligibility, that of *meaningful structure*, represents at the same time a reality and a norm, precisely because it defines at the same time the real motor and the end toward which this totality is geared, which is humane society [...] We should not suppose that nature evolves progressively toward legal, geometrical or causal structures; however, the hypothesis of a history dominated by the tendencies toward increasing meaningful and coherent structures, to reach a finale and transparent society composed only by such structures, is one of the principal *positive* hypothesis in the study of human realities" (Goldmann 1959, p. 111).

struggle. Politics of conviviality in diversity tends to eradicate difference as opposition and negation of the other. However, this progress toward more peaceful means of solving the socio-environmental conflicts emerging from the environmental crisis does not authorize an organismic vision of society. Reducing history to a general evolving ecological order neglects the fact that social change occurs by conflictive interests and antagonistic political forces. Contrasting positions mobilize the *environmental question* in the field of political ecology, where a dialogue of knowledge/*savoirs* open the paths to construct a sustainable world through confronted interests in the social appropriation of Nature (Redclift 1987; Martínez-Alier 1995b; Leff 2003, 2004).

If "ecological dialectics" cannot offer a renewed basis for the understanding of historical change and to ground the construction of a sustainable world order; if the techno-economic rationality that rules the world negates the underlying ecological order, to what extent can historical materialism—the science of capital founded on the theory of value—account for the environmental crisis and the transition toward sustainability? The environmental crisis has emerged as the expression of a "second contradiction of capital" (O'Connor 1991) responsible for the degradation of Nature and the ecological conditions of sustainable and equitable development. Ecology has emerged as the negated condition of production, providing a new material basis to construct an alternative (more organic and equitable) social order; it opens new horizons and possibilities to gear social change under principles of environmental rationality. In this perspective, Nature must regain its place in the production process as ecological potential and a fundamental condition for the sustainability of any mode of production. Nevertheless, the social order cannot be reduced to the biological organism, nor power and knowledge subsumed in the laws of evolution.

Critical environmental theory has set an end to dialectical naturalism and opened a new inquiry on the ecological basis for sustainable production. This theoretical shift leads us to scrutinize the ontology of historical materialism in general, and the epistemic structure of the labor theory of value in particular, to unground the conditions for life's sustainability.

Bibliography

Althusser, L. (1969), *For Marx*, London, Allen Lane.
Bachelard, G. (1938/1972), *La formation de l'esprit scientifique*, Paris, Librairie Philosophique J. Vrin.
———. (1972), *L'engagement rationaliste*, París, Presses Universitaires de France.
Baudrillard, J. (1974), *Crítica de la economía política del signo*, Mexico, Siglo XXI Editores.
———. (1983), *Les stratégies fatales*, Paris: Éditions Grasset & Frasquelle.
Bookchin, M. (1971/1990), *Post-scarcity anarchism*, Montreal/New York, Black Rose Books, Second edition, 1990.
———. (1990a), *Remaking society: Pathways to a green future*, Boston: South End Press.
———. (1990b), *The philosophy of social ecology. Essays on dialectical naturalism*, Montreal/New York: Black Rose Books.
Canguilhem, G. (1971), *La connaissance de la vie*, Paris: J. Vrin.
———. (1977), *Idéologie et rationalité dans l'histoire des sciences de la vie.* Paris: J. Vrin.
Castoriadis, C. (1998), *The imaginary institution of society*, Cambridge: MIT Press.
Engels, F. (1968), *Dialectique de la nature*, Paris: Editions Sociales.
Foucault, M. (1966), *Les mots et les choses*, París, Gallimard.
———. (1969), *L'archéologie du savoir*, Paris, Gallimard.
———. (1980), *Power/knowledge*, New York: Pantheon.
Funtowicz, S. & J. Ravetz (1994), "Emergent complex systems", *Futures*, 26 (6): 568–582.
García, R. (1986), "Conceptos básicos para el estudio de sistemas complejos", in E. Leff (Coordinador) (1986), *Los problemas del conocimiento y la perspectiva ambiental del desarrollo*, Siglo XXI Editores, México.
———. (1994), "Interdisciplinariedad y Sistemas Complejos", in E. Leff (Coordinador), *Ciencias sociales y formación ambiental*, GEDISA/UNAM/PNUMA, Barcelona.
Goldmann, L. (1959), *Recherches dialectiques*, Paris, Gallimard.
Harvey, D. (2003), *The new imperialism*, Oxford: Oxford University Press.
Heidegger, M. (1999), *Contributions to philosophy (from enowning),* Translated by Parvis Amad & Kenneth Maly, Bloomimngton & Indianapolis: Indiana University Press.
Illich, I. (1973), *Tools for conviviality*, New York: Harper & Row.

Janicaud, D. (1994), *Powers of the rational: science, technology and the future of thought*, Indiana: Indiana University Press.
Kosik, K. (1970), *La dialectique du concret*, Paris, François Maspero.
Lacan, J. (1971), "Subversión del sujeto y dialéctica del deseo en el inconsciente freudiano", *Escritos*, México, Siglo XXI Eds., Third edition, 1976.
———. (1976), "Subversión del sujeto y dialéctica del deseo en el inconsciente freudiano", Escritos, México, Siglo XXI.
Leff, E. (1980), "La teoría del valor en Marx frente a la revolución científico-tecnológica", en Leff, E. (Editor), *Teoría del valor*, México: UNAM.
———. (1994), *Ecología y capital: racionalidad ambiental, democracia participativa y desarrollo sustentable*, México: Siglo XXI Editores.
———. (1995), *Green production: towards an environmental rationality*, New York: Guilford.
———. (1998a), "Murray Bookchin and the end of dialectical naturalism", *Capitalism, nature, socialism*, Vol. 9 (4):67–93.
———. (2000a), "Pensar la complejidad ambiental", en Leff, E. (Coord.), *La complejidad ambiental*, Colección "Aprender a aprender", México: Siglo XXI Editores (2nd revised edition, 2003).
———. (2001), *Epistemología ambiental*, Sao Paulo: Cortez Editora.
———. (2003), "La ecología política en América Latina: un campo en construcción", *Polis*, Revista de la Universidad Bolivariana, II (5) 2003, Santiago de Chile, pp. 125–145.
———. (2004), *Racionalidad ambiental: la apropiación social de la naturaleza*. México: Siglo XXI Editores.
———. (2014), *La apuesta por la vida: imaginarios sociales e imaginación sociológica en los territorios del Sur*, México: Siglo XXI Editores.
———. (2018), *El fuego de la vida: Heidegger ante la cuestión ambiental*, Mexico: Siglo XXI Editores.
Lukács, G. (1923/1971), *History and class consciousness*, Merlin Press.
Marcuse, H. (1937), "Philosophy and critical theory", in *Negations*, USA/Great Britain/Australia, Penguin Books, 1968.
Martínez-Alier, J. (1995b), "Political ecology, distributional conflicts and economic incommensurability", *New left review* I/211.
Marx, K. (1965), *Œuvres, Économie I*, Paris: Gallimard.
Morin, E. (1977), *La méthode. La nature de la nature*, Paris, Editions du Seuil.
———. (1980), *La méthode. La vie de la vie*, Paris, Editions du Seuil.
Nietzsche, F. (1968), *The will to power*, New York: Vintage Books.
O'Connor, M. (1991), "Entropy, structure and organizational change", *Ecological economics* 3:95–122.

Redclift, M. (1987), *Sustainable development: exploring the contradictions*, London: Methuen.
Sartre, J. P. (1960), *Critique de la raison dialectique*, París, Gallimard.
Steiner, G. (1975/2001), *Después de Babel*, México: Fondo de Cultura Económica, 3a edición.
WCED (1987), *Our common future*, Oxford: Oxford University Press.

5

Marx's Theory of Value, Technological Change and the Forces of Nature

Introduction

Marx's value theory (the labor theory of value) is at the core of historical materialism, as the fundamental concept of the social structure that explains the organizing mode of production and the mobilizing force of the capitalist economic process. While the theory of value cannot be considered to be the "essence" of Marxist thought, it is one of its basic cornerstones: it condenses the social structure that forges the motor of history in social ground once the social relations of production that oppose capital to labor were established, turning political economy into an objective science, based on the concept of socially necessary labor time. In the work of Adam Smith, the theory of value was still caught within the mesh of representations and similarities typical of the classical episteme, which led to tautologies regarding the relationship between labor and commodities (Foucault 1966). David Ricardo conceived of labor as the generating principle of value, but the value was calculated in terms of salary or model commodities. With Marx, socially necessary labor emerges as the basis for value, as the structural and quantitative principle that allows us to attain objective knowledge of the capitalist process as the historical transformation of the conditions of production,

E. Leff, *Political Ecology*, https://doi.org/10.1007/978-3-030-63325-7_5

grounded on the ontological regime of the technique (Heidegger 1935/1977), that mobilizes and in turn is supported by technical change. The theory of value—subordinated in historical materialism to class struggle as the "motor of history"—is the scientific basis in Marx's understanding of capital accumulation and the development of its productive forces (Althusser et al. 1965/2016).

Value Marxist Theory: Technical Reduction to Simple-Direct labor as a Quantum of Value

In Marx's structural epistemology, "the concrete is concrete by virtue of being the synthesis of multiple determinations." The theory of value condenses the core of interdeterminations of complex factors and processes within the capitalist social structure that drives the development of the forces of production and which, in its constant interaction with the social relations of production and class struggle, leads to the transformation of the social structure from which it derives its laws.

For Marx, all commodities are the result of specific acts of labor that transform distinct objects of labor (raw materials, natural resources, energy sources) to produce a variety of interchangeable use values that have a particular utility for human needs. What gives these different acts of labor the same unit of measure is that they can be reduced to human energy, to "muscles, nerves, and brain." Productive labor is, for Marx, simple and direct labor force. It generally results from manual efforts upon the objects of production leading to their material transformation. The generalization of this type of labor arises from technical progress itself, which through industrial development transforms different forms of labor (in terms of diversity of movements and complexity of physical energy and mental processes) reducing them to simple and repetitive manual labor. The determination that makes labor time into a unit on which one can establish the equivalencies of commodity exchange is the result of the historical movement that has generated technical progress capable of producing the empirical principle of a quantitative theory of the dynamics of the capitalist mode of production. In this sense, Marx stated in his early writings:

> The use of the quantity of labor as the only measure of value without considering its quality supposes that [...] labor has been equalized by the subordination of man to machines or through the extreme division of labor.[1]

The theoretical basis for the theory of value lies in the concept of *abstract labor*. In Marxist theory, this concept is the result of an epistemological breakthrough and a theoretical elaboration leading to the construction of the concept of the social structure and the productive nucleus of economic phenomena. However, at the same time, it is the result of a historical process that makes simple and direct labor the productive principle of value. In this way, Marx avoids both the methodological individualism of vulgar economics and the rationalist idealism that derives historical reality from thought. Thus he stated:

> This abstraction of labor in general is not the mental outcome of a concrete totality of labor [...] labor has been converted, not only as a category but also as reality itself, into a means for producing wealth in general.[2]

This implies that within Marxist thought, the general categories of historical materialism—mode of production, social formation, social relations of production, development of the forces of production—coexist with temporal concepts such as those that constitute the theory of value—abstract labor and socially necessary labor as the principles of capital accumulation—whose temporal Nature depends on the transformation of the reality to which they correspond. Therefore, Marx considered,

---

[1] Marx, "Misère de la philosophie," in *Oeuvres Economie,* vol. I, Paris: Gallimard, 1965, pp. 28–29.
[2] Marx, "Introduction Générale à la Critique de l'Économie Politique," in *Oeuvres Economie,* I, p. 259.

> Despite their validity for all times, the most abstract categories are within their abstract determination, the product of historical conditions, and retain their full validity only for them and within their limit.[3]

Marx thus attacks the fetishist empiricism of vulgar economics and proposes a quantitative science based on the empirical conditions produced by history, those that emerge from the social relations of production:

> Exchange value appears thus as the natural determination of use values in society, as a determination that concerns them as things and based on which they substitute one another in the exchange process according to determined quantitative relations; they form equivalents just as pure chemical substances combine according to determined relations and form chemical equivalencies.[4]

Thus, Marx constitutes political economy as a quantitative social science with the same epistemological legitimacy as the natural sciences. Although abstract labor in its empirical manifestation as simple and direct labor is the source of all value, in fact, its quantitative determination does not emerge from the application of indeterminate labor time. Abstract labor that produces value is *socially necessary labor time*. The "social" and "necessary" Nature of labor supposes that use values produced by a given labor time represent a real "utility" in a commodity market. However, also, it implies that labor time that determines exchange value depends on the development of productive labor forces, whose productivity is modified by technology.[5] In this regard, we must analyze how technical change affects socially necessary labor.

Once industrial development reduces all labor to the application of simple and direct movements, each step of technical progress imposes

[3] Marx, "Introduction Générale à la Critique de l'Économie Politique," in *Oeuvres Economie*, I, pp. 259–60.

[4] Marx, "Introduction Générale à la Critique de l'Économie Politique," in *Oeuvres Economie*, I, p. 285.

[5] In this sense, Marx wrote: "By an increase in labor power or productivity of labor we generally mean a change in the processes that reduce the socially necessary time for the production of a commodity such that a lesser quantity of labor acquires the power to produce more use values," *Le Capital*, in *Oeuvres Economie*, I, p. 852.

certain average conditions of intensity on the application of labor power so that equal values are produced in equal amounts of time. Thus the value contained in any commodity is weighed by the average intensity required in its manufacture.[6] However, at the same time, technical progress introduces changes in labor productivity. Therefore a fundamental theoretical problem arises to establish the amount of socially necessary labor that results from technical innovation.

The Indeterminacy due to Technological Change of Social Labor Time

The first problem to consider once technological change effects in a differential manner the labor time needed to produce commodities in different industrial branches, is to determine which technical conditions define socially necessary labor in any capitalist formation in which technology is heterogeneous, and labor productivity is differentiated, not only among different productive spheres but also within one industry producing one specific use value. There is a lack of consistency along Marx's theoretical construction regarding this question: in some passages socially necessary labor is determined by the most productive techniques, while in others the measure of value is set by the average technical conditions at a given point.

When Marx analyzed the effect of the introduction of the steam engine in textile production, he considered that use values produced in technically inferior conditions reduced its value content; the labor time that produced them, and that continues to produce them, was devalued because "the product of its individual labor-hour represents no more than half a social labor-hour and gives no more than half its value."[7] Nevertheless, in other passages of *Capital*, Marx attributes the

[6] Marx explains: "All individual labor power is equal to all others in that it has the character of an average social force and functions as such (so that) in commodity production it does not use more than the average necessary time or the socially necessary time." in *Le Capital, Oeuvres Economie*, I, p. 566.

[7] Marx, *Le Capital*, I, in *Oeuvres Economie*, I, p. 566.

establishment of socially necessary labor to average technical conditions, not to the most productive techniques:

> The individual value of each piece, produced in (technically) exceptional conditions, will fall below the social value [...] Now, the value of an article is not its individual value but its social value, and this is determined by the time it costs, not in an individual case, but on the average.[8]

Regardless of which interpretation of socially necessary labor he opts for, the individual capitalist who introduces new technology reduces the labor time necessary to produce his commodities, leading to higher rent and relative profits over his competitors during the period while the diffusion of the technology throughout industry generalizes the increase of relative surplus value. However, this is not a satisfactory theoretical criterion for determining how much socially necessary labor is the determinant of value. If the most productive technology establishes socially necessary labor, then its consequence is the devaluation of commodities produced in inferior technical conditions. However, if average technical conditions fix socially necessary labor, then it depends both on the process of technological innovation and its diffusion on the industrial branch that make up the forces of production applied to a given product.

The clarification of this theoretical problem is fundamental in assessing the progressive elimination of the law of value. If the most productive technology determines socially necessary labor, then the innovation of automated technologies in any branch of industry would devalue all the articles produced mechanically or manually. However, if the average technical conditions determine socially necessary labor, then the diminishing effect of the law of value would depend on the degree of automation of the productive processes until its full generalization.

Marx sought to solve this theoretical problem by postulating that "labor for exceptional productivity counts as *complex labor,* or it creates, in a given time, more value than average social labor of the same type."[9] This theoretical subterfuge presents two problems: On the one hand,

[8] Marx, *Le Capital,* I, in *Oeuvres Economie,* I, p. 854.

[9] Marx, *Le Capital,* I, in *Oeuvres Economie,* I, p. 856.

there is no indication that all technical progress increasing labor productivity implies at the same time complex labor, in which case the empirical conditions that allowed Marx to posit simple and direct labor as the determinant of value formation would disappear with technological development. On the other hand, only the reduction of complex labor to simple labor would permit an evaluation of the quantity of value produced. Nevertheless, to paraphrase Maximilien Rubel, "the reduction of complex labor to simple labor is not based on experience, contrary to what Marx wrote in the *Critique ...* and in *Capital.* The laws governing this reduction were never formulated in *Capital.*"[10]

Thus, socially necessary labor, as the empirical and quantitative determinant of value formation, becomes an abstract principle whose effects are objectified through demand in market prices, establishing the labor time necessary for producing each commodity. Thus the competition of capitals in the market reflects the simple and quantitative unit in which the variable value of commodities derived from diverse productive processes in different times and technical conditions, affects the labor force's productivity.[11] Marx states that

> The law of value determines how much of its available time society can spend on the production of each type of commodity. In the workshop division of manufacturing, the proportional number set primarily through practice, and later through reflection, as a rule, determines *a priori* the mass of workers assigned to each particular function; in the social division of labor it acts *a posteriori* as a fatal, hidden, mute necessity, visible only in the barometric variations of market prices, which dominate and prevail over [...] the irregular arbitrariness of commodity producers.[12]

Consequently, with the development of productive forces and division of labor, the quantitative principle of the capitalist economic process becomes an invisible essence, perceptible only through its effects on the

[10] Marx, "Notes, *Le Capital,*" in *Oeuvres Economie*, I, p. 566, p. 1636.

[11] In this vein, Marx added: "To apply a similar measure we must have a comparative scale of different work days; it is competition which establishes this scale" (Marx, "Misère de la Philosophie," in *Oeuvres Economie*, I, p. 28.

[12] In Book III of *Capital,* Marx will say that values "are disguised behind production prices, and in the final analysis they determine them" (Marx, *Le Capital,* in *Oeuvres Economie*, I, p. 1592.

movement of market prices. Political economy, like any other science, becomes based on concepts that represent the hidden structure of the matter that determines and regulates its empirical manifestations—as the unconscious in psychoanalysis or the structure of atomic nuclei in physics. The epistemological peculiarity of Marxist science lies in this transformation of a principle that is at the same time theoretical and empirical, where direct and simple labor time produced in a given historical moment progressively loses its empirical support with economic progress. This has an adverse effect on the consistency and primacy of the law of value, which in the first moment over-determines the laws of supply and demand by generating the substance around which market prices are brought into balance, subordinating its theoretical hierarchy in a second moment, becoming an effect determined by individual capital competition in the market and the laws of supply and demand.

It is, therefore, necessary to analyze the theoretical implications of considering the law of supply and demand as that which over-determines value formation. Marx indicates that in order for a specific labor time to produce value it must at the same time produce use value, a demandable good. In this sense, all commodities for which there is no demand automatically lose their value. Nevertheless, the fundamental question that arises from this problem is whether or not demand can be established independently of the law of value.

The Theory of Surplus Value and the Laws of Supply and Demand

What Marx contributed to the theory of supply and demand is the fact that both are products of the very dynamic of capitalist accumulation and not of the free play of productive market factors or a subjective principle based on human desires, needs or preferences based of the principle of "rational choice." Conversely, in Marxist theory, the laws of value and surplus value determine both the supply and the demand for commodities. The development of the forces of production as a result of capitalist competition and the search for new investment sectors for the recapitalization of surplus value determine the quantity and the diversity of supply. This process modifies the structures of demand, employment and income distribution such that surplus value can be realized through the exchange of commodities and be newly invested in ensuring the expanded reproduction of capital.

With the theory of surplus value, Marx demonstrated that the determinant law of supply and demand is not reduced to an economic problem, such as the free play of factors of production, because it is class struggle within capitalism's social structure that determines the quantum of surplus value from the exploitation of labor force. With the law of value, he sought a quantitative measure of the economic process produced through this structure, not merely through the interplay of economic categories such as wages, income, production costs and distribution of wealth. For these reasons, the laws of supply and demand, although they can later destroy a given quantity of value constituted by the application of labor time, they cannot determine the constituent unit of value that depends on technological progress and class struggle. As long as technological change is a process not directly dependent on the laws of supply and demand, these laws cannot be the determinant principle of socially necessary labor and value formation.

The determination imparted by technical conditions upon socially necessary labor time arises again when Marx treats the concept of relative surplus value. Technological progress appears here as a process determined by the very dynamic of capitalist accumulation, which allows the production of growing relative surplus value from the workforce once proletarian struggles have limited the possibility of increasing absolute surplus value through the extension of the work day.

Increased productivity in industries that produce wage goods reduces socially necessary labor time for producing the input required to reproduce the labor force. By decreasing the value of the labor force, the necessary labor time for its reproduction is reduced and, all else being the same, the capitalist can appropriate a more significant part of the value produced during the day work. In this way, the production of relative surplus value (through the reduction of the cost of wages or necessary labor time) is necessarily connected to the effect that the reduction of socially necessary labor time has on the production of value. By devaluing capital and the commodities it produces, technological progress increases the relative surplus value extracted from the labor force, thus overcoming the law of diminishing returns.

Both processes lead to increased profits for an innovating capitalist. Nevertheless, for theoretical purposes, we should separate the increase in earnings produced through the increase in relative surplus value from that produced through the devaluation that arises when one capitalist introduces a technological innovation. Marx conflates both processes when he assumes that

> The capitalist who uses a perfected technique appropriates in the form of surplus labor a greater part of the work day than his competitors. On a small scale, he does what capital does on a large scale and in general in the production of relative surplus value.[13]

Thus, seeking a solution to the unresolved relationship between technological progress, socially necessary labor time and value formation, Marx resorts to the theory of surplus value. This theoretical turn arises from conflating the concepts of socially necessary labor time and necessary labor time. A capitalist who uses a perfected technique devalues the commodities produced by his competitors with less productive means. However, this does not allow him to appropriate more surplus-labor, because the necessary labor time is reduced only with the generalization of technical change in the production of wage goods. Marx himself said: "this extra surplus value disappears when the new technique is generalized."[14] While the innovative capitalist will make more profits as long as his technical innovation is not disseminated, relative surplus value is produced through the generalization of increased productivity of wage goods, that is, a decline in the value content and price of the average consumption basket. By blurring the specificity of these two processes, Marx conflated the theory of value and the theory of surplus value.

[13] Marx, *Le Capital, in Oeuvres Economie*, I, p. 856.

[14] Marx, *Le Capital*, in *Oeuvres Economie*, I, p. 856.

Surplus Value and the Organic Composition of Capital

The problem of calculating socially necessary labor time or the amount of value in a commodity gets even more confused when one considers not only that value is a product of the living and direct labor that the machine extracts from the worker; but also that all commodities contain a proportional part of the value contained within fixed capital, that is, raw materials, intermediate goods, and equipment consumed in the production of any given use value.[15] Raw materials and intermediate goods that are part of the composition of a new product transfer to it their original value, which is added to that produced by the socially necessary labor in the productive process; its value is affected, like that of any other commodity, by the changes produced by technological progress in the amount of socially necessary labor time necessary to produce them.

The case of machinery and equipment is different, because the value they transfer to the product depends not only on the value they contain but also on the rate of utilization and the period during which they conserve their productive function before being replaced by more productive capital. Marx assumes that "the time of capital reproduction corresponds to the time necessary for its consumption,"[16] so two techniques that contain the same amount of value but different durability because of their material constitution as use values transmit the same value to the product. If the organic composition of this capital is proportional to its durability, both techniques will produce the same surplus value, which allows their recapitalization once the productive life of physical capital is completed. This could be a reasonable hypothesis for an abstract theory of capital, but in reality, competition means that replacing one type of equipment with a more productive one depends on a balance between the cost and production rhythm of technological innovation as well as on the profits derived from monopoly over more productive technology, versus losses from quickly reinvesting capital. Therefore, the time it takes to reproduce capital—its economic obsolescence and its revaluation in the

[15] We can see in this process the "productive consumption of nature."

[16] Marx, *Grundrisse*, vol. 3, Paris: Ed. Anthropos, 1968, p. 305.

form of technological innovation—does not correspond to its natural productive lifetime.

If equipment is replaced through technological innovation before its natural wearing out is completed, this does not logically imply that the value is transmitted to the commodities that it produced during its useful life was equal to the total value transmitted by equipment functioning during the largest share of its natural wearing out with a competitor incapable of introducing technological innovations. The value transmitted by a machine to the commodities it produces depends not only on its value but also on the time it takes to produce and incorporate a technological innovation that will determine the useful time of value transmission, which is not the same as the "normal" operating time of the machinery in question. In any case, whether due to market competition or to the process of economic obsolescence and technological innovation, the law of value faces indetermination. The proportional part of the value transferred by any equipment to the commodities it produces does not depend only on the amount of time it functions as a result of market competition. It is also affected by technological innovation that is not quantitatively determined by the amount of manual or intellectual labor applied in scientific discovery and technological development. In this sense, Marx writes:

> The incessant progress of science and technology bestows upon capital a potential for expansion that is independent, within certain limits, of the magnitude of wealth of which they are composed [...] The progress of the productive potential of labor that is generated without the contribution of the capital operating at that time, but that it benefits from when it changes its skin, lowers its value more or less during the interval of time during which it continues working under its old form.[17]

As soon as capital accumulation applies scientific-technological progress as a requisite of capital's expanded reproduction, it is impossible to calculate the value contained in the capital incorporated within new technology or the value transmitted by the commodities it produces. The

[17] Marx, *Le Capital*, I, in *Oeuvres Economie*, I, p. 1112.

introduction of these new means of production devalues the machinery and equipment still operating as well as the value of the commodities they produce. So the value transmitted to a commodity by the old equipment depends not only on the labor time it contains and which it extracts from the labor force. The value of the machine at the moment when the new technology appears on the market cannot be re-evaluated based on the socially necessary labor time for producing the new machinery, but rather on its *productivity*, which has no quantitative relationship to labor cost or labor time necessary for its production. Its value is impossible to calculate being that it no longer depends on direct, living labor time applied in the production of consumer and production goods. Its value depends instead on intellectual labor, which cannot be reduced to direct, simple labor without inverting the over-determination of the law of value on the law of supply and demand. This "paradox" lead Marx to analyze value formation in the capital circulation process and value realization, in the perspective of the expanded reproduction of capital and the scientific-technological revolution.

Capital Expanded Reproduction, Technological Innovation and Value Formation

Following our argumentation, a technological change driven by capital accumulation has generated an increasing indetermination in the concept of value as formulated by Marx in the first book of *Capital*. The second and third books, devoted to the analysis of capital circulation and competition—published by Engels from Marx's drafts—do not reconstruct the concept of value; however, capital turnover cycles and competition among individual capitals introduce new conditions and determinations to capital formation and realization—the revaluing of value—that affect the consistency of its concept to encompass the different moments of the capital reproduction process. In this perspective, problems already raised with the concept of value extend to consider the confluence of the moments of production, circulation and revaluing of value—the reinvestment of realized value in new elements of the productive process—opening the question of conceptualizing intellectual labor as productive labor in scientific-technologic innovation.

For Marx, all labor applied in the circulation process is non-productive labor, meaning that it does not accrue value to commodities. Thus, the "determination" from the circulation process to value formation would seem to be limited to bring to its moment of "realization" the socially necessary labor time to produce any commodity by being consumed, as the accomplishment of the effective demand of a quantum of abstract labor embodied in the productive process. Nevertheless, circulation time conditions capital's reproductive dynamics over the mass of already formed value in a determined time, over the exploitation rates of the labor force over the capital's realization crisis.

The reduction of circulation time and the acceleration of the turnover of capital-money time increase the mass of surplus value, even when the exploitation rate remains invariant, broadening capital's space for its expansion process. The reduction of circulation time generates a greater quantity of productive capital available, capable of absorbing more labor force in a determinate time horizon, generating a larger quantity of value and surplus value. Thus, Marx states that

> A difference in the turnover period produces a difference in the annual surplus value rate, even if the mass of surplus value produced remains constant. However, besides that, there are differences in the capitalization and accumulation of surplus value, as well as in the quantity of surplus value produced during the year, when the rate of surplus value remains constant.[18]

To be sure, the rhythm for revaluing the realized value through consumption depends on the quantity of capital-money available for reinvestment in new production means; however, the reposition of equipment and the expansion of productive capacity depend on the rhythms of technological innovation, as well as the rates of depreciation and the conditions for the obsolescence of the already installed productive processes. In any case, any acceleration in the process of realization of surplus value acts as a mechanism that increases the rhythm for revaluing value. In this

[18] Marx, *Le Capital,* I, in *Oeuvres Economie,* I, p. 698.

sense, the effect of reducing the circulation time for any commodity can be compared to the effect of reducing the socially necessary labor time.[19] This parallelism between the effect of labor and circulation time at its turn is reflected in the confluence between the processes of formation and revaluing of value, as well as in the relation of technological innovation to capital accumulation.

In the process of value realization and revaluing, one strategic function of circulating capital is to orient part of available capital-money to the production of scientific-technological knowledge, that when incorporated in more efficient means of production allow increasing productivity, and thus, the production of value and surplus value. Technological innovation becomes an intrinsic necessity for the organic accumulation of capital to counter the tendential law toward the decrease in the profit rate, but also the effect of competition among individual capitals. Thus, the revaluing of value in capital's accumulation and expanded reproduction process is accomplished by assigning part of realized capital-money to the production of scientific-technological knowledge materialized in new means of production, where capital-money is invested as new productive capital of increased productivity.

Marx recognizes the importance of the scientific-technologic innovation process to raise the efficiency of the productive labor process in capital's accumulative dynamic; however, the analysis of this process was not included in his study of the *moment of circulation* and *turnover of productive capital.* Moreover, the production of scientific knowledge and its incorporation to new technological means of production appears as an external process to the object of capital, that is, to the capital's reproduction in its moments of production, realization, and revaluation of value and surplus value. In this sense, Marx states that "the capitalist [...] sells the product at its value and finds in circulation the material means of production to start again and renew the process without discontinuity."[20]

[19] Thus, Marx stated: "Through his operations, a mercantilist [...] can reduce the selling and buying time for many of his producers. He resembles then a machine that diminishes the unproductive expenditure of labor force or allows to increase production time" (Marx, *Le Capital*, I, in *Oeuvres Economie*, I, p. 570).

[20] Marx, *Le Capital*, I, in *Oeuvres Economie*, I, p. 507.

To be sure, the process of technological innovation does not correspond in time with the moment of transformation of the commodities-value in new capital-money, and then into the new elements of productive capital in the circulation process; However, Marx does not integrate the technological innovation process—nor the forces of Nature mobilized by technology—to that of capital turnover in terms of value when considering that "the natural forces that are offered freely can be integrated into the production process where they will act with more or less efficiency in a degree that will depend on the methods and scientific progress that cost nothing to the capitalist."[21]

Various passages in books 2 and 3 of *Capital* reveal Marx's theoretical insight and analytic penetration in regard to the role of scientific-technological innovation in value revaluation and capital reproduction, as well as to the limitations imposed by capital to the incorporation of the productive potentialities of the sciences for the development of the social forces of production:

> Progress in industry revolution regularly most of the labor means. These are not replaced in their primitive, but a renewed form. In one side, the mass of fixed capital invested in a natural form, and destined to function as such for a particular average period, offer a reason not to introduce (but progressively) new machines and thus constitutes an obstacle to a more rapid general introduction of perfected labor means. In the other side, when decisive innovations come into play, competition obliges to replace the old labor means, while still usable for new ones. Catastrophes and crisis are the main reasons that induce this early renewal of exploitation means on a larger social scale.[22]

Thus, technological innovation of the means of production is conditioned by the process that conditions the realization of value, which in turn affects the conditions for capital's revaluation in the contradictory process of the valuing and devaluing of value that characterizes capital accumulation: technological progress, which in one side increases the

[21] Marx, *Le Capital*, I, in *Oeuvres Economie*, I, p. 931.

[22] Marx, *Le Capital*, I, in *Oeuvres Economie*, I, p. 601.

exploitation of surplus value from the labor force, induces at the same time a devaluation of fixed capital. Marx explains it in the following terms:

> The constant improvements that devalue the machines in their actual installations make them lose a part of their use value, of their value in brief. The effect of this process is violent, mainly at the beginning of the introduction of the new machines, insofar that they haven't still attained a certain degree of maturity and have aged before having had time to reproduce their value [...] when tools, equipment, buildings, and fixed capital, in general, have reached a certain degree of maturity, such that they can maintain their fundamental structure at least a sufficiently long time, a similar depreciation occurs as consequence of the improvements in the methods for the reproduction of fixed capital.[23]

Marx condenses these two mechanisms that act in the devaluation of fixed capital, or better said, these two aspects of the process that allows counteracting the tendency toward the decrease in the profit rate produced by the increase of capital's organic composition in its accumulative process. The first aspect emerges from the devaluation of the installed fixed capital by the introduction of a technological innovation—a process that goes hand in hand with the devaluation of commodities produced by the devalued equipment—as the introduction of the new machines implies a reduction of socially necessary labor time for the production of commodities in this industrial branch; the second aspect consists in the reduction of socially necessary labor time for the production of the means of production. The extent and reach of this revaluation process will depend on the timing for the substitution of the previous means of production, that is, before its physical ware, or before the necessary functioning time for the realization of the invested value in the means of production under operation.

These effects of technical change over capital's revaluing process are not produced as a consistent linkage to the productive process through labor value theory, as technological progress is unconditioned by the dynamics of capital reproduction itself. The conditions for capital accumulation stimulate the production of scientific knowledge by planning research

[23] Marx, *Le Capital*, I, in *Oeuvres Economie*, I, pp. 926–927.

programs oriented to the incorporation of science to technological innovations to increase the exploitation of surplus value, by devaluing fixed capital under operation. As Marx stated, capital's crisis conditions the moment for the introduction of technological innovations, giving an impulse to the development of productive forces. In this undetermined context, technical progress produces this double support-effect to capital reproduction: (a) the rise in the surplus value rate as result of the devaluation of labor force that generates the increase in productivity of labor processes; and (b) the devaluation of capital that reduces its organic composition, generating the conditions for its revaluation and reproduction.[24]

Thus, technological progress produces the double effect of devaluing fixed capital and at the same time the value of the labor force. However, if these two processes generally coincide, they do not necessarily implicate one another; it is possible to imagine a technical change that demands more manual and intellectual labor time in the elaboration of new means of production, while its effect might be a more than proportional increase in the productivity of consumer commodities. Conversely, it is possible to improve the construction of machines and equipment that reduce their value, without having to increase necessarily its productivity. In this sense, while the technical and the organic composition of capital are intimately related, they maintain their specificity.[25]

[24] In this respect Marx explained: "The development of labor productivity in one production branch, iron or carbon, or machines or buildings that in turn might depend on the progress in the domain of intellectual production, in particular of the sciences of nature and its applications can lead to a decrease in value, and cost, of the means of production in other branches of industry, for example in the textile industry or in agriculture.

Consequently, the commodity that leaves an industrial branch as the product enters another branch as a means of production. The higher or lesser reduction in its price depends on labor productivity in the sector where it comes out as a product; at the same time, it leads not only to a decrease in the price of commodities where it participates as means of production but also depreciates constant capital of which it is one of its elements. Consequently, it works as a condition for the rise of the profit rate" (Marx, *Le Capital*, I, in *Oeuvres Economie*, I, p. 909).

[25] Marx clarified: "The difference between the technical composition and the value composition of capital appear in each industrial branch in the fact that for a constant technical composition, the value relation of the two fractions of capital might vary or remain constant, while its technical composition is modified [...] a difference in the volume of constant capital can also indicate a change in the mass of production means functioning for a certain quantity of labor force; this difference can also come from the fact that the value of the means of production employed will vary from one sector of production to another" (Marx, *Le Capital*, I, in *Oeuvres Economie*, I, p. 939).

However, the difficulty to integrate the technological innovation process to capital's reproductive dynamic implies, beyond the consideration over the cost involved in the production of knowledge, the need to conceptualize intellectual labor in relation to value formation and capital revaluation. Marx did not integrate the technological innovation process to the circulation circuit; neither did he develop it as part of the production process. Marx uses the term "complex labor" to refer to the more qualified technical labor of maintenance, repair, and improvement of productive equipment, that has the effect of raising the productivity of simple and direct manual labor applied to the productive process. However, this complex labor is qualitatively distinct to intellectual labor that generates knowledge, methods, and techniques to be incorporated into the new means of production. With the increasing scientifically based production, a more significant part of complex labor in the productive sector of machinery and equipment consists effectively in labor applied to the scientific-technological innovation process. Thus, the own conditions for capital reproduction have generated a new productive sector of *intellectual goods*—scientific and technological commodities; intellectual property rights—which productivity does not depend directly on determinate labor time.

Thus, the unresolved problem in *Capital* of the reduction of complex into simple labor, reaches new dimensions the moment that intellectual labor—specially scientific-technological labor—acquire specificity, relative autonomy, and considerable proportions as wage-labor in the production of commodities, bringing up the need to conceptualize the integration of scientific production and technological innovation to capital's valuation. Therefore, the integration of the process of production of knowledge to capital's reproductive dynamic is not limited to a circulation sector where a portion of realized value is assigned to scientific research as part of the production costs and technological innovation absorbs a portion of the produced surplus value. As a condition of capital's revaluation process, intellectual labor is not a simple category for the distribution of economic rent, but a *moment* of capital's revaluation, a condition of the process of capital's production and reproduction.

Intellectual Labor and Scientific-Technological Productive Forces

The valuation of intellectual—scientific-technological—labor is a theoretical condition to evaluate the transmission of value from the means of production to the commodities produced; if socially necessary labor time for the production of different commodities can be considered as an effect of labor productivity once technological innovations have materialized in new means of production, it is impossible to estimate the value that they transmit to the product without conceptualizing the value they contain; and this value cannot be reduced to manual labor time needed for its construction, which does not represent but a minimum portion of the complex intellectual labor time on which depends the productivity of productive forces. Thus it becomes imperative to conceptualize *intellectual labor as productive labor* to be able to think the revaluing of value in the process of capital reproduction driven by scientific-technological innovation.

The conditions of capital reproduction and accumulation are responsible for inducing a qualitative change in the technological innovation process different from the first methods introduced to increase the productivity of labor. In the first stages of capital formation, the division of labor and its reduction to a simple and direct expenditure of labor force as the constitutive principle of value creation, as well as simple cooperation that emerges from the concentration of workers around the first machines that use the direct forces of Nature, appear as the primary sources that increase the productivity of labor. In this sense, Marx could have considered that technical progress cost nothing to the capitalist.[26] However, at the same time, more abstract concepts, as the entropy law, born from the study of the steam machine, cannot be conceived as a creative act external to the determinations impinged on the necessity of

[26] "Such increase of production forces—a process that costs nothing [to the capitalist] is the division of labor and combined labor inside the production process [...] Another productive force that costs nothing, is scientific power [...] that can only be appropriated by the use of machines" (Marx, *Grundrisse*, in *Oeuvres Economie*, II, pp. 1745–1746).

raising the surplus value rate by making the production process more efficient (Leff 2000).

Marx circumvents the necessity to conceptualize intellectual labor as productive labor based on his conception of the relation of technical progress to the labor process and the productive process where the value is formed. Marx supposes that technical change consists predominantly of the reduction of productive labor to simple and direct tasks. Considering this is the central tendency of technical progress, Marx thought the principle that forms value as an abstract, undifferentiated and direct labor time, that is, a time where labor and production times coincide.[27] However, this tendency might only have prevailed in the first stages of capital accumulation; and not even then, if we consider the differentiation of labor time and production time in agriculture, where the direct forces of Nature without the direct intervention of human labor are responsible for production inside and outside the market (see Chap. 6). As technical change progressed, it shifted to the direct application of scientific knowledge to production.

Under the technical conditions that prevailed before the technological applications of science to production, it was possible to increase the productivity of labor through the division and fragmentation of productive tasks and their reduction to simple and repetitive functions. That is why the historical conditions of technical progress make the simple manual, and direct labor appears as the essence of productive labor. The division of labor and the association of simple tasks to which the labor process is reduced thus appear as an increased productive force of capital, as an external potency to labor force itself that costs nothing to the capitalist. What each worker loses in qualification by reducing his complex abilities to simple gestures appear as a force that concentrates in front of him as capital. In this stage, the magnification of capital's productive power as the effect of the division of labor does not emerge from a social process of intellectual production articulated to the production of commodities,

[27] "There is no capital valuation as productive capital as long as it is performed in that part of its production time that exceeds labor time [...] the more production and labor times coincide, productivity increases as well as the valuation of a given productive capital in a given time. Thus the tendency of capitalist production to reduce the excedent of production time over labor time to a minimum (Marx, *Le Capital*, I, in *Oeuvres Economie*, I, pp. 563–564).

but as technical progress that springs from disarticulation of individual technical abilities, turning every worker into just a piece in the chain of production.

With the advent of manufacture, the mechanization of production generated the need of complex labor in front of simple manual labor. A new class of technicians was formed whose function consisted in supervising the production process to maintain the cadences of manual labor, to provide necessary repairs to the equipment and to execute complex tasks in the productive process. However, this stage did not imply either an organic articulation of the production of knowledge to the production of commodities. Technical innovations were still produced by isolated inventors and not as a process induced by capital's dynamic. Scientific work was not yet wage-labor paid out of variable capital, so intellectual labor was not considered in terms of productive labor.

It is big industry that generates the need for intellectual production in the form of wage-labor, separated in time and space from direct labor tasks, but organically articulated to the process of capital valuation. With the automation of production, its productivity will depend more and more on the potency of natural laws incorporated into production processes through technology tan from the productive power of manual labor. From then on, production of commodities will depend more from the mediated effect of intellectual labor through the incorporation of science and technology as direct productive forces of the means of production, tan from simple labor time applied directly to production.

The Law of Tendential Fall in the Rate of Profit and the Revaluation of Value

The process of capital accumulation poses a real, as well as a theoretical problem, for value formation, as the increase in the technical and the organic composition intensifies the value of fixed capital and reduces the relative part of variable capital, that is, of labor force exploited in the production process. This indetermination in value formation increases as the accumulation process follows the tendency to mobilize technical progress that materializes in productive processes that reduce direct human labor as the principal source of value.

Marx offered a theoretical response to these "contradictions" of capital accumulation in his study of the *law of the tendential fall in the rate of profit*. There he states that this tendency toward the fall in the rate of value production would be offset by the increase in the mass of labor force being exploited by capital in its expanded reproduction process, and by the proportional increase of raw material transformed by a capital of larger dimensions and increasing productivity.[28]

Notwithstanding the undervaluing of the direct forces of Nature appropriated by capital, particularly in the rural and agricultural sector, the theoretical problems posed by the scientific grounding of production induced by capital accumulation go beyond these responses to the contradiction between value formation and the tendency toward the fall in the rate of profit. Scientifically based production generates a qualitative transformation of labor and the productive process that affects the principles on which the concept of labor is founded. Basically, the problem of the relative decrease in variable capital in relation to constant capital—the relative reduction of real labor time and of the number of workers in relation to an increase of fixed capital, both physically and in terms of value—is not solved by the increase in real wage-labor time applied to production or through the evaluation of energy transmitted by labor

[28] Marx explains: "the tendential fall in the rate of profit, or the relative decrease of capitalized surplus labor compared with the mass of materialized labor put in movement by living labor, does not exclude in any way the increase in the total mass of labor [...] exploited by social capital, nor [...] the total mass of surplus labor that it appropriates [...] The number of workers employed by capital, that is, the total mass of labor that it puts to function, the total mass of surplus labor that it absorbs, the quantity of surplus value that it creates, the quantity of profit that it produces, can therefore increase, and increase progressively, despite the progressive fall in the rate of profit [...] Thus, the same laws of social capital engender a mass of increasing absolute profit and a decreasing rate of profit" (Marx, *Le Capital*, I, in *Oeuvres Economie*, I, pp. 1005–1007).

Marx elsewhere remarks: "Insofar as labor productivity develops, the value of raw materials constitute an element of value that constantly increases the value of the product-commodity; this doesn't happen only because [the value of raw materials] is totally incorporated in its value, but because in every proportional part of the total product, the part represented by the use of machines and the one formed by labor newly added are in constant decrease. Facing this tendency to fall, there is a proportional increase of the other part of value represented by raw materials, save that this increase is cancelled by a proportional decrease in the value of raw materials due to a growing productivity of labor employed to produce this raw material" (Marx, *Le Capital*, I, in *Oeuvres Economie*, I, pp. 922–923).

force. The problem lies in the possibility of evaluating the value transmitted by machinery and equipment as means of production, in relation to the new added value generated by the labor force of different degrees of qualification, as well as by the socially necessary labor time for the reproduction of labor force that determines the capitalized surplus value. If we maintain the principles of value formation, capital expansion will depend on its capacity to increase the relative rate of surplus value, the limits of which are not only conditioned by the productivity of labor, but also by class struggles that establish the value of labor force; but also by the contribution of technology and the productive forces of Nature, where production time does not coincide with direct labor time, and where its determination in terms of value has vanished.

The compensation of the reduction in value, both from the transfer of value form the means of production and raw materials, as well as from new productive labor to commodities produced, faces theoretical problems. The hypothesis of the transmission and addition of value from raw materials and from equipment to consumers' goods is questionable as long as the problem of the value of the productive forces from Nature is not solved. With the direct application of science and the technological organization of production, the agricultural production of raw materials for industry and the recycling of its products raise the problem of reducing productive labor to a quantum of simple and direct labor that form value. Conceptualizing the intellectual labor incorporated as productive labor becomes imperative. Furthermore, the contribution of ecological processes to the offer of natural resources and the production of raw materials with traditional techniques in non-capitalist social formations raise the problem of conceptualizing its products in terms of value (see Chaps. 6 and 7).[29]

Consequently, to apprehend the process of revaluing of value, it is necessary to consider the transformations induced by capital's reproductive dynamics in the labor process and to conceptualize the determinations of intellectual labor, insofar that science and technology became the principal potency in the development of capital's productive forces, as well as the contribution and the conditions from ecological processes to capital's

[29] Leff 1995.

production and extended reproduction. Therefore we can conclude that there are no sufficient and consistent conceptual bases for establishing a quantitative theory of value once scientific knowledge and technology have become the capital's dominant forces of production. This does not exclude the possibility of considering a *qualitative theory of value*, although such a theory still poses conceptual problems.

From Quantitative to Qualitative Theory of Value: The Capitalistic Appropriation of Nature

Capital has the intrinsic need to rely on technological progress in order to increase the production of relative surplus value and to overcome the adverse effects of the increased organic (value) composition of capital on the rate of profit. This valuation–devaluation–revaluation of capital occurs in a contradictory process tending toward the replacement of labor-value formation as the principal determinant of capital dynamics. In general, any revaluation of capital incorporated within a new technology involves the introduction of fixed capital with lesser value and with an increased capacity to produce surplus value, although with a decreased capacity to extract value from the labor force per unit produced.[30] To be sure, the very expansion of capital provokes an increase in the labor force incorporated into the production process and is opposed to the declining tendency of value produced in individual cases. Needless to say that the disappearance of value formation as a quantitative principle determining

[30] "Given the general bases of the capitalist system, the development of the productive powers of social labor always rise to a given point of accumulation to become, from then on, the most powerful mechanism [...] The development of the productive potential of social labor triggered by such progress crystallizes in qualitative changes [...] in the technical composition of capital, that is, that the amount of equipment and materials increases more than the labor forces needed to make them function [...] These changes in the technical composition of capital are reflected in its value composition, in the progressive increase of constant capital and the decrease of variable capital. However [...] this change in the value composition of capital does not fully reflect the changes in its technical composition. The reason is that progress in labor potential, manifested through growth in equipment and materials put into motion with a lesser sum of labor, also diminishes the value of most products that function as means of production" (Marx, *Le Capital*, I, in *Oeuvres Economie*, I, pp. 1134–1135).

capitalist development does not eliminate the social relations of exploitation upon which the capitalist mode of production is founded.

The resolution of capital's internal contradictions in the world's more industrialized nations involved the socio-cultural transformation of pre-capitalist formations such that the latter have been either displaced or forced to adapt their traditional practices to the capitalist mode of production resulting on the extended proletarianization of urban and rural workers. Maintaining the rate of profit thus depends in large part on the exploitation of labor with different qualifications, on the resources of "underdeveloped countries" and "traditional peoples," and on the unvalued assets from Nature. Thus, although the scientific-technological revolution tends to devalue productive equipment and reduces much of the direct labor it sets into motion, commodities produced with more productive technologies contain the value produced in the elaboration of the raw materials, and intermediate products produced either with traditional techniques or with modern techniques that use a considerable quantity of labor. That is why the dominant tendencies within Marxism subscribe to value formation as the determinant of the capital accumulation process and internationalization of capital. However, this does not support a quantitative theory of value, nor exempts the importance of considering the importance of the "natural" and "scientific" forces of capitalistic production.

The main problem with a qualitative theory of value arises with the qualification of the conditions of value production as a determinant based on complex and differentiated labor time, technological processes and natural forces that constitute capital's actual productive forces. Its purpose raises and faces the problem of finding a homogeneous measure of these differentiated material and energy sources, production processes and production times that cannot be reduced to a standard measure of socially necessary time. Marx assumed that technological progress appears as a process external to the very production of value,[31] or alternatively, as a historically uniform process that therefore does not affect value

[31] Thus, Marx stated in the *Grundrisse*: "We have introduced the development of the forces of production as an exterior element" (Marx, *Grundrisse*, in *Oeuvres Economie,* II, p. 145).

relations.[32] However, from the moment that the development of the forces of production is conceived as being independent of value formation, the organic cohesion of the theory of value as a determinant of capitalist accumulation is broken.

This theoretical disarticulation also occurs as a result of the separation between manual and intellectual labor as determinants of the development of the forces of production. Although Marx admits the existence of complex labor in addition to simple labor, which together accounts for collective labor within a hierarchy of labor forces, intellectual labor always appears as a property extracted by capital from the proletariat and then concentrated as a capital asset to exploit labor power.[33] However, Marx never explains the necessary connection between the value produced through the exploitation of the proletariat and the intellectual labor that increases the exploitative capacity of capital. Being that science appears as an "independent productive labor force"—since scientific labor and its crystallization in the development of productive forces is considered non-productive labor in the capitalist sense; that is, it does not produce value—it is impossible to articulate the innovative process that gives labor its socially necessary character or to see technological progress as an effect of value formation. Nor is it possible to similarly incorporate to this theoretical scheme the forces of Nature that science sets into motion through technological innovation in commodity production, nor the ecological conditions of an eco-technological sustainable production process.

Scientific labor acquires another perspective within the theories of surplus value and circulation. In the capitalist system, Marx writes, "the determinant objective of production is surplus value [...] Therefore only the worker who produces surplus value for the capitalist and whose labor furthers capital can be considered productive."[34] Furthering capital does

[32] "The progressive development of the social forces of production acts almost or entirely uniformly on the labor time necessary for the production of various commodities" (Marx, "Critique de l'Économie Politique," in *Oeuvres Economie,* I, pp. 289–290).

[33] "What piece workers lose is concentrated before them in capital [...] This excision that starts with simple cooperation [...] is completed with big industry which makes *science an independent, productive labor force* enrolled in the service of capital" (Marx, *Le Capital,* I, in *Oeuvres Economie,* I, p. 905, my emphasis).

[34] Marx, *Le Capital,* in *Oeuvres Economie,* I, p. 1002.

not merely mean extracting value in the production process but rather the capacity to reproduce the conditions of the labor force's exploitation. It is not enough to extract a quantity of value that can be recapitalized in the form of fixed capital once a machine or piece of equipment is used up. Capitalist accumulation and competition require that surplus value, in order to be recapitalized, crystallized within means of production with growing productivity,[35]That is, through technological progress.

In this sense, there is no more productive labor for capital than scientific-technological labor. More so than simple, direct labor, it allows the surplus value produced in the direct process to be recapitalized and reproduced in the capital cycle. Furthermore,

> Through the discovery of new, useful materials or new qualities of the materials already in use, the machine multiplies the investment spheres for accumulated capital. By showing the proper methods for reutilizing capital's excrements in the circular course of reproduction and social consumption, it converts, with no assistance at all from capital, these non-values into so many additional elements of accumulation.[36]

Therefore, if the production of value depends on direct, simple labor, the development of the capital's productive forces depends on scientific labor. Since capital accumulation itself determines a growing tendency toward the substitution of direct, living labor by the direct application of scientific forces in commodity production, the specific determination of value as the fundamental principle of the structural dynamic of capital tends to vanish.[37] This turn poses serious theoretical problems to grasp

---

[35] "Surplus value is thus convertible into capital because the net product in which this surplus exists already contains the material elements of new capital" (Marx, *Le Capital*, I, in *Oeuvres Economie*, I, p. 1084).

[36] Marx, *Le Capital*, I, in *Oeuvres Economie*, I, pp. 1111–1112. Aren't genetic natural resources and biotechnology—those "non-values"—the primary sources of the most powerful industry today in the era of globalized capitalism?

[37] Marx himself declared that "As long as time—the quantum of labor—is established by capital as the only determinant element of production, direct labor considered as the principle for the creation of use values, disappears, or at least is reduced quantitatively and qualitatively to a role that is certainly indispensable, but subordinate, in relation to scientific labor in general, to the technological application of natural sciences, and to the general productive force resulting from the social organization of the entirety of production" (Marx, "Principes d'une Critique de l'Économie Politique," in *Oeuvres Economie*, II, p. 301).

the "concreteness"—in the synthesis of multiple differentiated determinations—of capital dynamics in the stage of scientific-technological revolutions. Given that value produced by labor is the basis of capital, Marx writes: "if production could be carried out with no labor at all, neither value, nor capital, nor value production would exist."[38]

In the *Grundrisse* Marx had recognized that:

> The exchange of living labor against materialized labor, in other words, the determination of social labor as the opposition between capital and wage labor constitutes the last development of value relations and the system of production founded on value. Its permanent condition is the mass of direct labor time, the quantum of work applied as a decisive factor in the production of wealth. However, as big industry develops, the creation of real wealth depends less on time and the quantity of labor than in the action of other factors set in movement by labor, whose powerful efficacy has no comparison with the immediate labor time set forth in the production process; it depends over all of the general states of science and technological progress, of the application of science to production [...] When labor, in its immediate form, will have ceased to be the great source of wealth, labor time will cease to be the measure of labor, and exchange value will cease to be the measure of use value. Surplus labor of human masses will cease to be the condition for the development of general wealth [...] From then on, production based on exchange value collapses, and the immediate process of material production abandons its form and its miserable contradictions.[39]

Here Marx describes the turning point from a past historical moment, which produced the social conditions for thinking value formation as the basic principle of capital dynamics, to a future historical moment, a utopian moment, in which all labor will have disappeared. Thus a vacuum of thought emerges that impedes the comprehension of the principles that determine commodity production in the technological-ecological phase of capitalism and the transit to a sustainable mode of production founded on environmental rationality. This passage in the *Grundrisse* has opened

[38] Marx, "Principes d'une Critique de l'Économie Politique," in *Oeuvres Economie,* II, p. 250.

[39] Marx, "Principes d'une Critique de l'Économie Politique," in *Oeuvres Economie,* II, pp. 305–306.

the way to a series of "theoretical deviations" in Historical Materialism. With the elimination of the law of value, the development of the forces of production appears as the determinant factor in the transformation of social relations of production. Thus, Radovan Richta wrote that,

> Since progress in production resides in mechanization and widespread industrialization, capital constitutes its efficient and appropriate form of movement. Historically, we can find there a justification for the existence of capital as an external, transitory social form of the development of civilization [...] The relations of production are but a form of the movement of the forces of production (Richta 1969, pp. 30–34).

Other neo-Marxist authors, like Herbert Marcuse, following Martin Heidegger's position on technology (Heidegger 1935/1977), as the ontological regime that destroys the world in modern times, argued that social domination in the scientific-technological revolutionary era leading to automation is commanded by the primacy of technological reason and scientific rationality.[40] Furthermore, the scientific-technological revolution appears as a privileged mechanism for human liberation, taking the place of class struggle and the transformation of the social relations of production as determinants of societal progress.[41]

The problem expressed in this passage of the *Grundrisse* arises from the fact that here Marx confuses two determinations in one: he conflates the

---

[40] "Reason, as conceptual thought, as behaviour, necessarily leads to domination. Logos is the law, the command, the order for power over knowledge" (Marcuse 1968, p. 190).

[41] Thus, in an ambivalent view over the power of technology Marcuse stated that "Political change cannot become qualitative social change except by reorienting technical progress, that is since a new technology can be developed." (Marcuse 1968, p. 252). "In order to transcend technological reality, a prior condition is that it is first accomplished. By being realized it will constitute at the same time the rationality that will enable its transcendence" (ibid., p. 255). "If technological rationality reached perfection, it would translate ideology into reality, and at the same time, it would transcend the materialist antithesis of this culture" (ibid., p. 258).

Contradictions in Marcuse's "dialectical thinking" are shifted from social structure to "technological reason"; the conditions for its disappearance are the full realization of scientific-technological development. This implies that liberation depends upon the full development of the forces of production and the general automatization of labor processes, not on political practice aimed at transforming social relations of production, and with nature. This "dialectic" of history falls under *Gabor's Rule*: "Everything that is technologically feasible must be realized, whether it is judged to be morally good or condemnable" (Gabor 1964).

exchange of living labor for materialized labor with the opposition between wage labor and capital. The problem is not only the reduction and the progressive elimination of simple and direct labor as a determinant of value formation and the increasing application of science as capital's main productive force but the oblivion of the productivity of Nature itself and ecological sustainability as a condition for the productive process. Without a doubt, at present the production of wealth increasingly depends more on the use of "natural" forces through science and technology than on direct, living labor, leading to the "collapse of production based on the law of value." However, the scientific-technological potential is not just a by-product of proletarian labor appropriated by capital to exploit labor muscle; scientific labor is intellectual wage labor, exploited by capital.

Furthermore, the progressive elimination of living, direct labor as the basis of the theory of value does not imply the transformation of the capitalist mode of production into new productive rationality. General abstract labor is but one determination which, combined with a whole series of other determinations, constitutes the structural dynamic of capital's mode of production. The disappearance of the law of value does not put an end to the generalization of commodity exchanges, nor to the global exploitation of human labor and Nature, nor to the opposition between a wage class and a capitalist class whose power rests on the production, possession, and control of scientific knowledge. These are the multiple determinations that constitute the capitalist mode of production.

Technological progress first transforms and then progressively eliminates the quantitative determination of value; it then produces a tendency toward the progressive substitution of direct manual labor with indirect intellectual labor in commodity production, through which it appropriates the forces of Nature. At the same time, capital extends its tentacles capitalizing Nature, appropriating the ecological means of production, the environmental goods, and services of the planet. Through the change in the quality of the labor processes generated by scientific, technological and industrial developments, human labor continues to be the basis of production, together with peasants' labor and with Nature's productive processes that act without the direct intervention of man. These transformations in the labor process do not eliminate the social relations of production, the property-appropriation relations nor the exploitation and social control based on the power of the capitalist class over the process of production, including the ecological and technological means of production, peasant's labor, and knowledge.[42]

Technology, the instrument *par excellence* of calculation and control, paradoxically has made the epistemological project based on quantitative science *an impossible dream* for the social sciences. This does not mean that the science of history has failed, but rather that historical materialism demands new conceptual bases to orient political practice to gear social transformation toward the construction of a sustainable world.

Marx recognized the importance of technological progress for the expanded reproduction of capital. However, his analysis lacked the appropriate conceptualization of the determined and determinant character of intellectual labor, of scientific production and Nature's productivity in the contradictory process of valuation–devaluation–revaluation of capital as the effect of the development of productive forces and the strategies for the capitalist appropriation of the forces of Nature. Marx recognized the determinant character of scientific knowledge and its technological applications to raise the productivity of capital's productive processes. However he never characterized the "knowledge sector." He did not conceptualize intellectual work as productive labor or the productive forces of Nature and cultural creativity of the people, integrating the economic, technologic, ecologic and cultural determinations and conditions for the revaluation of capital, and its transition to a sustainable mode of production.

Marx's theory of value considers the effect produced by intellectual labor already incorporated in the means of production as they raise the productivity of manual labor, that is, in its effect on the socially necessary labor time that, as we have seen, is affected by competition among capitals and by their turnover cycles. However, intellectual labor is not considered as a determined and a determinant principle for the process of capital reproduction. This would not affect the theoretical fundaments of historical materialism if the role played by the production of knowledge were marginal to the valuation process of capital, or if it could be explained

[42] The appropriation of traditional knowledge through ethnic-bio-prospection is an "exemplar" of the strategies of the capitalistic appropriation of natural-cultural forces of production.

from the determinations of the law of value as exposed in *Capital.* However, in his study of capital turnover and competition, the concept of value becomes less and less concrete, and Marx himself admits the difficulties it faces to apprehend the capital's real movement as determined by the theory of labor-value.[43]

The productive processes of Nature and the productive power of scientific knowledge are determined by the "law of value" in a sense and to the extent that the impulse of capital accumulation drives technological progress and the metabolism of Nature. Scientific knowledge is not simply produced as the effect of an internal logic of scientific discovery (Kuhn 1962; Popper 1973) and there is no "technological reason" operating independently of social conditions that drive its productive powers, so it cannot become by itself the determinant principle of the economic process. Conversely, it is not possible to substitute the laws of economics for any scientific law governing natural and technological processes—that is, the second law of thermodynamics, the law of entropy—to determine the ensemble of productive processes in any given society. Thus, the need arises to rethink the ontological and the epistemological foundations of historical materialism capable of integrating the ideological formations, the social imaginaries and the production of scientific knowledge with the conditions of ecological productivity and sustainability under a new social and productive rationality.

The economic process founded in labor time has been displaced by an economy based on capital's capacity to appropriate scientific knowledge as a means of production and a power device to exploit the forces of Nature. Capital accumulation is not only based on the over-exploitation of cheap labor from Third World countries but also on the appropriation of its natural resources—the undervaluation and free appropriation of the biosphere, of its ecologic goods and services, including the privileged access to genetic resources from its biodiversity through "intellectual property rights," and to fossil fuels that maintain an overcapitalized

[43] Thus, in face of the difficulties to explain the equalization in the rate of profit in productive sectors with different organic compositions and turnover times, Marx considers: "It would seem that here, the theory of value is incompatible with the real movement and the empirical phenomena of production, and it might be necessary to renounce to understand them" (Marx, *Le Capital*, I, in *Oeuvres Economie*, I, p. 945).

agriculture and growing urbanization that induce the entropic degradation of the planet.[44]

To value the complex ecological, cultural and technological processes—the *negentropic potential*—that supports a sustainable economic process, it is necessary to shift from the actual economic metrics that reduce the world's ontological and axiological diversity to the objective, quantitative uniform market values, building a more qualitative theory capable of integrating economic, technologic, ecologic and cultural processes; a plural axiological and epistemological scheme capable of expressing the antagonisms between the techno-economic ontological regime and environmental rationality with the multiple cultural rationalities that determine the processes of appropriation of Nature, incorporating the ecological conditions for sustainability to productive processes.

The complementarity of Nature's objective and subjective values in the construction of an environmental rationality calls for a new theory capable of integrating the ecological conditions for sustainability and the cultural values assigned to Nature—through imaginaries and identities forged in the interrelations of material and symbolic processes—expressed in the communal and environmental rights of indigenous and peasant peoples for the reappropriation of their bio-cultural patrimony of productive resources. This implies the need to renew the conceptual principles of historical materialism to incorporate the power of technology harnessed by environmental ethics and the forces of Nature signified by cultural meanings and values as fundaments of a sustainable production process. Thus, historical materialism is reconstructed from the perspectives opened by eco-Marxism, political ecology, and environmental rationality toward a democratic, diverse and sustainable eco-socialism.

---

[44] The intellectual property rights over the genetic resources of biodiversity and its strategies to invade the tropical regions of the Third World with its transgenic products express the power of the capitalist ecological scientific-technological era.

Final Remarks

The scientific-technological revolution, as a process determined by the dynamics of capital, has yielded to the dissolution of the empirical principle that founded the theory of value, that is, simple and direct labor as the quantitative determinant of commodity production. This fact has two significant repercussions for Marxist thought. The first involves the organic relationship between social theory and social practice; the second involves the epistemological specificity of the science of history. These two aspects of Marxist thought have always constituted an organic totality. The epistemological changes generated by the effects of technology in the process of capital reproduction and in the sustainability conditions for production pose the need to re-elaborate the concepts of historical materialism in order to incorporate the productive function of intellectual labor and the forces of Nature in the process of capital's reproduction/ transformation toward a sustainable economic paradigm. Notwithstanding the above, reconceptualizing productive labor to integrate its manual and intellectual functions, as well as the forces of Nature—the entropic limits and negentropic potentials of Nature—inasmuch as this qualitative theory of value might give account of an ecological economy, it will never preempt the historical and social conditions for the production of knowledge that mobilizes the reproduction of capital and its transition toward ecological socialism through another ontological basis for sustainable production. Knowledge of the socioeconomic determinations of the production of knowledge in its productive function cannot preempt the need to know the historical determinations over the production of knowledge in their theoretical function to apprehend the Real through the history of metaphysics and the constitution of the ontological regime of scientific rationality—the *logocentrism of science* (Derrida)—as the mode of theoretical appropriation founded in the objectification of the world: of Nature and human beings.

Perhaps Marx was not convinced that the internal laws of capital would lead directly to the dissolution of the capitalist mode of production. However, by presenting the dynamics of capital as the effect of quantitative and objective laws, he opened the way for Lukács (1923/1971) and others to consider the rise of class-consciousness as a result of these laws. Thus revolutionary practice acquired an objective character, determined by the internal laws of capital. With the disappearance of the law of value as the determinant quantitative principle of social transformations, political practices were no longer the effect of an automatic mechanism. Social change is not the effect of teleological determinism. From the stands of Historical Materialism, history geared by class struggle generates social structures with their temporal, tendential laws. These structures are not transformed merely as the effect of their internal laws but through the class struggle that develops within them and by power strategies in knowledge (Foucault). Social practices transform social reality and modify their internal laws. Therefore there are no absolute laws that determine praxis, nor can there be any pre-established laws of Nature that determine the development of the productive forces, nor can praxis determine a priori natural laws.

Scientific-technological revolution is transforming the labor process, turning the "forces of Nature" into the dominant forces of production of wealth. Laws of Nature geared by historical conditions of scientific discovery and technological innovation have become fundamental determinants of production. However, Marx and Engels did not conceptualize the workings of Nature in the metabolism of the biosphere—the negentropic potentials and the entropic limits for the development of the productive forces. Time involved in natural and technological processes; time that forges identities and social actors, progressively replaces labor time as the determinant of commodity production. No essential value-law does determine natural, theoretical and social processes. This implies the need to inquire about the power-knowledge strategies that govern social-economic processes, in order to deconstruct economic rationality forged by a mechanistic quantitative episteme, and orient the construction of environmental rationality based on the productive potentials of Nature, technology, and culture.

The environmental crisis is setting a limit to the expansion of capital opening a historical transition from economic to environmental rationality. It is no longer the case of assessing the growth of capital, its limitations, and crisis through the theory of value, but of thinking and constructing environmental rationality to enable a sustainable economic process. The scientific-technological revolution has operated a profound transformation of the labor process liberating the forces of Nature as the predominant productive forces in the production of economic wealth and turning on the equilibrium of ecological systems as a condition for a sustainable economic process. The complex times that merge and emerge in the processes of ecological productivity, technological innovation, and social creativity, embedded in cultural identities and existential ontologies of the people, have displaced labor time in the industrial sector as the determinant principle of the production of use values. The productivity of Nature enhanced by scientific knowledge and technological innovation, integrated with the metabolism of the biosphere, under the control of ecological resiliency and the care of cultural creativity, have become fundamental conditions for a sustainable economic process. The production and distribution of wealth will depend on the strategies for the production of knowledge and the thermodynamic and ecological conditions for the sustainable social appropriation of Nature.

Bibliography

Althusser, L., Balibar, E., Establet, R., Macherey, P. & Rancière, J. (1965/2016), *Reading Capital*, translated by Ben Brewster and David Fernbach, London: Verso.

Foucault, M. (1966), *Les mots et les choses*, París, Gallimard.

Gabor, D. (1964), *Inventing the future*, Harmondsworth: Penguin.

Heidegger, M. (1935/1977), "The question concerning technology", in *Heidegger. Basic writings*, trans. D. Farrell Krell, New York: HarperCollins, pp. 283–317.

Kuhn, T. (1962), *The structure of scientific revolutions*, Chicago: The University of Chicago Press.

Leff, E. (1995), *Green production: towards an environmental rationality*, New York: Guilford.

———. (2000), "The scientific-technologic revolution, the forces of nature and Marx's theory of value", in *Capitalism, nature, socialism*, Vol. 11 (4):109–130.

Lukács, G. (1923/1971), *History and class consciousness*, Merlin Press.
Marcuse, H. (1968), *L'homme unidimensionnel*, Paris: Ed. Minuit [*One-dimensional man*, Boston: Beacon Press, 1964].
Marx, K. (1965), *Œuvres, Économie I*, Paris: Gallimard.
Popper, K. (1973), *La logique de la découverte scientifique*, Paris: Ed. Payot
Richta, R. (1969), *La civilization au carrefour*, Paris: Ed. Anthropos.

6

Revaluing Nature: From Exploitation of Peasantry in Capitalism to Emancipation of Indigenous Peoples and Sustainability of Life on Earth

Value Theory and the Question of Peasants' Poverty and Persistence

Historical materialism—the science of capital—and economic theory at large have seen value theory vanish as the conceptual grounds that sustain the rationale of the economic process. Notwithstanding this fact, capital has remained well and alive, expanding in the world order and invading the planet. Capital accumulation extends its arms to the whole biosphere, grounding its feet in the planet by dispossession of the livelihoods of traditional peoples, the peasants and indigenous peoples that inhabit the world resisting the techno-economic rationality that expands over their life-territories. Marxism has failed to provide a satisfactory explanation for peasant poverty and the persistence of the peasantry in capitalism. Julio Boltvinik (2010) responds to this "lack of knowledge" postulating that both questions are functionally linked by their submission to the

Published originally in Leff, E. (2016), From the environmentalism of the peasantry and indigenous peoples to the sustainability of life, in Boltvinik, J. and Archer Mann, S. (Ed.), *Peasant's Poverty and Persistence in the 21st Century. Theories, Realities and Policies*, COLMEX/CROP/Zed Books.

E. Leff, *Political Ecology*, https://doi.org/10.1007/978-3-030-63325-7_6

functioning of agrarian capitalism. He draws his arguments from Marx's value theory to explain why peasants' poverty persists as a prerequisite for the possibility of capitalism in agriculture. Thus, if poverty is produced by capitalism, the "poorness of the poor" is constructed through Marxist discourse. The comprehension of the persistence of poor peasantry is "locked in" the persistence of Marxism, albeit an extension of its conventional approach that failed to account for the seasonality of the labor process in the overall capitalist agricultural production process. Once peasantry is defined as people living and producing "outside" the capitalist system, differentiating them from rural proletarian workers—those who reproduce their labor force by earning a wage working in capitalistic agriculture—Boltvinik sees the persistence of peasantry as a result of the functional interdependence of the peasantry—petty commodity producers—with the capitalist mode of production, disclosing entrenched relations that had remain hidden in their articulation through simple commodity exchange (Fossaert 1977).

The corollary of this questioning into the capitalist reasons and determinations for the persistence of poverty is the will to complete what is lacking in Marx's Theory of Value by offering a "general theory of value." In practical terms, this would be accomplished by adding to seasonal wage labor the required income to complete the total annual cost of reproduction of rural labor force, either by increasing consumers' prices, through trade protection and/or, by subsidizing peasants' production; in the way, it is argued, that developed countries do. It is pretended that these pragmatic reforms to theory and economic policy would not only contribute to explain but should also solve the problem of persistent peasant poverty in the South.

Beyond the intention to contest Boltvinik's account of the functional interdependence of those two forms of production in terms of its more pragmatic and political aims, it is crucial to challenge the proposed pretension of reforming value theory to incorporate the full cost of reproduction of peasants' labor force—including its seasonal and often very long periods of non-labored days—to offer a "general theory of value" that would translate in practical terms as subsidies to peasants' traditional agriculture. We must explain the persistence of peasantry, not only through the mechanisms that articulate it functionally to capitalistic agriculture but through other reasons, such as the social resiliency of their cultural imaginaries and traditional practices as their attachment to a dwelling life-territory. This ontological-historical-anthropological-social-ecological perspective challenges the absolutism of economic reasoning—its internal logic and structural determinism—opening critical thinking to a more complex understanding of human history in the face of the environmental crisis of capitalism. The environmental question has produced a shift from traditional Marxism to eco-Marxism in the perspectives opened by political ecology and environmental rationality (Leff 1993a, 1995, 2004).

The Poverty of Theory: Seasonality of Labor and the Historicity of Marxism

The functional articulation of the capitalist mode of production (CMP) with peasants' simple commodity production explains why the capitalist does not have to pay the full cost of reproduction of peasants' labor power, as its subsistence economy provides a substantial part of the primary means of survival—for their endosomatic consumption and physiological needs—that is, for the reproduction of available labor power to provide seasonal work to capitalist agriculture. The explanation for the reasons why the peasants are forced to offer and sell their seasonal labor in the market is because their subsistence economy is insufficient. Moreover, the reason it is insufficient is not because peasants are intrinsically and traditionally poor, but because of colonialism first, and afterward capitalism, induced an impoverishing process that entailed the pillaging of their resources, the degradation, and deterioration of the productivity of their ecosystems, the appropriation of their cultural patrimony of natural resources, the dispossession of their territories, and the colonization of their knowledge. As a consequence, indigenous peoples and peasants have been displaced to marginal lands that render the lesser productivity that would justify the prevalence of differential rent as an explanation of their persistent poverty.

The destruction of their means of production and their livelihoods goes beyond the state of survival in the most eroded fields due to population pressure and the degradation of their socio-environmental conditions. Their functional interdependence does not fully attribute their poverty to the CMP or by differential rent, but by the *historical process of entropic degradation of their environment and their livelihoods*. In this context, the explanation of the capitalistic determination of their present "state of poverty" requires a historical analysis of this process of dispossession—from the enclosure of the commons as the dual expropriation of the direct producers and the earth in the primitive accumulation of capital—and a critical deconstruction of Marxist concepts to understand the condition of peasantry and indigenous peoples—of that "labor force" whose labor conditions are interlinked with the workings of Nature.

Thus, it is one thing to elucidate why peasants' seasonal work is functional and profitable, subsidizing capitalist agriculture by merging with the unvalued forces of Nature that contribute to the production process in the sequential or seasonal moments of the "productive" labor process. It is quite another to extend this explanation to the causal reason for the poverty and persistence of peasantry. *Toute proportion gardée*, it would be tantamount to justifying the persistence of Nature because of its functionality in the agricultural production process, that is, to circumscribe the ontology of Nature to the theory of value.

Differential rent does not offer either the full explanation of the peasantry poverty and persistence. Land rent has become hybridized by the intensity of technological intervention and complexified with capital profits in overcapitalized agriculture.[1] Take for instance the case of transgenic productivity. Seasonal labor is still needed at certain moments of the production cycle. However, under these conditions, the "reserve army" of rural workers exceeds by far the employment furnished by

[1] The determination of differential rent by natural conditions is displaced by the technological alteration of soil's fertility. As Marx himself acknowledged, "rent could not be the constant measure of the degree of fertility of a piece of land because the modern application of chemistry is altering every minute the nature of soils and geological knowledge is starting to change in our days all the old estimations of relative fertility […] Fertility is not a quality as natural as it might seem; it is intimately linked to today's social relations […] Rent, instead of *grounding man to nature*, has linked the exploitation of the Earth to competition" (Marx, "Misère de la philosophie," in *Oeuvres Economie*, I, Paris: Gallimard, 1965, pp. 125, 123).

transgenic latifundia and other forms of capitalist agriculture. Unemployed workers do not always have the possibility of completing their income in seasonal work in neighboring plots; often they take flight to neighboring countries to engage in the over-capitalistic production of crops that then compete and displace the traditional communal lands' production, as is the case with Mexican peasants abandoning their *ejidos* and *milpas* to work in the capitalistic production of transgenic maize in the United States. Unwillingly they become "organic proletarian agents" of the disruption of their traditional livelihoods and the risks involved in polluting their rich eco-cultural patrimony of genetic resources (Alvarez-Buylla and Piñeyro 2013). Thus, the resistance and possible *rexistence* of peasantry must be conceptualized in the context of the complex determination of capital dispossession and extermination of the peasantry in the present geopolitics of "sustainable development" (Porto-Gonçalves 2002; Leff 2002a).

In brief, the problematic of peasant persistence and peasants' poverty lies in the actual exploitation and strategic needs of capitalism and the resistance strategies of the peasantry and indigenous people. But the problem for understanding the workings of these processes arises from the "poverty of theory" to account for the complex ontological conditions that explain the global domination of techno-economic rationality in modernity, the rootedness of the peasantry and indigenous peoples to a territory and the condition of humanity and of life in the planet Earth (Leff 2014, 2018). If Marx constructed the theory of the social structure that is accountable for the exploitation of labor and, as a consequence, of Nature—in short, for the production of poverty and land degradation—once capitalism developed in the modern world produced by the history of metaphysics, historical materialism falls short of means to explain the condition of the peasant's persistent poverty.

The rootedness of the peasantry and indigenous peoples to their lands and territories is the result not only of a synchronic but also of a diachronic determination that overflows the historical process of articulation and subsumption of the traditional modes of production to the increasingly dominant capitalist mode of production. It involves the relations of the ontological regimes that constitute these socio-economic formations in the way they depend on the ecological conditions and in the way they enact thermodynamic processes of their territories. In this perspective, even genetic structuralism approaches (Goldmann 1959) prove insufficient to grasp the attachment of people to their territories from an ontological-historical-anthropological-ecological perspective.

Cultural resiliency and rootedness in Nature might prove to have ontologically deeper roots and to be more transcendent than the logic of peasants' persistence and survival subjected to the capitalistic squeeze of their conditions of existence. Even Marx's understanding of the metabolic rift of capitalism regarding the alienation of the production process from the natural conditions of their existence is insufficient to assess the questions posed by the principle of sustainability. Exploitation and erosion of life conditions under capitalist production, as well as for all socio-economic formations, have to be envisioned through the entropic degradation produced by the economic process (Georgescu-Roegen 1971).[2] The sustainability of life calls for a broader theoretical perspective: a questioning not only of the persistence of the dominant and subsumed modes of production, but of the kind of knowledge and wisdom involved in the emancipation of peasants, indigenous peoples and human beings in the construction of sustainable productive processes and the persistence of life on the planet (Leff 2014, 2018).

In this respect, a fundamental question remains unanswered in the domains of traditional Marxism and eco-Marxism regarding peasants' and indigenous people's poverty, their historical and structural causes, and their responses to their persistence and survival. To transcend the domineering structure of capitalism, to construct a sustainable economic world order, and to reconstruct the sustainable livelihoods of the Peoples of the Earth and their sustainable futures the theory of value is insufficient. The main shortcoming of the theory of value is not that it fails to include the discontinuity of labor in seasonal production processes, such

[2] A metaphor from Nature might be useful in understanding the capitalistic draining of peasants' territories. The inscription and institutionalization of the CMP are analogous to the introduction of eucalyptus on ecosystems where it overgrows by absorbing water and nutrients from the soil, drying the land, sucking the blood and squeezing the lives of the people. Thus, capital is implanted and extends its roots deep in the conditions of the sustainability of ecosystems and subsistence of their peoples; capital pumps up all the energy necessary for its continuous growth and expansion, consuming and exhausting the conditions for the sustainability of life and the reproduction of peasants, of Nature, and eventually of the capitalist system itself.

as those of agriculture. The basic problem is that *Nature* is not valued and that Nature does not determine value or surplus value. Marx himself stated:

> The value of labour-power is determined, as in the case of every other commodity, by the labour-time necessary for the production, and consequently also the reproduction, of this specific article. Insofar as it has value, it represents no more than a definitive *quantity of the average social labour objectified in it.* Labour-power exists only as a capacity of the living individual [...] For his maintenance he requires a certain quantity of the means of subsistence. Therefore the labour-time necessary for the production of labour-power is the same as that necessary for the production of those means of subsistence [...] However, labour-power becomes a reality only by being expressed; it is activated only through labour. However, in the course of this activity, i.e., labour, a definite quantity of human muscle, nerve, brain, is expended, and these things have to be replaced. Since more is expended, more must be received. If the owner of labour-power works today, tomorrow he must again be able to repeat the same process in the same conditions as regards health and strength. His means of subsistence must, therefore, be sufficient to maintain him in his normal state as a working individual [...] the owner of labour-power is mortal. If then his appearance in the market is to be continuous, and the continuous transformation of money into capital assumes this, the seller of labour-power must perpetuate himself 'in the way that every living individual perpetuates himself, by procreation' [...] The labour-power withdrawn from the market by wear and tear, and by death, must be continually replaced by, at the very least, an equal amount of fresh labour-power. Hence the sum of means of subsistence necessary for the production of labour-power must include the means necessary for the worker's replacements, i.e., his children, in order that this race of peculiar commodity-owners may perpetuate its presence on the market. (Marx 1978, Vol. I, pp. 274–275)

In Volume II of *Capital*, Marx adds:

> Working time is always production time, the time during which capital is confined to the production sphere. However, it is not true, conversely, that the entire time for which capital exists in the production process is necessarily therefore working time [...] What is at issue here are not interrup-

tions in the labour process conditioned by the natural limits of labour-power itself [...] What is involved is instead [...] an interruption conditioned by the nature of the product and its production, *during which the object of labour is subjected to natural processes of shorter or longer duration*, [...] while the labour process is either entirely or partially suspended [...] Between seed-time and harvest, the labour process is almost wholly interrupted [...] In all these cases, additional labour is added only occasionally for a large part of the production time [...] therefore, the production time of the capital advanced consists of two periods: *a period in which the capital exists in the labour process, and a second period in which its form of existence—that of an unfinished product—is handed over to the sway of natural processes, without being involved in the labour process.* (Marx 1978, Vol. II, pp. 316–317, *my emphases*)

Moreover, in the *Grundrisse*, Marx reflected:

The fixed capital here allegedly acts quite by itself, without human labour, like, e.g., the seed entrusted to the earth's womb [...] The time required here for the product to reach maturity, the interruptions of work, here constitute conditions of production. Not-labour time constitutes a condition for labour time, in order to turn the latter really into production time. The question belongs only with the equalization of the rate of profit. Still, the ground must be cleared here. The slower return—this is the essential part—here arises not from circulation time, but rather from *the conditions themselves in which labour becomes productive*; it belongs with the *technological conditions of the production process* [...] *Value*, hence also surplus value, *is not = to the time which the production phase lasts, but instead to the labour time*, [both] *objectified and living, employed during this production phase. The living labour time alone [...] can create surplus value, because (it creates) surplus labour time.* (Marx 1973, pp. 668–669, *my emphases*)

What these citations of Marx disclose goes well beyond the fact that non-seasonal and not-working time that constitute conditions of the production process and for the reproduction of agricultural labor power is not adequately valued in the theory of value. They also reveal—and this is the crucial point of debate that poses the environmental question to Marxism—that the natural processes involved in the production of commodities, in the value of the labor force in general and in the reproduction of peasants' labor force in particular, are not valued. The problem is crystal clear: Nature contributes to production, but only socially necessary labor time—the labor-time necessary for production, and consequently for the reproduction of labor force—determined by technological progress, contributes to value formation and the establishment of the rate of surplus value! The problem does not lie only in the fact that labor time fails to coincide with production time, but more fundamentally in the fact that neither Nature's contribution to production nor the destructive effects of production on Nature is valued. While the poor peasant survives through the articulation of modes of production—many times in extreme conditions of poverty as capital does not pay for dignified, just, and egalitarian standards of living in the reproduction of its needed labor power—Nature does not get paid at all for its contribution to the overall productivity of capitalist agriculture, nor to the global economy. Simply stated, *Nature has been externalized by the economy*; Nature contributes to production but does not determine value, not in the way that the concept of value and surplus value theory are structured in Marxist theory of value.[3]

Recognizing this theoretical fault, Marx hinted at capital's "metabolic rift" as the alienation of capital from its natural conditions. To overcome it Marx wrote in Vol. III of *Capital*:

> Liberty in this sphere (in the field of natural necessity) can only consist in this, that socialized man, the associated producers, will govern the human metabolism with nature in a rational way, subjecting it to his collective control instead of being dominated by it as a blind power; in order to enact it with a minimum expense of energy and in more dignified and appropriate conditions to its human nature. (Marx 1981, Vol. III, p. 959)

[3] To be sure, it is not value theory that exploits and externalizes Nature, but the ontological condition of capital, that becomes the object of critical theory. Thus, Marx unveiled the will and the mechanisms of exploitation constituted and instituted in economic rationality, particularly of the CMP. However, Marxism does not deconstruct nor escapes metaphysical thought that constructed the objectified reality from where critical reflection emerges and is inscribed. Cf. Marcuse (2005).

Eco-Marxism addressed this condition imposed by capital accumulation and reproduction over Nature as capital's *metabolic rift* (Foster 2000) and as *the second contradiction of capital*, disclosing the fact and how capitalism destroys the ecological conditions for the reproduction of capital (O'Connor 1998). However, displaying and internalizing this forgotten contradiction will not transcend the constraints of the Marxist theory of value to envision and to construct a sustainable world order. Not only the uncertainty of ecological processes challenges the purpose of rational control over human metabolism with Nature. The environmental question calls for a deconstruction of economic rationality and the transition towards an environmental rationality that beyond any will to power of the rational, is grounded in the social imaginaries of indigenous peoples and peasant societies, in the self-understanding of their conditions of life and livelihoods and in the collective control of the ways in which their modes of production, of their practices of appropriation and transformation of Nature, enact the metabolism of their life-territories.

In the course of these theoretical developments, peasantry has survived and persists through complex socio-cultural-political strategies of resistance and emancipation, beyond their adaptation to the conditions of capitalist agriculture. However, how long can we expect Nature to hold onto the biosphere's life-support systems when its resiliency mechanisms are being eroded by capitalism? How can peasants survive without a territory that supports their livelihoods?

Energy-Value, the "Simple Reproduction Squeeze" of Peasants' Labor and the Unvalued Forces of Nature

From the standpoint of ecological economics and eco-Marxism, economics and Marxism have been left without an adequate theory of value (Altvater 1993).[4] Environmental economics is the economics of the

[4] "But in the course of economic development, the scarcity of goods in relation to human needs is compounded by the fact that they are no longer at wide enough disposal; *to scarcity comes to be joined shortage*. Paradoxically, the scarcity of resources is what makes rational economic calculation

environmental externalities of the economic system, that is, of the natural processes that contribute to production but are not accounted for by the principles of value formation, but instead, are the buried and ignored ecological costs of the economic process. Ecological economics is, in no small extent, the economics of energy flows throughout the economic process, the economics of the metabolism of matter and energy governed by the entropy law. Following the proposal of Howard Odum, ecological economics has attempted to construct an energy value theory. From Podolinsky to Neurath, and through the school of ecological anthropology—from Leslie White to Richard Adams and Roy Rappaport—there have been numerous attempts to complement Marxist labor-value theory, with a theory of energy-value (Martínez-Alier 1987).

The idea of combining an energy theory of value with Marx's theory of labor-value seems to have been an original contribution of Podolinsky. Thus, he intended to determine the minimum conditions for human survival with the analysis of energy flow efficiency (Martínez-Alier 1995, p. 72). However large the impact of this principle in ecological economics and in determining the rationality of cultural organization through the energy flows in agriculture in ecological anthropology (Rappaport 1971), these attempts to complete Marx's theory of value with thermodynamic and ecological measures have failed. Though, in principle, complex matter and energy flows should be more natural to reduce to basic energy unit measures than it is to transform complex labor to the abstract, simple and direct labor power that defines value as socially necessary labor time, it is impossible to internalize a homogeneous ecological measure of matter, energy, and time within a unit of value. As Albert Puntí (1988) showed, "same quantities of energy coming from different sources have different times of production" (cit. in Martínez-Alier 1995, p. 53). In brief, the approaches of ecological economics lack a quantitative unit of matter-energy-time that could fit into the structural dialectics of value to surplus value, and that would enable one to establish a rate of exploitation of Nature as a surplus ecological value.

both possible and necessary. However, its very success in growth and expansion have led to a *shortage* in the vital quality of nature, which in turn undermines the principle of *scarcity* and thus economic rationality itself" (Altvater 1993, p. 6).

Consequently, the limitations of the theory of value to determine peasant's poverty and the persistence of peasantry are not solved by adding the non-worked days to account for the total value of the reproduction of rural labor force, as the failure to consider the contribution of Nature to value formation is not resolved. A quantitative theory of value is untenable once its conceptual strength has collapsed with technological progress—the indeterminacy of value by the average productivity of technologies or by the more productive technology, and ultimately by the displacement of direct labor by scientific and technological knowledge in the production process—as one is left with the impossibility of defining a unit of value (socially necessary labor-time) to produce exchange value by any homogeneous unitary measure of matter-energy or labor-time value.[5] Explaining peasants' poverty in terms of value theory would require a re-elaboration of the theory of labor-value and of capital, not just by adding seasonal labor, but by rendering an account of the contribution of Nature's processes to value and price formation, not merely in the cycle of the simple reproduction of capital, but in its expanded reproduction, a purpose that has been abandoned by eco-Marxism as an impossible task.

This does not rule out the fact that peasant's labor subsidizes capitalist agriculture through the "simple reproduction squeeze" as well as the fact that agricultural production is undervalued in order to fuel industrial development and urban growth. Moreover, peasant's poverty is not only produced through the "simple reproduction squeeze" in the reproduction of its labor force, but is also affected by the "metabolic rift" caused by exhausting the soil fertility by transferring its elements to urban consumer's commodities, the transfer of value—of these ecological distribution processes—from rural areas to the cities, and from poor to rich countries, that are not captured by the theory of value. Thus, Marx wrote that,

> Big landholdings reduce agrarian populations to an always decreasing minimum, opposing to it an always growing industrial population gathered in the big cities. The conditions thus created an irreparable rupture in the metabolism determined by the laws of life, wherefrom the exhaustion of

[5] Cf. Chap. 5.

> the earth's resources that commerce extends well beyond national frontiers (Liebig) [...] Big industry and big mechanized agriculture act together. If originally the first tends to ruin the labor force, man's natural force, while the second exploits the natural force directly from the earth, they end by joining up in their progress: the industrial system in the countryside weakens rural workers, while industry and commerce supply the means to agriculture to exhaust the Earth. (Marx, *Le Capital*, III, in *Oeuvres Économie* II, p. 1424)[6]

The ontological reduction of all entities to the realm of objective calculation—from the use of land measuring since the origin of sedentary societies to land surveying devices in the capitalistic rationalized agrarian production—explains the constructed mechanisms of capital accumulation—the alienation of modern culture from Nature; the separation of urban life from rural livelihoods—that are not accountable by a quantitative value theory. The scale of production, the forms of property and technological development play an essential part in capitalist agriculture by determining the economic surplus that can be appropriated and reinvested for the expanded reproduction of capital. Today, natural forces of production are hybridized by biotechnology to fuel capitalist agriculture. Thus, a "post-Marxist" theory of differential rent is envisioned as a process of *ecological distribution*, whose workings go beyond the fact that the more productive plots are established in the best preserved and the most resilient land.[7] The intervention of technology in the structure, the chemistry and the metabolism of the biosphere and the geosphere complexify the possible transformation of the flows of energy—of the entropic degradation and negentropic potentials—into economic value measures. As land is fertilized with fossil fuels, which are sold at prices below their environmental costs, it is impossible to include the costs of reposition of exhaustible resources in the economic calculations of

---

[6] In his analysis of the agrarian question, Kautsky expressed the view that to the transfer of value from rural to urban areas, "corresponds an increased loss of nutrients in the form of corn, meat, milk and so forth, that the farmer must sell to pay the taxes and interests from debt and rent [...] Even though such transfer does not signify the exploitation from agriculture in terms of the law of value [...] it does imply a material exploitation as the impoverishment of the earth's nutrients" (Kautsky 1899/1998, Vol. 2, p. 214).

[7] Cf. Chap. 2.

capitalist production and its effects in terms of ecological distribution and accumulation by dispossession (Harvey 2003).[8]

Moreover, how would differential rent account for the "conservation" strategies of the "green economy" and "sustainable development" where peasant's labor is involved and marginally paid for the preservation of their bio-diverse territories to compensate for the ecological damage inflicted by the global economy, but which in terms of value theory appears as non-productive labor? Following the above, poverty is not a homogeneous state of being, nor can it be reduced to its capitalist determinations. The peasant's "simple reproduction squeeze" and the exhaustion of Nature is operated via diverse and complex mechanisms to keep moving the "treadmill of production" (Gould et al. 2008).

In what follows, the category of peasantry will be applied not only to people working the land (the *campesinado*) as a social class, but also the indigenous peoples that inhabit rural areas. The distinction between them is not only an ethnic question and a matter of their miscegenation with the peasantry, but an issue regarding the degree to which they maintain (or not) their cultural identities and traditional practices, and their degree of integration into the capitalist system and the global economy. Peasants' ways of survival, of persistence and forms of *being-in-the-world*, are the result of history. Capitalism drove and accentuated the separation of the rural peasantry from industrial proletarians and urban dwellers through the dualist construction of a society that divided Nature from culture. The CMP structured the dialectical, functional, and systemic relationship between rural and urban areas. No wonder we keep going back to the origins of capitalism to understand what triggered this condition of the peasantry in modernity all the way up to the present day.

However, if capitalism has not expelled peasants completely from rural areas and agricultural production, can this be explained merely by Marx's theory of differential rent or by some diffuse "obstacles derived from some natural features of agricultural production" (Mann and Dickinson

[8] An example in the recent past is how low-priced Mexican oil has fueled the artificially high productivity of capital-intensive monocultures in the United States, including maize that then displaced traditionally produced maize at the smaller scale and lower agrochemicals intensities. This vicious circle produced unequal competition with Mexican peasants' production that has remained poor because of these complex capitalist determinations.

1978)? Is value theory a conceptual tool to understand peasants' poverty? Alternatively and more precisely, Can we assert that the theory of value is responsible for misvaluing the work of the peasantry?

The fact is that by construing the theory of value as the key to assert the fundamental contradiction of capital to industrial labor, Marxism explained the structural condition that reduces the value of labor and increases surplus value. However, it just did not value the "workings" of Nature—the generativity of *physis*; ecological productivity and resiliency—and of the creativity of peasantry that acts together with Nature to lower the value of the reproduction of the labor force of industrial workers and peasants themselves. Although the theory of value is the critical theory of capitalism, it is nevertheless inscribed in the techno-economic ontology of modernity—in Heidegger's world of *Gestell*, of the disposition of all entities as objects and the calculative appropriation of the world (Heidegger 1935/1977).[9] However, if technological change inscribed in the ontology of capital determines the production of surplus value, it does not operate outside the world where the "ontology of Nature" determines the productivity of subsistence commodities for the basic needs of the proletariat nor the reproduction of peasantry. By ignoring the participation of the forces of Nature in the overall production process, the theory of value cannot account for the ecological conditions for the expanded reproduction of capital. The "productive consumption of Nature" in the theory of value and surplus value has become the unvalued destructive (entropic) consumption of Nature.

In this "ecological perspective" the major problem is not only that of discontinuous labor time in agriculture and the value of agricultural labor power, but of the contribution of Nature in establishing the value of the means of subsistence and of any commodity, in a way that would allow capitalism—and any other mode of production—to internalize the ecological conditions for its expanded reproduction in space and time. This

[9] Marxist critical thinking has remained inscribed within metaphysical calculative thinking. From that standpoint, it uncovers the exploitative nature of the capitalist system, it views in the proletariat the objective subject of history, it claims justice for the poverty of dispossessed peasantry, squeezed by the law of value, oppressed by the techno-economic rationality of modernity; but it overlooks the ontology of life that sustains their livelihoods, the resilience and the productive metabolism of the biosphere, humanity, and the capitalist system itself.

problem has no solution within the theory of value or the structure of the capitalist mode of production. It calls for the deconstruction of economic rationality and the construction of *another productive rationality*. Beyond a change of economic paradigms, it implies a shift from economic rationality to environmental rationality, a turn from the techno-economic ontology of modernity to the ontology of life (Leff 1995, 2004, 2014, 2018).

Peasants' Persistence/Resistance/Rexistence: The Struggle for Life

The enigma of peasants' persistence not only contradicts the Leninist prognosis of their disappearance from the face of the earth (being either proletarianized by capitalist agriculture or absorbed by industry or tourism). Peasants are inscribed in another existential ontology: not economic ontology but the ontology of life. As shown by Chayanov (1974), the family peasant unit does not seek to maximize profits or to obtain the average rate of profit. While capitalist agriculture profits from this condition of traditional petty commodity production, peasants have not only persisted by becoming functional to capital; instead, in a more extended period, indigenous peoples/peasants have survived by resisting dispossession by colonization and capitalistic domination. They have resisted by *cultural resilience*, by the deep cultural rooting of their ontological existence in their ethnic spaces and by their innovative adaptations to displacement from their native territories; by emancipation from subjugation of their traditional knowledge and practices, and from erosion of the ecological conditions of their habitats; from the reinvention of their identities and reconstruction of their life-territories.

From the standpoint of the environmental question, the critical issues that arise are not only the conservation of biodiversity for Nature's sake or the persistence of peasantry in the process of modernization and continuous capital accumulation; but the survival and the destiny of the living planet and human life. In this perspective, the theory is not only concerned with the disappearance of traditional cultures and life forms by capitalist entropic use and its destructive consumption of Nature. The environmental question focuses on critical thinking in the sustainability of life on the planet. In this perspective, the outstanding issue arises from the ineluctable *entropic degradation of Nature induced by the economic process*. Economic growth feeds not only on the non-valued conditions of life and on scarce natural resources but transforms all matter and energy that flows through the process of production (and consumption) in degraded energy, ultimately in the form of unrecyclable matter and irreversible heat. These are the workings of the economic process on Nature, on the complex structure of ecosystems, on the life support systems of the biosphere. This is the fatality of the economic process that triggers the *entropic death of the planet.*

The persistence of peasantry, their ways, and means of survival, their emancipation strategies, cannot be ascribed to their condition within the worldwide domination of the capitalist system and its economic rationality. The becoming of their being is deployed in the *resistance/rexistence* of their cultural being in the perspectives of environmental rationality and the construction of their sustainable life-worlds, opening the possibility of building a sustainable global economy. The reversal of unsustainable capitalism and the transition to a sustainable mode of production is not solved by an adjustment of value accounting to equalize economic distribution to seasonal work of the "persistent peasantry" nor by the restructuring of value theory or the ecological reform of economics to internalize negative environmental externalities. Environmental rationality challenges eco-Marxism, ecological economics, and political ecology to devise an alternative mode of production based on the conservation and enhancement of ecological conditions and the productive forces of Nature, embedded in socio-cultural-ecological relations of production—in the web of life—and embodied in different cultural beings.[10]

In brief, a sustainable economic process cannot be achieved—as environmental economics intends—by "internalizing externalities," having first coded those "externalities"—ecological breakdown, pollution, biodiversity, climate change, greenhouse effect gases, environmental goods and services—in economic terms: say, having capitalized Nature and imposed

[10] Cf. Chap. 7.

the commodification of Nature; nor can we make the economic process more ecological by restraining economic behavior to the ecological conditions for the reproduction of Nature (Passet 1979) so as to attain a "steady state economics" (Daly 1991). Georgescu-Roegen's "bioeconomics" is the more radical critique of economic rationality after Marx, underlying and expressing the "second contradiction" of capital, the ineluctable self-destruction of the ecological bases of capitalism by the way that the economic process activates and magnifies the workings of the entropy law (Georgescu-Roegen 1971). However, this "bioeconomics" remains a critique of the thermodynamic unsustainability of the prevalent economic rationality rather than economics based on the thermodynamic principles, in the negentropic potentialities and the ecological conditions for sustainable production.

If capitalist-induced *entropic degradation* is what is driving the ecological destruction of life support systems and cultural resiliency, then the future persistence of peasantry will not depend on its functional utility for capitalism, but on envisioning and constructing a sustainable mode of production, one based in the *negentropic potentials of life*. That is, a mode of production based on "managing" the ecological conditions of the biosphere and internalizing the material and the symbolic conditions of human existence. Thus, the labor process must be envisioned and oriented toward enhancing and magnifying the principle of life: from the emergence potentiality of *physis* and the process of photosynthesis to the eco-technological productivity of the biosphere.[11]

A *sustainable negentropic paradigm of production* is built on the complex articulation of three orders of productivity: ecological, technological, and cultural. Ecological productivity is based on the ecological potential of different ecosystems. It can be enhanced through scientific research, green technologies, and sustainable management practices to define and guide the cultural-economic value of the techno-ecological output of the

[11] Photosynthesis is the negentropic transformation of radiant solar energy into biomass from where the complex forms of life have emerged (Schrödinger 1944). Entropy works in the metabolism of food chains and in all industrial metabolisms throughout the economic process to maintain the complex organization and evolution of life. Notwithstanding this ineluctable entropic condition of all thermodynamic processes, negentropy can be conceived as the overall process that forms and maintains the life support systems of the planet. Thus, sustainability must be constructed in the balance of negentropic/entropic processes in the metabolism of the biosphere.

different production processes. Ecological technologies and practices include the management of high-efficiency photosynthesis, secondary succession—the regeneration, growth, and evolution of natural ecosystems after the slash and burn of traditional shifting agriculture—selective regeneration of valuable species in ecological processes, and multiple associated cropping, agro-ecological, and agroforestry practices (Leff 1986b, 1995).

This alternative paradigm of production is articulated in a spatial and time frame of non-modern cultural imaginaries and ecological practices. A rich diversity of peoples/cultures and their different territorialities open a new theoretical perspective that incorporates the relations of cultures to their territories. As Milton Santos argued, different temporalities cohabit in geographical space. This alternative conception of historical time has important political implications for social movements, such as the actuality of ancestry invoked by the Afro-Colombians of the Pacific and the imaginary of *Vivir bien* ("living well") of Andean peoples, opening up the question of co-evolution of peoples/cultures and Nature/territories. Their territorialization processes, based on ecological conditions and cultural practices, clearly differ from the logic of space occupation and territorial division of labor in the capitalist world. Thus, ultimately what is at stake in political ecology are conflicts over territories; not only the clash of interested parties for the appropriation of land and natural resources, but the confrontation of alternative modes of production and patterns of space-construction, of ways of territorializing and inhabiting the world, of "living sustainably in a sustainable world" (Leff 2014, Chapters 3, 6).

The social agents of this historical transformation are not the industrial proletariat or urban citizens, but the inhabitants of rural territories: the dwellers in ecosystems; the peasants and indigenous peoples of the Earth. These are the agents for building a *negentropic society*. This statement brings us to the point of inquiring where and how do these people stand on Earth, what are their attachments to their territories, what are their imaginaries of sustainability, and what is their potential for becoming the social actors in the construction of a sustainable world.

The peasants' struggles of the twentieth century (Wolf 1972b), from the Mexican agrarian revolution to present social movements, such as *Via Campesina* and the "Landless Movement" (*Movimento dos Trabalhadores*

Rurais sem Terra), with their distinctive categories of peasantry and revolutionary strategies, focused on their attachment to, their claim to, and their appropriation of land as a basis for survival and the maintenance of their traditional cultural organization. Today, environmental struggles do not only claim a piece of land, but a *territory*, a space to be restored and reconstructed from their "deep" cultural identity roots—i.e., "profound" México (Bonfil Batalla 1987)—as the habitat where they can experience their *habitus*, where they can deploy their imaginaries and practices in order to preserve livelihoods, envision their sustainable life-worlds. A territory is conceived as a central multidimensional space for the organization of communities' ecological, economic, and cultural practices; space where cultural identities are being reconfigured, where the new political actors and environmental movements are emerging to reappropriate their bio-cultural heritage and reconstruct their life-territories. Territorial conflicts are the core of political ecology in Latin America.[12]

The new socio-environmental indigenous and peasant movements for the reappropriation of Nature emerge in this post-Marxist and post-structural perspective. A new peasantry is arising from the ecological ground that has been marginal though functionally linked to the capitalist system precisely in those regions that remained at the margins of the market, occupied by traditional populations that harbor the most significant natural wealth of the planet in water and biodiversity.[13] Thus, facing the colonizing and exploitative character of the new geopolitics of globalization and sustainable development, a series of critical and creative responses from different Latin American peoples emerge in the conflictive field of political ecology. An emblematic example is that of the struggles of *Seringueiros* in the Brazilian Amazon Region who, from their

[12] In words of Arturo Escobar, "The territory is conceived as a multidimensional space, fundamental for the creation and recreation of the communities' ecological, economic and cultural practices […] We can say that in this articulation of cultural identities and appropriation of the territory underlies the political ecology of the social movement of the black communities. The demarcation of collective territories has led activists to develop a conception of territory that emphasizes articulations between settlement patterns, space use and use-meaning practices of resources" (Escobar 1999, p. 260).

[13] It is most interesting to note the coincidence and overlapping of preserved biologically diverse regions of the world, on the one hand, and the persistence in these territories of cultural diversity on the other (Boege 2008).

syndicalist claims to the land where they had worked as proletarian rural workers of the rubber industry in the late nineteenth century, have come to establish their *extractive reserves* as a strategy of sustainable production and livelihood (Porto-Gonçalves 2001a, 2004).[14]

The radical imaginaries root ecosystem dwellers in their territories are manifest as strategies of resistance/*rexistence*. We can trace this political existential ontology in the discursive strategies of present socio-environmental movements in Latin America, as the one expressed in the political discourse of the Afro-Colombian movement of the *Process of Black Communities* (Proceso de Comunidades Negras) of the Colombian Pacific Coast, in their claims for the reappropriation of their rich patrimony of biodiversity, the reinvention of their identities and the construction of their future (Escobar 2008). Thus, the Process of Black Communities in the Colombian Pacific claimed their rights to:

> 1. *The reaffirmation of identity* (the right to being black) [...] from the perspective of our cultural logic and worldview in all of its social, economic and political dimensions [...] 2. *The right to the territory* (as the space for being), as a vital space and a necessary condition for the recreation and development of our cultural vision [...] as a habitat and space where black people develop its being in harmony with nature. 3. *Autonomy* (the right to the exercise of being/identity) [...]. 4. *Construction of an autonomous perspective for the future* [...] an autonomous vision of economic and social development based on our culture and traditional forms of production and social organization [...]. 5. *Declaration of solidarity* [with...] the struggles for rights of black populations throughout the world [...] for alternative life projects. (Escobar et al. 1998, cited in Escobar 2008, p. 223)

These emancipation processes are spreading in the claims of indigenous peoples for autonomy, territory, and dignity. They are being legitimized in the constitution of cultural and environmental rights, mobilized by their movements of indigenous, peasant and Afro-descendant communities in defense of their rights to a territory and their traditional

[14] Chico Mendes, leader of a new peasants' socio-environmental movement in their struggle against the hegemonic economic forms of exploitation of Nature, instituted the strategy of *extractive reserves* as a new "agrarian reform" countering the strategies of "sustainable development."

practices in many Third World countries and in all latitudes of the Latin American region, from the Amazonian cultures to the Zapatista movement in Mexico, from the Mapuche in the further south of the continent, to the Seri or Comcaac in the arid north of Mexico. The fire of life enlightens the environmentalism of the peoples of the Earth and the persistence of the peasantry of the countryside that supports the *rexistence* of a sustainable human world and life itself on Earth. These are the social points of anchorage of emergent environmental rationality.

Thus, the persistence of peasantry and traditional peoples can be theorized beyond its functional interdependence with capitalist agriculture, as supporters for building other possible worlds, if we trust that humanity has not been engulfed or swept away by the progress of capital expansion or drowned by the entropic death of the planet; if we can still envision sociologically and enact politically a sustainable future grounded in the immanence and under the conditions of life, as the *social enownment of Nature*.

Bibliography

Altvater, E. (1993), *The future of the market*, New York: Verso.

Alvarez-Buylla, Elena y Alma Piñeyro (Coords.) (2013), *El maíz ante los transgénicos. Un análisis integral sobre el caso de México*, México: CEIICH-UNAM/UCCS.

Boege, E. (2008), *El patrimonio biocultural de los pueblos indígenas de México. Hacia la conservación in situ de la biodiversidad y agrobiodiversidad en los territorios indígenas*, México: INAH/Comisión Nacional para el Desarrollo de los Pueblos Indígenas.

Boltvinik, J. (2010), "Poverty and persistence of the peasantry", Background paper for the International Workshop on this Issue, CROP/El Colegio de México.

Bonfil Batalla, G. (1987), *El México profundo, una civilización negada*, México: Grijalbo.

Chayanov, A. (1974), *La organización de la unidad económica campesina*, Buenos Aires: Nueva Visión.

Daly, H. (1991), *Steady-state economics*, Washington: Island Press.

Escobar, A. (1999), *El final del salvaje. Naturaleza, cultura y política en la antropología contemporánea*, Bogotá: CEREC/ICAN.

———. (2008), *Territories of difference. Place, movements, life, redes*, Durham/London: Duke University Press.

Fossaert, R. (1977), *La société, Tome 2, Les structures économiques*, Paris, Seuil.

Foster, J.B. (2000), *Marx's ecology: materialism and nature*, New York: Monthly Review Press.

Georgescu-Roegen, N. (1971), *The entropy law and the economic process*, Cambridge, Mass.: Harvard University Press.

Goldmann, L. (1959), *Recherches dialectiques*, Paris, Gallimard.

Gould, K.A., D. Pellow & A. Schnaiberg (2008), *The treadmill of production. Injustice and unsustainability in the global economy*, Boulder: Paradigm publishers.

Harvey, D. (2003), *The new imperialism*, Oxford: Oxford University Press.

Heidegger, M. (1935/1977), "The question concerning technology", in *Heidegger. Basic writings*, trans. D. Farrell Krell, New York: HarperCollins, pp. 283–317.

Kautsky, K. (1899/1998), *The agrarian question*, Swan: London.

Leff, E. (1986a), *Ecología y capital: hacia una perspectiva ambiental del desarrollo*, México: UNAM

———. (1986b), "Eco-technological productivity: a conceptual basis for the integrated management of natural resources", *Social science information*, 25 (3): 681–702.

———. (1993a), "Marxism and the environmental question: from critical theory of production to an environmental rationality for sustainable development", *Capitalism, nature, socialism*, Vol. 4 (1): 44–66, Santa Cruz, California.

———. (1995), *Green production: towards an environmental rationality*, New York: Guilford.

———. (2002a), "La geopolítica de la biodiversidad y del desarrollo sustentable: economización del mundo, racionalidad ambiental y reapropiación social de la naturaleza", in Ceceña, A.E. & Sader, E., *La guerra infinita: hegemonía y terror mundial*, Buenos Aires: CLACSO-ASDI, pp. 191–216.

———. (2004), *Racionalidad ambiental: la apropiación social de la naturaleza*. México: Siglo XXI Editores.

———. (2014), *La apuesta por la vida: imaginarios sociales e imaginación sociológica en los territorios del Sur*, México: Siglo XXI Editores.

———. (2018), *El fuego de la vida: Heidegger ante la cuestión ambiental*, Mexico: Siglo XXI Editores.

Mann, S.A. & J.A. Dickinson (1978), "Obstacles to the development of a capitalist agriculture", *Journal of peasant studies*, Vol. 5 (4): 466–481.
Marcuse, H. (2005), *Heideggerian Marxism*, edited by Richard Wolin & John Abromeit, Lincoln & London: University of Nebraska Press.
Martínez-Alier, J. (1987), *Ecological economics. Energy, environment and society*, Oxford/Cambridge: Basil Blackwell.
———. (1995), "Distributional issues in ecological economics", *Review of social economy*, Vol. LIII (4): 511–528.
Marx, K. (1973), *Grundrisse,* Penguin Books, UK
———. (1976, 1978, 1981), *Capital* (three volumes) Penguin, UK.
O'Connor, J. (1998), *Natural causes: essays in ecological Marxism*, New York: Guilford.
Passet, R. (1979), *L'économique et le vivant*, Paris: Payot.
Porto-Gonçalves, C.W. (2001a), *Geo-grafías. movimientos sociales, nuevas territorialidades y sustentabilidad*, México: Siglo XXI Editores.
———. (2002), "O latifúndio genético e a r-existência indígeno-camponesa", *Geographia*, Ano 4, No. 8, Niterói: Universidade Federal Fluminense, pp. 7–30.
———. (2004), *Geografando nos varadouros do mundo: da territorialidade seringalista à territorialidade seringueira*, Brasilia: IBAMA.
Puntí, Albert (1988), "Energy accounting: some new proposals", *Human ecology* 16 (1):79–86.
Rappaport, R.A. (1971), "The flow of energy in an agricultural society", *Scientific American*, 224 (3):116–132.
Schrödinger, E. (1944/1969) *¿What is life? The physical aspect of the living cell*, London/New York: Cambridge University Press.
Wolf, E. (1972b), *Las luchas campesinas del siglo XX*, México, Siglo XXI Editores.

7

Marxism and the Environmental Question: Towards an Environmental Rationality for Sustainability

The Ontological Grounds of Historical Materialism

After the dissolution of the theory of value as a cornerstone of *historical materialism* and the core concept of the science of *capital*—of the development of the forces of production as the driving force of history–, it might seem "anti-paradigmatic" to intend to ground the environmental question in the eroded soil of Marxist theory of value and nonsense to try to reformulate historical materialism from an environmental perspective, beyond its oblivion of Nature and the failed attempts to ground it in natural dialectics. However, ¿is it not that contradiction has its profound meaning in the counter-discursive character of thinking? If the collapse of really existing socialism and the power strategies of the global order

Paper delivered at the XIII Coloquio de Antropología e Historia Regional, organized by El Colegio de Michoacán, August 7–9, 1991. Published originally in Leff, E. (1993). Marxism and the environmental question: From critical theory of production to an environmental rationality for sustainable development, *Capitalism, Nature, Socialism*, 4(1), pp. 44–66. This revised text incorporates a synthetic version of an article unpublished in English: Leff, E. (1980). Alfred Schmidt y el fin del humanismo naturalista, *Antropología y Marxismo*, No. 3 abril–septiembre de, pp. 139–152.

E. Leff, *Political Ecology*, https://doi.org/10.1007/978-3-030-63325-7_7

based on market economy have in a way undermined the foundations of historical materialism and put into the proof of history the Marxist political economy, it remains that Marxism has provided the most critical analysis of the root causes of the environmental crisis arising from capital accumulation and economic rationality. However, the science of capital has failed to integrate ecological processes and cultural values into the general conditions for the expanded reproduction of capital and its transition toward a sustainable mode of production and social rationality grounded on the conditions of life.

Marxism offers a theoretical basis to demystify the dominant neoliberal discourse and to clarify the current conflict between the conditions for the expansion of capitalism (based on investments, markets, and profits) and those of ecological and social sustainability. The discourse of modernity and post-modernity has marginalized Marxism as a critical theory of history, of society and of the environment, of theory capable of understanding the natural and human forces driving the conditions and the destiny of life in the planet (Habermas 1987). In the a-critical discourse of "sustainable development" and "natural capital," the environmental question is reduced to emergent issues to be considered under the "new economic order" and the transition to a liberal democracy under the ruling ontological regime of capital. Although these issues have generated diverse critical social responses, manifest in the radical ecological discourse and in the new socio-environmental movements they still lack a theoretical framework and a strategic program to build an ecologically sustainable world order.

The ecological vacuum of conventional economic theories and historical materialism has given rise to several strands of analysis and reflection regarding the effects of economic growth on environmental degradation and the destruction of the biosphere's ecological web of life that supports the sustainable supply of natural resources. The theory of value and economic theory, in general, face the difficulty of quantifying natural wealth and cultural meanings, as well as assessing complex and long-term economic, social and ecological processes, in terms of present market values or prices. The irrationality of capital accumulation becomes manifest from its effects on the depletion of natural resources and environmental degradation, as well as the inability of economic rationality to control and internalize negative impacts on Nature and society with its theoretical concepts and policy instruments.

Thus, the environmental crisis does not only manifest the most profound critique of the ecological irrationality of economics; it comes to question the ontological grounds of historical materialism. Even though authors have devoted their reflections to rescue the ecological grounds of Marx's thinking and historical materialism (Benton 1966; Burkett 1999; Foster 2000), the question is to see if logical arguments under the scrutinity of the consistency of Marx's theoretical construction support this claim. Marx intended to ground his thinking in a materialist ontology, from his Epicurean formation in his early days to his will to turn Hegel's idealist dialectics on its social factual material feet and in class struggle; since his earlier writings Marx and Engels confronted all forms of idealism and disclosed the roots of social contradiction from the material condition of the capitalist society and the alienation of exploitative social relations as relations among things—advancing on Heidegger's notion of total world objectification in the era of *Gestell*—and the fact that Marx conceived metabolism (*Stoffwechsel*) between the economic process and industrial transformation and its exchanges with Nature. However, the ontology that prevailed in Marx's views of the transcendence of social man throughout history is the dialectics between the development of productive forces and social relations of production. In modernity, the driving force of history is the dialectics of capital and economic value as its determinant essence, and not the creative evolution of life arising from Heraclitus's *physis* to negentropic potentials of the web of life in the biosphere (Leff 2020). At the bottom of Marx's conception of the driving force of history is the class struggle. However, class consciousness and social conflict have been framed under the logic of capital accumulation and the development of productive forces driven by the powers of rational thinking, of scientific discovery and technological innovation: of *human logos* over the emergent, generative, diversifying forces of *physis*.

With these questions in mind, we can begin to reconstruct a Marxist theory of production grounded on the potentials and conditions of Nature: a productive rationality that accounts for the incorporation of natural processes in the general conditions of production and the construction of an environmental rationality based upon the principles of eco-technological-cultural productivity, ecological resiliency, territorial rights and environmental justice. Whereas Marxism offers a historical, economic, and social perspective on the study of environmental problems and a theoretical paradigm that can be reworked in a way to incorporate the environment into the productive process, critical environmentalism offers Marxism knowledge as well as political and ethical principles to assess and judge the socio-environmental and ecological conditions to build sustainable productive practices and livelihoods, and a sustainable world order. Beyond the effort to unearth an underlying ecological ground from the natural dialectic of orthodox Marxism and the appropriation of a balanced ecology from collective workers, historical materialism needs to be renewed from the principles of the ontology of life and the existential conditions of the Peoples from the Earth.

For this purpose, historical materialism, like the critical theory of history and political economy, must rework the categories of Nature and culture, placing them in the very center of the productive process: as social relations of production and into the dynamics of the development of productive forces. It must also question the capitalist and "socialist" models of economic growth based on the maximization of profits and short-term economic surplus. A new theory of production must be constructed incorporating the environment into the productive process: not as externalities of the productive system nor only as a general condition for sustainable production, but rather *as the potential for alternative productive rationality.* The principles of environmental rationality provide a new basis for the sustainable development of the forces of production that integrate cultural, technological, and ecological processes, thus generating productive processes that are more equitable and sustainable by being grounded in the ontology of life and on the political power of social-environmental movements in the construction of life-territories and a sustainable world order (Leff 2004, 2014, 2018).

2. Deconstruction of Economic Value Theory and Construction of Marx's Concept of Nature

Marx constructed the concept of objective value—of labor-value as "socially necessary labor time," as a "social law" and as a theoretical principle to found "critical thinking" over "vulgar economics," that is, of the theoretical construction that alienated man from Nature in modern times. However, if classical economics masked the fundamental exploitative social relation of the capitalist mode of production, Marx could not avoid externalizing Nature from the process of value formation. This "paradox" opens an epistemological question and to deconstruct the ontological regime that configured the mode of comprehension of the social structure and on the material processes that mobilize history, including the objectivity of the social process that reifies reality by considering it to be a relation among things and that externalized Nature from the theory of value. In this context emerges the inquiry over the concept of Nature in Marx, and its place in the scheme of comprehension of historical materialism. In the epistemological configurations derived from the ontological regime that Heidegger named *Gestell* (Heidegger 1935/1977)—where the totality of beings are disposed as objects to be appropriated through measure and calculation, both the concept of labor-value and the concept of Nature are *En-Framed* as the grounding categories of historical materialism. In this scheme, history and the capitalist mode of production lose their natural ground; Nature is captured by the lens of an objectifying rationality, of an ontological regime that orients social praxis in the teleology of history driven by the economic process: in a genealogy of modes of production that entail different evolving modes of appropriation of Nature, in the dialectical transcendence of history driven by the development of productive forces and social relations—the class struggles confronted in those social structures—that destine life in the planet.

In a hermeneutical analysis of *The Concept of Nature in Marx*, Alfred Schmidt (1971) engaged in unearthing a fundamental category in Marx's historical materialism. The concept of Nature is constructed in Marx's philosophical writings as an ontological category, more so than as a scientific concept supporting his theory of the capitalist mode of production. The concept of Nature in Marx expresses a trans-historical ontological category that would comprise the world totality and its historical becoming from a materialist or realist perspective, breaking up with Kant and Hegel's transcendental idealism. Here, praxis as labor that transforms matter occupies the central place. The concept of Nature is conflated as the whole of reality. Thus, Schmidt states that

> The work here presented […] can be described as an attempt to present in its main aspects the mutual interpenetration of nature and society as it takes place within nature, conceived in its broadest sense as the total reality comprising both moments […] Marx showed that society itself was a natural environment […] in the 'metaphysical' sense that Mar's theory is a theory of the world as a whole […] Nature was for Marx both an element of human praxis and the totality of everything that exists […] The concept of "nature" is identical to the "whole of reality" in the Marxist view. (Schmidt 1971, pp. 16, 27, 29)

Assigning an ontological basis to the concept of Nature, Marx conflates the ontology of the Real with the category of the Natural. By the same rate, the ontological regime of "Nature" understood as the emergent potency and becoming of *physis* is conflated with the order of techno-economic rationality of modernity where the capitalist mode of production is configured—both the theory of capital as its scientific mode of knowledge (Althusser et al. 1965/2016)—and its inscription in the broader mode of understanding the historical process where life's creative evolution (Bergson 1907/1998) that springs from *physis* and the rationality of modernity born from the *human logos* converge (Leff 2018). Thus, the way is foreclosed to an ontology capable of grasping the Real constructed by different levels of materiality—different ontological regimes—as well as the relations between the realms of the Real and the Symbolic that constitute the social rationality of capital, as well as other different social formations. The "reciprocal interpretation of Nature and society" is not considered here as the articulation or the inter-determination among natural and social processes, but is reduced to the internalization of the world by human praxis that constitutes a "natural whole." From this conception of Nature, Schmidt constructs his phenomenological interpretation of Marxism. By so doing, he disarticulates the concepts of use value and exchange value to imagine a metaphysical and ahistorical relation between Nature and society, between man and his productive practice, that reduces the Real, Nature and History to the abstract existence of man.

To be sure, we are not dealing here with an existential ontology that unearths Marx's concept of Nature, but with an intent to rescue Marx's ontology of historical praxis that, while opposing the mechanistic formalism derived from Descartes and Newton and Kant's *a priori* rationalism, falls prey of the metaphysical realist concept of Nature. Feuerbach made the transit from Kant and Hegel's transcendental idealism to the humanist subjectivism that influenced young Marx's writings.[1] Human praxis became the principle of reality to man in the constitution of his life-world, his knowledge, and transformative praxis. However, Marx work did not stop at this phenomenology of history; his fundamental contribution was to dis-cover the economic structure that *enframes* human praxis under the capitalistic regime and the determinations of the historical process where all social praxis is deployed. The epistemological principle of the primacy of the Real over Thought, of praxis over consciousness, was the point of inflection for a materialist ontology of history: of the natural, physical-biological; of the social, historical-symbolic. This lead in differentiating the forms of materiality as the expression of the Real in the reality of the world, opening critical thinking to overcome the idealist, subjectivist and naturalist conception of history. When Marx refers in *Capital* to the submission of man to the "natural laws" over which he has no dominance, he refers to the objective laws of history. Marx deconstructs the ideological effect of the mechanistic determination of the laws of Nature in naturalizing historical determinations, to view the laws of value and surplus value as the proper laws of history. Thus, Marx states that:

> The act of vision [...] is a physical relation between physical things. However, the value form and the value relation of the products of labour has absolutely nothing to do with their physical nature. It is only a determinate social relation among men that take here and for them the fantastic form of a relation among things. (Marx 1965, p. 606)

Marx's concept of praxis opens the possibility to overcome Feuerbach's naturalist monism, that is, the abstract character of a general ontology that relates Nature and consciousness of Nature with the vital process of

---

[1] For Feuerbach, "Metaphysics or logic is only real, immanent science, when it is not separated from the so-called *subjective spirit* [...] The reality, the subject of reason is man only. It is the *man* who thinks, not the Ego or Reason" (cit. in Schmidt 1971, p. 26).

society, where the relation between the Real and Knowledge of the Real is reduced to the knowledge of Nature as a practical consciousness of the world.[2] The metaphysics of Nature that underlies the philosophy of praxis does not consider the specificity of scientific knowledge as a particular mode of cognitive apprehension of the Real—and of its technological applications–, that while being *enframed* in a historical mode of comprehension, distinguishes itself from other phenomenological and existential modes and forms of knowing that are configured within the practical-transformational character of social praxis under other cultural ontological regimes. This distinction in the forms of knowledge is essential considering the increasing relevance of scientific knowledge as a productive force for capital, and also of the emergent human rights of indigenous peoples to conduct their praxis from other ontologies—i.e., the cognitive modes under the "living-well" imaginaries of the Quechua and Aymara peoples.[3] Thus, by positing Nature as an ontological category that conflates the natural and the social, this "Marxist view of Nature" acted as a smokescreen that Marx himself had to overcome in order to think a proper concept of Nature within capital's ontological regime, articulated to the theory of the capitalist mode of production, as well as the strategic role of the technological application of the scientific knowledge of Nature to the process of capital accumulation.

The concept of Nature in Marx is not an omni-comprehensive ontological category that underlies the whole of Marx's work, nor the critical concept of the transcendental dialectic of history. Different meanings of Nature are encrypted in the ontological principles and the theoretical discourse of historical materialism and the science of capital. Thus, diverse natural processes underlie the concepts of value and differential rent, insofar as they affect the socially necessary labor time for the production of commodities, surplus value and the rate of profit. Nature takes the

[2] "Nature in its broad sense is the sole object of knowledge [...] it includes the forms of human society [...and] it only appears in thought and reality in virtue of these forms" (Schmidt 1971, p. 29). Schmidt relies on *The German Ideology* and on the *Paris Manuscripts*, where Marx stated that "nature, taken abstractly, for itself, rigidly separated from man, is nothing to man," to affirm that "Just as in Marx's view there is no purely immanent succession of ideas such as 'intellectual history' might investigate, so also pure historically unmodified nature does not exist as an object of natural-scientific knowledge" (ibid., pp. 30, 50).

[3] See Chap. 11.

name of entropy and negentropy to analyze thermodynamic effects on the biosphere and the environment of the economic process in present times. From a pluralistic epistemological perspective, all monistic approaches to unify Nature and society, the ontology of life and economic rationality appear as ideological premises. For historical materialism there is no society in general, nor Nature in general, but as empirical objects or metaphysical categories incapable of being articulated in scientific discourse. For the materialistic theory of history, all modes of production articulate the environment where they develop by appropriating Nature, enacting by their internal "logic" Nature's metabolism. Nature itself is organized within different orders of materiality; it is apprehended theoretically by different scientific disciplines and through technological knowledge is incorporated as a productive force in the economic process. Thus, the over-determination of the economic process mobilizes the theoretical and technical articulation of Nature to the capitalist production process (Leff 1986c, 1994, Chap. 1).

Marx did not think only the transcendental dialectics of History and World as the unification between Nature and society through the labor process. Marx's materialism is not founded on the principle that "everything is matter," or in thinking the world as "a metaphysical determination by which all beings appear as labour material." Historical materialism as the science of the capitalist mode of production and its transcendence in history searches to understand the social structure that converts Nature into objects for labor, in raw materials and natural use values as inputs for the production of value and surplus value. Marx's materialism is not a vision of the world as a relation among things; this is rightly the core of Marx's critique to that false metaphysical view that has generated the alienation of social subjects. Marx unveiled the ideological effect produced by the capitalist production process as the "commodity fetishism" that makes appear reality as a relation among things. In *Grundrisse* Marx denounced:

> The crude materialism of the economists, by which man' social relations of production and the determinations that keep things going are seen as being subsumed under these relations as natural properties of the things, is just as much a crude idealism, even fetishism, as it ascribes to things social

relations as determinations immanent to them, and thus mystifies them. (cit. in Schmidt 1971, pp. 227–228)[4]

Marx confronts naturalist metaphysics by demarcating historical materialism from a naturalist vision of history, from the naïve realism of Engels' *Dialectic of Nature*, and from a monist ontology that would reduce the world complex and relational ontology to an abstract dialectic between subject and object of labor, to a transcendental phenomenology that would guide the teleology of history through the evolution of the labor process that would end up liberating man of his alienation with Nature. However, this demarcation from objectified thinking that Marx brilliantly deconstructed was insufficient to overcome the metaphysics of Nature that accompanies the transcendental dialectic inscribed in the labor process, and that will root its creed in the "deep understanding" of the dialectical naturalism of the emergent ecologism.[5]

The conversion of Nature in objects of labor and its products in commodities, and its generalized exchange dictated by the socially necessary labor time—its value–, is not a materialist philosophical principle nor a dialectic of the labor process in general through human history, but the effect of a determined social structure, of the rationality of the capitalist mode of production. Schmidt sees in human labor the "manifestation of a natural force," not considering the historical and economic determinations of capital accumulation that modify the labor process as well as the forms and rates of exploitation depending on the conditions for the revalorization of capital's value through class struggle: that of the capitalist to

[4] Thus, Marx anticipated the structure of Heidegger's ontological order of *Gestell* announced in *The Question of Technique*, of the objectified world disposed for appropriation through calculation, viewed here as the effect of the structure of Capital, of the social relation that transforms Nature into natural objects to be appropriated in theory and in practice by the calculation in value of labor force and the scientific appropriation of natural forces put at the disposal of the production of surplus value (see Chap. 7).

[5] Thus, Schmidt stated that: "Nature becomes dialectical by producing man as transforming, consciously acting Subjects confronting nature itself as forces of nature. Man forms the connecting link between the instrument of labour and the object of labour. Nature is the subject-object of labour. Its dialectic consists in this: that men change their nature as they progressively deprive nature of its strangeness and externality, as they mediate nature through themselves, and as they make nature work for their purposes […] the dialectic of the labour-process as a natural process broadens out to become the dialectic of human history in general" (ibid., p. 61). See the critique of dialectical naturalism in Chap. 4.

raise the rate of profit and to search for new areas for the production of surplus value; that of the proletariat to diminish the daily working hours and improve his quality of life.

Schmidt prevents himself from identifying his Marxist hermeneutics with the dialectic of Nature or an evolutionist conception of history. He thus affirms that social history must not be conflated in natural history, nor should we subsume social relations in natural laws, as is the case with social Darwinism where "human history is an appendix of natural history, and its laws of motion are phenomenological forms of biological laws." Nevertheless, a certain ecological view moulds his understanding of the capitalist production process. For Schmidt, the key to the connection of the social exchange with Nature lies in the concept of *Stoffwessel* (*metabolism*) stating that "with the concept of organic exchange Marx describes the social process following the model of natural phenomena" (ibid., p. 99). Schmidt finds in the notion of *Stoffwessel* a central concept of *Capital*, a pivotal keyword to assign to Marx an ecological conception of society.[6]

Schmidt searches to respond to the problems raised by the dialectic interrelation between Nature and society by digging into Marx's ontological and epistemological principles. In his exegesis of Marx's writings, Schmidt finds in labor the critical term that links the legality of matter with human ends, in a way that if "the laws of nature exist independently of and outside the consciousness and will of men [...] men can only become certain of the operation of the laws of nature through the forms provided by their labour-process" (ibid., p. 98). Thus, Schmidt reduces the determinations that guide social praxis and the conditions for the

[6] Schmidt considers that "The whole of nature is socially mediated and, inversely society is mediated through nature as a component of total reality. The hidden nature-speculation in Marx characterizes this side of the connection." Schmidt's specular identification between Nature and society leads him to state that: "in the direct labour-process, i.e. the metabolism between man and nature, the *material side* triumphs over the historically determined form; in the process of exchange, which depends on the labour-process, the *historically determined form* triumphs over the material side" (Schmidt 1971, pp. 79, 92). By subsuming the social forms of appropriation of Nature in the abstract moments of the exchange between matter and value, a false door was opened in the imaginary field of a Marxian ecological anthropology—as the "ecology of the peasant's mode of production" (Toledo 1980)—that wishes to regard social formations as the articulation of the formal-abstract determinations of economic value with the material conditions drawn from an empirical analysis of material-energy flows in ecosystems enacted by labour processes.

appropriation of Nature dictated by the laws of Nature through the labor process oriented in the naturalness of human ends. Although he admits that "Labour's purposes [...] are limited objectively by the material at mens' disposal, and by its laws, subjectively by the structure of men's drives and needs" (ibid., p. 100), through the labor process man subjects natural laws to his ends. Schmidt's humanist ideology—that of the sovereignty of man's consciousness and will—obscures Marx's construction of the social structure underlying the capitalist mode of production that prevails in the forms of expropriation of Nature that determine the *Stoffwechsel* of the labor process with Nature.

To be accurate, the "ends" of the labor process depend on the subjective needs and desires of man, and on the laws of the material he disposes of to satisfy them. However, the subject is not the principle of his wishes and needs, nor the laws of Nature are immanent and static, nor is science by itself a freeway for the emancipation of man. Marx produced knowledge of the historical process and the social structure that conditions human desire and transforms it into a growing demand of commodities, that operates as a will to know that drives and orients the production of scientific knowledge: not to submit matter and Nature to the "ends of man," but to the logic of capital. Schmidt sees history as a process oriented by the abstract ends of man; the dialectic between Nature and society is turned into the means to achieve man's will. The teleological activity of man becomes the transcendental process that allows the subject to gather the split moments of the object and subject of labor, of society with Nature. In this vein Schmidt asserts:

> Nature can only combine with itself after the emergence of organic life, and specifically of man as a self-conscious active Subject because as Marx said, it is in labour that nature sheds a part of itself and confronts itself through the division into 'material of nature' and the purposeful 'force of nature.' Man's existence for himself consists in his ability to exploit nature's Mechanism and Chimism in order to realize his purposes [...] Man's finite-teleological activity does not break out of the natural context; it does not need a transcendental principle to explain it, however much, because it is historical activity, it negates nature. Purposes at first foreign to nature do not merely make use of nature but have themselves natural causes. (ibid., p. 106)

Thus, Schmidt identifies the evolution of organic life with human history. Use value and exchange value lose their specific determinations; the exchange of commodities can be reduced to matter exchanges, to a natural metabolism because what is "valid for the stages of transformation a natural object, viewed in isolation" is valid also "for the relation of man and nature in the history of society in general." This conception of the relations between Nature and society appears as the reflection of the "contradictory unity of the moments of knowledge in Marx [...] mediated through historical praxis, where both epistemological realism and socially applied subjectivism make themselves felt in his thought" (ibid., p. 107).

Praxis became a foundational argument in Marxist philosophy to think the building and transforming of the world. However, the concept of praxis is not reduced to the mediation between an objective realism and a subjective humanism. Social praxis in our times of environmental crisis and the construction of a sustainable future is not guided by the emergence of an ecological consciousness arising from biological evolution but is inscribed under the ontological and epistemological regimes that configure the rationality of modernity. What characterizes the *epistemic* order in modernity, is the fact that things are no longer perceived in their relations of differences and similitudes: the word ceases to designate the thing; the Real appears as the effect of a process of production driven by the structures and the principles of life, language and history (Foucault 1966). Human praxis is deployed within those structures as a complex of social—productive and discursive—practices.

Marx made his critique of Feuerbach's materialism for having conceived reality as an object given to intuition and not as the result of human activity, as praxis. Schmidt disregards the effect of capitalist production in its mode of *objectification of reality* and in codifying the *materiality of knowledge*. Going beyond the phenomenological intentionality of the subject, for Schmidt, if sensualist knowledge prevails along the pre-capitalist period, the industrial society would have produced the conditions for objective knowledge and an increased subjectivity of reality. He thus states: "With the ever-increasing reduction of nature in modern times to the level of a moment in social action, the determinations of objectivity entered progressively and increasingly into the subject. This displacement of emphasis within the labour relationship towards the subjective side was conceptually expressed by the principle that only what was made by the Subjects was in a strict sense knowable" (ibid., p. 122). The materialist primacy of praxis is reduced to the phenomenological consciousness of the subject. It is the subject that produces reality through labor, orienting reality toward his ends, and not the capitalist mode of production that determines the subjectivity of individuals. This subjective consciousness of reality that emerges from the separation between object and subject in the labor process leads to the dilution of the social in the natural. Thus, the humanism of the young Marx of the *Paris manuscripts* opened the doors to the utopia of a historical teleology, where man and Nature would be reconciled. Dialectic materialism, taken as a general ontology of the Real, is inscribed in the impossible purpose that intends the positivist unification of knowledge.

However, history is not a particular case of a general ontology grounded in the category of Nature, nor can the ontology of Nature be subsumed in the trans-historical ontology of human labor and production. The realist epistemology in which Marx's critical theory is grounded avoids the trap of conflating "Nature" and "society" in the identity between humanism and naturalism, dissolving the primary difference between the Real and the Symbolic, that fundamental contradiction between *physis* and *logos* that the environmental crisis has revealed. Conversely, Schmidt envisions

> A society which still received its nourishment through the metabolic interaction with nature, but that at the same time was organized in such a way that it could renounce the ruthless exploitation of the latter would allow the realist moment in Marx's epistemology to emerge still more strongly. The fact that nature is also a being-in-itself, existing independently of the manipulatory intervention of humankind, the truth of materialism, which does not see things as already aprioristically modified, but allows them to have their say, as it were, comes into its own. (ibid., p. 155)

However, the construction of such an ecological society will not emerge just as the becoming of a historical teleology guided by the evolution of Nature or driven by a transcendental subject of history. The transition to environmental rationality will be the result of a social praxis devoted to the deconstruction of the naturalist metaphysics that underlies social theory and our understanding of the conditions for the sustainability of human existence, that is, through a strategy and a politics of knowledge. In the dialectic of history, capitalism has exacerbated the difference between Nature and society; society has been alienated from its original organic constitution, while capitalism has imposed its rational dominance over Nature. In this sense, Schmidt considered that.

> Measured against the concrete determinacy of the labour-process as a specific capitalist phenomenon, there is something peculiarly unhistorical and natural-like about the forms which preceded it [...] Only with the transition to capitalism does the mastery of nature take a new quality; only at this point does the labour process, which Marx initially stated was in its general determinations for all stages of society, becomes a strictly social process. Now, as he said, those general determinations no longer suffice, and therefore, precisely in their abstractness, turn to be characteristic of the particular stages of pre-bourgeois production [...] What the critique of political economy was concerned with, and what it wishes to explain, was instead the typical phenomenon of bourgeois society, the "separation of these inorganic conditions of human existence from that existence, a separation that is only fully completed in the relation between wage labour and capital." (ibid., pp. 169, 174)

This clarification is essential, yet requires further questioning to deconstruct Marx's concept of Nature. First of all, finally in this appendix Schmidt puts his finger on the most sensitive point, the turning point of the general abstract ontological category of Nature where one can sense that labor is the process that links man to Nature in their dialectical transformation throughout history, to the scientific concepts that lead to understanding how specific social formations, and particularly the capitalist mode of production, through the structure and praxis "ordered" by the law of value, do not only establish interactions between the labor process and the workings of Nature through *Stoffwechsel*, but actually impinge on the metabolism of the biosphere. This is the turning point in the history of Nature and of human history in the planet, where the ontology of life's creative evolution is dominated by the ontology of capital, where the imposition of the scales, rhythms, and forms of expanded reproduction of capital overflows and breaks with the ecological equilibrium of the biosphere, overwhelming humanity with the environmental crisis.

However, idealizing pre-bourgeois societies and the traditional social formations for having established a natural relation and material interchange with Nature must also be overcome. The concept of Nature must be reconstructed as a fundamental ontological order—the ontology of life—under which man will have to transcend his alienation with Nature. The significance of the concept of Nature that emerges through Schmidt's hermeneutics of Marx, as well as in the organismic vision of history—the ecologic episteme that wills to unearth an ecologic ontology in Marx—is encoded in the objectivism that has filtered into the ontology of Nature and the techno-economic rationality that dominates the modern world driving the metabolism of the biosphere toward the entropic death of the planet. Marx's critique of economic reason is still trapped in his understanding of "the natural," in the complicity that has been established between science and economics that naturalizes history as the economization of the world, in a conception of civilizing progress over-determined by the economic process, in a transcendental dialectic that under the capitalist mode of production subordinates use value to abstract exchange value and the market logic, alienating Nature in an objectified world driven by capital's will to power.

Bolívar Echeverría clarifies the point in question:

> For him [Marx], to construct a proper world, modern life needs to rely on a peculiar economic device, that consists in the subordination, subjection or subsumption of the 'social-natural' process of reproduction of human life under a "social-artificial" process, only transitorily necessary: that of the reproduction of the market value of things in the modality of "valuation of value" or "capital accumulation". In the base of modern life a mechanism acts repeatedly and tirelessly, that subordinates systematically the "logic of use value", the spontaneous meaning of concrete life, labour and human enjoyment, to the production and consumption of "earthly goods", to the abstract "logic" of "value" as a blind substance indifferent to any concretion, that only needs to validate itself with a margin of profit in quality of

"exchange value". This is the cruel reality of alienation, of the submission of the realm of the human will to the hegemony of the purely "thingly" "will" of the world of commodities inhabited by capitalist economic value. (Echeverría 1998, p. 63)

The capitalist mode of production exploits Nature to the benefit of the market and the extraction of surplus value, by disposing of the potentials of Nature and human beings as objects for economic expropriation. However, despite capital's legacy of destruction and degradation on Nature, it has not yet annihilated the self-organizing and productive processes emerging from natural sources of life nor the existential resistance and cultural creativity from the peoples of the Earth. The productive forces of Nature and the ecological organization of the biosphere, under the dominance of the destructive forces of capital, lie there as a potential reserve waiting to be *enowned*, embodied and embedded by new productive rationality. If throughout history men have survived through diverse modes of *appropriation/ transformation* of Nature; if capitalism imposed its accumulative logic as a mode of *expropriation/ degradation* of Nature; after the environmental crisis a new era has opened up in the evolution of life in the planet for the *social enownment of Nature*: a mode of extracting matter and energy, of enacting and transforming the metabolism of the biosphere, respecting the conditions of life in the planet. *Enowning* Nature will imply en-knowing Nature: learning to live within the conditions and inhabiting Nature in the immanence of life on Earth.

That will be a turning point in history opening the horizons of sustainable, meaningful life on this planet. However, we must ask: To what extent did Marx anticipate these historical events? In which ways did Marx prepare human thinking to understand the conditions of sustainability? How ecological is the legacy of Marx's historical materialism?

3. How Ecological is Historical Materialism?

The emergence of the environmental crisis reverted to criticism of the prevailing paradigms of economics that besides triggering the crisis as the result of the commitment to economic growth, they did not foresee its consequences. Marxism, either in the practices of real socialism or in its theoretical revolutionary discourse, was not exempt from its lack of ecological consciousness. This fact mobilized theoretical reflection on the ecological grounds of historical materialism. The inquiry diverged between those who tried to trace and discover the ecological basis of Marxism and those who intended to construct its missing ecological foundations. Among the leading proponents of an ecological foundation of Marxism is John Bellamy Foster (2000), who has argued that orthodox Marxism is already constructed on ecological ground. If Alfred Schmidt tried to prove that the *concept of Nature* underlies Marx's thinking, not too far from it, Foster has found the key to open the door to Marx's ecology in his understanding of the *ecological rift*,[7] drawn from the concept of *Stoffwechsel*—of matter interchange and ecological metabolism—that establishes the unifying linkage of human labor, social praxis, the development of productive process and the transformation of social relations of production, with Nature.

Foster has pretended from start to end of his work that "it is possible to interpret Marx differently, one that conceives ecology as central to his thinking" (ibid., p. vi). He thinks that "Marx's world-view was deeply, and indeed systematically, ecological [...] and that this ecological perspective derived from his materialism" (ibid., p. viii). For Foster, Marx's materialism stems from realist—non-atomistic, non-teleological, non-mechanistic, non-deterministic—ontology, derived from Epicurean lineage through Lucretius, and anchored in Darwin's evolution theory of the living and on Liebig's science of nutrient circulation in the soil and animal metabolism. The radical question that arises is not only to acknowledge that mans' transformative practice affects soil metabolism or even the evolution of life, but to disclose if the motor of history, driven by class struggle, is inscribed in modernity within the techno-economic ontology that drives the development of productive forces already rationalized and empowered by science, or, if human praxis is inscribed in the ontology of life, emerging from the "fire of life," inscribed within the

[7] Ecological rift expresses the fact that the way that capital operates on Nature is not by appropriation, but instead as *expropriation*, i.e., appropriation without an equivalent (in Marx's terms, also without exchange and without reciprocity). On Marx's distinction between appropriation and expropriation, see John Bellamy Foster and Brett Clark (2018), The expropriation of nature, *Monthly Review* 69(10), pp. 1–27.

immanence of life; and if claiming Marx's ecology implies that this understanding of the environmental question was evident in his mind and was the ontological substratum of his historical materialism. Foster argues that

> One consequence of Marx's new, practical materialism, however, was that the focus of materialist thought shifted from nature to history, without denying the ontological priority of the former. It is true that Marx tended to see his materialist conception of history as rooted in a materialist conception of nature, which together constituted the realm of natural history (in its Baconian sense that included human production). Nevertheless, his emphasis in his social critique was overwhelmingly on the historical development of humanity and its alienated relation to nature, and not on nature's own wider evolution. (Foster 2000, pp. 113–114)

The question here is not only to acknowledge Marx's awareness and concern over the metabolic rift caused by capital accumulation but to clarify whether in Marx's materialism the driving force of history is the social dialectics involved in the development of the productive forces of humanity—what Louis Althusser (1969) referred to as the economic over-determination in capitalistic contradictions[8]—or in the material evolution of Nature. Foster considers that where Marx and Engels "emphasized ecological contradictions" in the *Communist Manifesto*—viewed as the contradictions between the countryside and the city–, "they did not seem to believe that they were developed to such an extent that they were to play a central role in the transition to socialism" and would have "a distinguishing feature of the latter dialectic of the construction of communism" (Foster 2000, p. 140). However, it is in *Capital* that "the materialist conception of Nature in Marx reaches its full integration with his materialist conception of history." There,

> In *Capital*, Marx employed the concept of "metabolism" (*Stoffwechsel*) to define the labour process as "a process between man and nature, a process by which man, through his actions, mediates, regulates and controls the metabolism between him and nature." Yet, an "irreparable rift" had emerged in this metabolism as the result of capitalist relations of produc-

[8] Cf. Althusser, Contradiction and overdetermination, in *For Marx* (1969), and Leff (1992b).

> tion and the antagonistic separation of town and country. Hence under the society of associated producers, it would be necessary to "govern the human metabolism with nature in a rational way." (Foster 2000, p. 141)

Without denying that ecological dynamics underlie the contradictive relations between town and country, this exegesis of Marx's natural materialism is far from disentangling the complex ecological processes involved in the capitalist entropic expropriation of Nature in capital's metabolic rift. Acknowledging the ecological costs and environmental degradation generated by capital production and extended reproduction process, or claiming the need to refurnish Nature with natural elements robbed by capital—in a similar vein as in our days Georgescu-Roegen (1971) envisioned the entropic costs of the economic process or that Herman Daly (1991) claims a "strong sustainability" as a norm imposed to the economic process to allow the renewal of natural resources and to maintain ecological equilibrium in a steady-state economics[9]—does not lead directly to turning this dialectical contradiction into its opposite: a new production paradigm supported by the negentropic potentials of Nature. The ecological rift and entropic effect of human appropriation of Nature through labor have been the rule since the primary difference between the Real, and the Symbolic was grounded on Earth. Capitalism has exacerbated this process opening the abyss to the entropic death of the planet. However, humanity still has to deconstruct the techno-economic rationality of capital's regime and replace it with the theory and praxis of ecologically rational reappropriation of Nature and human governance of the metabolism of the biosphere. This turning point in history and historical materialism will not occur without envisioning environmental rationality to open the possibilities for that unprecedented event.[10]

---

[9] Similarly, Justus von Liebig postulated the natural-material "law of replenishment" (or "law of compensation") necessary for ecological reproduction, Cf. Justus von Liebig, *Letters on Modern Agriculture* (London: Walton and Maberly 1859), 254–255, cit. in Foster (2018).

[10] Foster considers that "we can hope to restore the essential metabolism through a revolutionary overturning of the capitalist integument and the creation of new, co-evolutionary material reality. This is Marx's core ecological message […] The future lies with the development of the twenty-first-century socialist/ eco-socialist movement, to be rooted in a diverse, all-inclusive environmental working class. What is needed is the revolutionary reconstitution of the interdependent social metabolism with nature, bringing it under the rational control of human beings" (Foster 2018).

4. Historical Materialism and the Environmental Crisis

How did Marxism react to the environmental crisis? How does Marxism respond to the environmental question? How does Marxism deal with the environmental problems of our time? Although Marx could not predict the irruption of the global environmental crisis half a century ago and the extent of global ecological imbalance that has increased since, he did anticipate the effects of the capitalist mode of production on the destruction of planetary resources and the loss of soil fertility. He stated in *Capital*:

> Every progress in capitalist agriculture is a progress not only in the art of exploiting the worker but the art of exhausting the soil; every progress in the increase of the fertility of the soil during a specific time is a progress towards the destruction of its more lasting sources of fertility [...] The capitalist production only develops the technique and the combination with the process of social production by exhausting the two sources from where all wealth emerges: the earth and the worker." (Marx, *Capital*, I, in *Oeuvres Économie*, I, pp. 998–999)

However, Marx's account of the effect of capital accumulation on the destruction of Nature did not generate an adequate theoretical response through his critical concepts. Nature was the object of reflection by Marxism insofar as it was integrated into the ideological superstructure of traditional social formations—in their cultural imaginaries, myths, and practices—or as a general ontological category of Nature and objectively as the metabolism between society and Nature through the labor process. In order to establish itself as a social science, Marxism had to demarcate itself from the naturalist ideology of its time, leading to the fertile development of the theory of social determinations over Nature, which included the way that capitalism gave impulse and oriented the production of scientific knowledge on Nature and the incorporation of the forces of Nature to the productive process through technology; however it excluded the natural contribution and conditions for the creation of

However, nowhere did Marx transform this message into the theoretical construction of that co-evolutionary material reality, inscribing the production process in the material immanence of life.

wealth under the capitalistic regime and as a bio-cultural patrimony for the sustainable livelihood of the peoples of the Earth. Only after the ecological crisis that irrupted in the world in the last half-century was a reflective process set off within Marxism concerning the ontological regime, the epistemological bases, and the theoretical concepts of historical materialism to respond to the environmental question.

The emergent environmental discourse during the 1970s was regarded with skepticism by some neo-Marxist theorists and precursors of political ecology, as "false consciousness" (Enzensberger 1974), stating that the economic crisis was the fundamental cause that triggered the need to open up new ecological productive sectors (the anti-pollution industry) to provide an outlet to reinvest profits in order to avoid a crisis of overproduction (Granou 1973). The environmental question itself, the underlying condition—the need to preserve natural resources and ensure ecological balances—for all sustainable economic and productive processes—in traditional, capitalist or any future society—to ground the economy on the productivity of Nature and the creative evolution of life, was almost completely ignored.

The unmasking of environmentalists' "false consciousness" still conceals one of the basic beliefs of Marxist analysis of the transcendence of capitalism toward an ecological socialism: that the transparency of social relations and the society–Nature relationship would come about only as a result of the elimination of scarcity that characterizes the pre-capitalist and prevails in the capitalist mode of production. Therefore, although the transition to socialism would be led by class struggle, its material possibilities depended upon the development of the forces of production and, in particular, on the forces of Nature released by the scientific-technological revolution. Thus, Marxism embraced the will to unleash the free development of productive forces, both under capitalism and during the transition and building of socialism (in the latter case under different forms of ownership and control of the means of production), exacerbating the environmental degradation of the planet. This ideological inertia has ignored the ecological limits of growth and the ecological bases for a lasting and sustainable development of the forces of production, which were recognized by an emergent eco-Marxism only in the late 1980s.[11]

The crisis of Marxism is not only the result of the collapse of really existing socialism and capital's self-destructive development of the forces of production. It can also be attributed to certain "blind spots" in the construction of historical materialism as the science of history. Although Marxism understood capital as a relationship of social exploitation and promised the rationalization of the development of the forces of production by eliminating internal crises of under- and overproduction, of value realization and profit-making, through the ownership of the means of production by the proletariat represented by a socialist state, it did not incorporate the forces of Nature as productive potential in its theory of social labor nor the ecological conditions for the extended reproduction of capital. The theory of capital was structured in the order of *Gestell* (Heidegger 1935/1977) under the "framework" of an objectified world where the sustainability conditions of the economic process grounded in the ontology of life were reduced under techno-economic rationality to a unitary measure of value.

Marx turned Hegelian dialectics and grounded the ontological contradiction in the social relations of production; he unearthed and clarified the principle of the exploitation of surplus value that structured the capitalist mode of production. However, he could not escape the epistemic frame of his time, and searched for a unit of value quantification based on the simple and direct labor to which capital reduces labor power. The concept of socially necessary labor time, upon which value is measured, is devised using capital's exploitative relationship with labor. However, this measurement of value excludes the productive forces and the conditions of reproduction of Nature, which in their ontological diversity and ecosystemic complexity are irreducible to homogeneous units. Marx's theory of capitalist production is thus questioned: not only as a result of ecological destruction and capital's inability to assign a sustainable value to natural resources but also because of the indeterminate effects

[11] Cf. E. Leff, *Ecología y Capital*, Mexico: UNAM, 1986 [*Green Production*, 1995]; J. O'Connor (1989). Combined and uneven development and ecological crisis: A theoretical introduction, *Race and Class*, 30(3), pp. 1–11; J. O'Connor, *Natural Causes. Essays on Ecological Marxism*, New York: Guilford, 1998.

introduced by technological progress in the calculation of socially necessary labor time as a measure of value.[12] Marx's theory of production does not incorporate the natural and cultural conditions for the production of value under capitalism, nor in its transition to a socialist society. The environment question thus invites Marxism to reformulate its vision of the dialectical transcendence of history based upon the destructive development of productive forces. The environmental question also poses the need to respond to present-day power relations under global change and world transformations in which new historical subjects emerge, creating new conditions for the labor process and the social appropriation of Nature that impinge on the metabolism of the biosphere.

The need to incorporate ecological conditions into the analysis of the economic process is giving an impulse to new theoretical developments and methodological approaches in a new branch of ecological Marxism. One first question that arises is whether or not the processes that determine the metabolism of the biosphere and the ecological conditions of productive processes are the object of ecological science external to historical materialism, understood as the critical theory of the capitalist mode of production; or if ecological potentials and conditions should be internalized in the conceptual apparatus of a new paradigm of a political-ecological economy. This problem opens up two options. The first is the possible articulation of two sciences, ecology and historical materialism, in which ecology would account for the structure and workings of the ecological base of natural resources, as a set of ecological potentialities and conditions, norms and constraints, for a sustainable production process external to capital's internal contradictions.[13] The second is to conceptualize the environment as the articulation of different productive processes (natural, cultural, economic, technological) with ecological processes as co-determinant forces and conditions of production, leading to a paradigmatic reformulation of the theory of capitalist production and any sustainable development processes (Leff 1986a, 1995).[14]

[12] See Chap. 5.

[13] Thus, René Passet proposed the notion of "economics under constraints" in *Ecological economics* (Passet 1979).

[14] For the epistemological principles of this productive paradigm see E. Leff, Sobre las relaciones sociedad-naturaleza en el materialismo histórico, in E. Leff, Ed., *Biosociología y articulación de las*

This reconstruction of the theory of production, integrating social and natural processes, goes beyond the recognition of a "Marxist concept of Nature" centered on the concept of metabolism (*Stoffwechsel*) as a phenomenological general mediation and material exchange in all processes of production and social appropriation of Nature (Schmidt 1971; Foster 2000). A new theory of production is needed to unveil the "second contradiction" of capitalistic production, related to the ecological conditions for capital's production and reproduction process (O'Connor 1988). In *Capital*, Nature appears less than as "second contradiction" of capital, as a "contradiction of second order": as the over-determined effect of the exploitation of labor rather than the dialectical movement of Nature (Leff 1992b). Beyond any general ontological concept, Nature must be incorporated to the development of productive forces from its modern scientific concept, as the negentropic potentials and the entropic limits of the metabolism of the biosphere.

General categories of Nature and labor do not capture the specificity of the relations between the capitalist mode of production and non-capitalist social formations and the environment; of particular modes of social appropriation of Nature and their environing ecological processes. The concept of "socio-economic formation" offered an initial framework to analyze the structure of the modes of production of pre-capitalist societies and their supra-structures. Authors like Maurice Godelier (1969, 1974, 1984) and Claude Meillassoux (1977a, b) offered fundamental studies on the relationship between cultural organizations (family, marriage and kinship relations; religious representations and ideological formations) and the natural environment in the development of their production process and resource use patterns. These contributions by Marxist theorists are important, but they provide an insufficient theoretical basis for analysing the ecological and cultural processes affecting today's sustainable development of productive forces. From an environmental perspective, we should see culture not only as super-structural values or as a structure that takes the place of the productive base in traditional, non-capitalist societies; instead we should analyze the ways in

Ciencias, México: UNAM, 1981; E. Leff, *Ecología y Capital*, Mexico: UNAM, 1994, Chaps. 1 and 2.

which the ecological conditions are reconfigured in the social imaginaries and cultural values of traditional peoples; the way these are embedded in their bio-cultural patrimony of resources (Boege 2008, 2018) and are enacted through productive practices, functioning as a principle of productivity in the sustainable use of natural resources (Leff 1993).

Another eco-Marxist current has been developed that emphasizes Nature's function in supporting or limiting production (Ely 1988). However, lacking a theory explaining the transition to a sustainable economic process based on environmental rationality, this approach does not incorporate natural processes into the productive process itself, leaving intact the paradigm of the capitalist mode of production. Therefore, there is a need to establish a concept of Nature that is appropriate for building a productive process based on the social appropriation and democratic and participatory management of the environment viewed as a resource base and means of production, as a life-support system and ecological condition of existence, which supports different modes of production and lifestyle patterns.

The environment is more than just one more element of the conditions of production or a cost of economic growth. We must conceptualize it as productive potential, as an integral part of the social forces of production, in a productive paradigm that is not economistic, yet pertains to the realm of political economy, since production under environmental rationality involves participation as well as conflict among political actors in the social appropriation of Nature. Political ecology embraces the dialogue of knowledge under the ontological regime of environmental rationality that guide the juridical procedures for the just resolution of contradictions between conservation and growth, between environment and development, between the self-destructive appropriation of Nature by capital and the aim of subsuming the appraisal of the environment under the concept of natural capital, and the construction of new life-territories grounded on the ontology of life: in the negentropic productivity of Nature and the cultural creativity of the people.

Thus, environmental rationality as the ontological regime that embodies the understanding of the conditions for sustainable living in the planet opens the ways for a more equitable and sustainable production. By legitimating the rights of existence and the territorial rights of the peoples of the Earth, environmental democracy opens the gates for environmental knowledge to flow and for the environmental movement to deconstruct the one-world ruled by market mechanisms and to create a new world constructed by different worlds, founded on diverse, sustainable productive rationalities; as the riverbed where flows the mainstream of the forces of Nature into multiple and diverse modes of production, life-territories, and forms of being-in-the-world.

In this sense, the environmental challenge to historical materialism and Marxist political economy goes beyond a synthesis of orthodox Marxism and the new ecology, or the incorporation of energy rationality into the metabolism of production. Eco-Marxism aims to integrate ecological processes into the development of the forces of production and the social relations of production into the web of life, leading to the democratization of an ecological society. Beyond accounting for the environmental externalities and the ecologic costs of the productive process, this eco-social theory incorporates Nature as a condition for sustainability of the productive process and as the guiding force for the construction of different life-worlds and modes of existence. Ecological Marxism requires the elaboration of new concepts concerning an equitable diverse and sustainable production process along with instruments for monitoring and assessing its effects in the metabolism of the biosphere.

Environmental rationality mobilizes the political ecology arena for greater autonomy and participation in self-management of the productive process of associated workers and communities, as well as in global decision-making by civil society; it opens new windows to envision new paths to make the transit toward new productive rationality based on the socialization of Nature.

5. The Ecological Limits and the Environmental Conditions of Sustainable Production

The growing dynamic of the economic process, the trends of the extended reproduction of capital in the total mobilization of all entities in the world and its effects in the metabolism of ecosystems, drives an increase in the productive (destructive) consumption of raw materials, a heightened rhythm of extraction and transformation of material inputs, the production of non-recyclable waste, and the degradation of the environment of planet Earth. The failure to value Nature transformed by the economic process—the *Stoffwechsel* between society and Nature—has resulted in ecological imbalances, a decline in soil fertility, and the depletion of non-renewable resources that are not accounted for in the value of capital and in price formation of natural use values, while the accumulation and reproduction of capital is reflected in the destruction of Nature and the entropic degradation of the planet.

The environmental crisis signals the explanatory limits of a theory in which natural use values are valued only insofar as they incorporate labor time or internalize the "scarcity" of natural resources through the market. Thus, social labor must be redefined to incorporate necessary labor for the renewal of the production conditions of Nature and the enhancement of the environmental potential to build a sustainable world order. However, the place of Nature in sustaining production and life-worlds is not limited to the intervention of man on Nature through labor. Nature manifests its forces and conditions for the sustainability of life. The environmental question challenges the theoretical status of social labor, the concept of value and Nature, not only to assess the extended reproduction of capital or the restricted areas of environmental protection and the capitalization of Nature, but for the construction of a productive paradigm based on an environmental rationality under which social labor and production practices are bound together with the forces of Nature and the metabolism of the biosphere to enable a sustainable and appropriation and transformation of Nature under the bio-thermodynamic and ecological conditions of life on Earth.

Marxism must be "enriched" with new concepts from bio-thermodynamics and ecological science to environmental sociology and political ecology to maintain its status as a critical, and at the same time a positive, theory of production. Ecological Marxism must respond to the challenge to offer a theoretical framework within which to analyze the connections among natural and social processes, the links between the "objective" and "subjective" conditions of sustainable production, as well as for the analysis of socio-environmental conflict and struggles for the appropriation of Nature; to orient the political practice of environmental movements, not only in relation to their contribution to a postmodern culture of difference, but also with respect to the political conflicts, class interests, and social strategies in the construction of a sustainable and lasting productive process. In the conceptual dialogue of Environmentalism with Marxism, new insights guide the critical analysis of the relationship between society and Nature—and between the economy and resources—for an eco-Marxist theory of sustainable production.

An analysis of Nature as a potential or a limiting norm for production brings us to the question of assessing the ecological conditions under current conditions of capital expansion and technological change. This leads to evaluate from the possibilities for recycling and diluting production and consumption waste, the impact of the expansion of the economic process over scarce resources and the global ecological equilibrium of the planet. Although the environmental crisis produces different impacts at local and regional levels, it has become increasingly global (food and nutrition crisis, water scarcity, poverty and social inequality across the world; ecological imbalance and climate change throughout the planet) with transnational and cross-class effects that manifest the global limits to economic growth.

The proposals for "zero economic growth" were widely rejected by Third World countries that demanded their right to develop, while the economic order resisted being deconstructed. The project to "dematerialize production" (von Weiszsäcker et al. 1997; Schmidt-Bleek 2008) failed because its strategies were not viable for de-growth of economic rationality through technological innovation and to halt capital expansion. In this context, environmental principles were put forth as a non-destructive alternative road to development, as "ecodevelopment strategies" (Sachs 1980) and "sustainable development" (WCED 1987), in order to preserve natural resources with social equity for future generations. In a more critical appraisal and response to the environmental crisis, an eco-technological productive paradigm was envisioned, one that transforms the limits to the development of destructive productive forces through the entropic appropriation of Nature by grounding the production process in the negentropic potentialities of Nature (Leff 1986b, 1994, 1995, 2019).

The assessment of the limits to growth leads to question the concept of scarcity within the framework of new strategies of capital (technological innovation, new products, market expansion, capitalization of Nature, green economy, biotechnology, transgenesis) in contrast with the resiliency limits of ecosystems, and in face of the resistance of the people to erosion of their bio-cultural patrimony, displacement from their territories and degradation of their livelihoods. These limits should be reconsidered in the light of different strategies or resource use: the productive potentialities of ecosystems and the creative innovations of the people to reconstruct their life-territories.

6. Ecological Marxism and the Environmental Conditions of Production

The general conditions of production, understood as everything that exists in Nature and society that is not produced according to the process of value formation and the laws of the market but rather which establishes the certain necessary conditions for capitalist production, should also be redefined in the context of extended capital reproduction. In this sense, ecological conservation principles that provide essential environmental services for the reproduction of Nature and the survival of humanity were incorporated into economic policies. These socio-environmental conditions include all those activities considered to be of strategic interest for the State that is unprofitable for capital, such as infrastructure and public services; conservation areas, and environmental norms for industry, subsidies for basic foodstuffs, and health services. However, these traditional public sector activities are being challenged by neoliberal economics through privatization policies that extend the laws of the market to all environmental common goods and services, conditioning and reducing the ecological and cultural conditions of production to the dominating ontological regime of the economic process—the imperative of the development of the productive forces driven by capital accumulation, the potency of technology and the power of the rational—throughout the planet, from North to South, and East to West.

Thus, the strategies of capital and the geopolitics of sustainable development have been focused on the re-elaboration—reframing, rephrasing and welding anew—of those conditions for production that are most difficult for capital to generate and regenerate, those processes which are ontologically excluded from and actually exterior to economic rationality due to their nature that rejects being objectified and quantified. These include qualitative processes where capital accounting and discount rates can not value time cycles—the resiliency and productivity of ecosystems; conservation conditions and regeneration times of natural resources; environmental services and common goods of the biosphere; health conditions, environmental quality, the meaning of life; local and global impacts; trans-generational effects of ecological risks and long-term ecological processes; the bio-cultural patrimony of the people and their existential rights. These socio-environmental conditions of production cannot be absorbed under the theories of a green economy and reduced to the concept of natural capital without exerting violence on Nature.

Historical materialism needs to work out a theory of those natural processes and socio-cultural values that provide the *ecological and existential conditions of sustainable and meaningful production*. By establishing a system of protected areas and a normative juridical system for the ecological ordering of the territory and of productive processes, the State has promoted the spatial distribution of productive activities, industrial and domestic waste management, recognizing territorial property rights of traditional peoples and even, as in the case of Bolivia and Ecuador, the "rights of Nature." Civil society has started to adopt energy-saving and pollution-reduction productive and consumer practices, eventually delinked from the dominant realm of capital and the market. By self-managing the productive activities of their life territories and livelihoods, indigenous peoples, peasant societies, and rural communities contribute to preserving the general conditions of the production of capital, as well as the conservation of resources for present and future generations. However, these ecological and existential conditions remain "external" and "limited" to capital's economic rationality, to the ontological regime that drives the development of the productive forces against the sustainability of life in the planet.

7. Environmental Externalities and Ecological Costs of Uneven Development

Ecological destruction, in its double expression, as over-exploitation of natural resources and environmental degradation, has been the effect of the capital accumulation process. The failure to value Nature favors the overuse of its resources. This fatality is inscribed in the ontological regime of capital. Environmental resources that are a condition of production appear as a cost for the widened reproduction of capital; these ecological and social costs are externalized to the people and the environment, to the overall conditions of life on the planet. Two moments of capital are combined in this process.

On the one hand, the extended accumulation of capital generates an expanded demand for natural resources that enter into the productive consumption of capital—following its economic laws—and the destructive degradation of the environment—following the entropy law (Georgescu-Roegen 1971). On the other hand, realization crises and technological innovation generate processes of "destructive creation" (Schumpeter 2014/1942), multiplying and accelerating the extraction of non-renewable resources and imposing productive patterns that are aimed at maximizing short-term profits without taking into consideration the conservation and regeneration of natural resources. This process leads to further deterioration of the environment and the quality of life of the people, increasing environmental risks throughout the planet.

Capital externalities thus become new production costs. These are being "internalized" by geopolitical strategies rather than being established by any economic mechanism of price formation. New concepts and power strategies have been devised to adjust environmental processes and cultural values to market mechanisms. Thus, the neoclassical approach has proposed the concept of "natural capital," to "internalize" Nature and the environment within the realm of capital. The task is not an easy one given the incommensurability in the mutually external and yet co-dependent economic and environmental processes (M. O'Connor 1989). There are limits on capital's (and the State's) ability to translate Nature into market (or planning) prices, such as Nature's capacity and potential to contribute to the production of use values; or to account for ecological processes such as the resilience-regeneration-productivity of ecosystems and the overall metabolism of the biosphere in the face of capital intervention; or the reduction of the complex chemistry of pollutants and meteorological dynamics of the atmosphere to carbon unit pricing accounting. Labor value, as a concept that binds together labor processes and capital cycles, cannot be easily tri-polarized in order to internalize the negentropic potentials, ecological processes and overall metabolism of the biosphere that contribute to and limit the production of use and exchange values. Nature cannot be disaggregated into discrete and homogeneous value units, like labor versus capital, nor can it integrate and blend complex ecological cycles into mechanistic capital cycles.

Therefore, it is not enough to regard Nature as a cost that is calculated in terms of natural capital. We need a political economy of the environment that is critical, in order to see poverty, unemployment, and the destruction of natural resources as effects of given relations of production. But we also need a *positive theory of production*, which can give support to a new productive rationality in which natural forces along with labor power and technological potentials integrate in the development of productive forces and contribute to an equitable and sustainable production and distribution of wealth, with global ecological balance being constructed through diverse, sustainable livelihoods and modes of being-in-the-world.

There has been a broad discussion of the impact of capital on different environments and socioeconomic formations. In contrast to the ideological discourse on global environmental problems and humanity's collective responsibility, political ecology highlights the difference between the environmental problems of rich and poor countries, as well as the uneven distribution of the environmental costs among nations, regions and social classes generated by the asymmetrical exchange between the North and the South, leading to over-extraction of resources (current capital's neo-extractivism) and the transfer of destructive and polluting processes to underdeveloped countries and tropical regions in the South.

The combined effects of environmental degradation and social polarization are manifest, as in the case of the decline in soil fertility and resources owing to deforestation, erosion, salinization, and desertification caused by the introduction of inappropriate technical models (e.g., capital-intensive agriculture, transgenic monocultures, and cattle-raising in tropical regions) and their impact on poverty, social inequities, and malnutrition in rural areas. Urban-industrial concentration, rural–urban separation, and regional migration, also cause socio-environmental severe problems (Faber 1988; J. O'Connor 1988). The intensive occupation of urban space and the ecological rift that results from uneven development have brought about the entropic degradation of the biosphere, with the over-exploitation and under-utilization of environmental resources, and a failure to take advantage of the environmental potential for ecologically sustainable development.

The adverse effects of uneven and combined development should be counteracted by an alternative mode of production capable of combining and integrating multiple strategies of sustainable societies and communities, and even of exploring possible complementarities of self-sufficient environmental economies as a transition toward a sustainable world order.

8. The Emergence of Political Ecology and Ecological Marxism

Political ecology and ecological Marxism are new areas of social theory and praxis that sprang from historical materialism inquiring the root causes and social effects of the ecological imbalance and environmental degradation generated by capitalist expropriation of Nature. At the same time, they seek to incorporate material principles for the ecological and thermodynamic reordering of the production process. Thus, a new thermodynamically based economics has emerged, providing a critical theory of the entropic decay produced by economic rationality (Georgescu-Roegen 1971). From other theoretical strands, new anthropological approaches analyze the rationality of traditional societies in terms of their energy flows (Rappaport 1971); the emerging discipline of ecological economics seeks to integrate ecological conditions with global economic processes and subsistence micro-economies. However, these contributions cannot replace political economy as a field where the ecological and energy bases, conditions, and potentials for an equitable, sustainable, and lasting production process raises the question of ecological distribution and power strategies over the appropriation of Nature.

Within this tendency, eco-Marxism has been inspired by the thermodynamic theory of open systems. In opposition to the unified, uniform, and modern quantitative rationality, based on the predictability, normative character and control of natural, cultural and social processes, eco-Marxism is internalizing concepts of opening, diversity, indeterminacy, coevolution, co-dependence, and dispersion (M. O'Connor 1989;

Martínez-Alier 1987, 1989). There are thus several ways of enriching Marxism in order to develop a critical analysis of production through the consideration of environmental issues. The following are fundamental:

First, the *constitution of socio-environmental formations* as production units: Marxism offers a holistic vision of the relationship between society and Nature based on social rather than biological over-determination and the centrality of production and social reproduction. From an environmental perspective, the objective is to incorporate ecological potentials and conditions into the capitalist production process and then analyze the articulation with environmental units of production, that is, with non-capitalist socioeconomic formations. In its most general form, these economic-environmental formations can be included within a broad typology of social formations depending upon different forms of land ownership, social access to resources, the property of the means of production, and market exchange of products and inputs (Fossaert 1977). Nevertheless, these typologies do not take into account the complex articulations with natural processes such as appraisal of natural resources, environmental goods and services, and ecological potentials in the sustainable production process. The constitution of socio-environmental formations in the transition toward a world order governed by environmental rationality poses the problem of the conflicting relations of different ethnic styles and community resource use practices in diverse self-sufficient local economies with the market economy (Leff 1993).

Second, the analysis of the interconnections of the global economy with local socio-environmental processes: Marxism is a social theory, based upon the material nature of the social relations of production, which establishes the relations of determination, causality, and conditioning with a series of natural and social processes. Marxism is a theory that de-naturalizes and de-subjectivizes social processes (Althusser et al. 1965/2016; Althusser 1970). In this sense, Marxism opposes naturalist, biological, and energy-centred approaches, as well as methodological individualism, which analyze the relationship between society and Nature from biological evolution, or the ecological perspective of the carrying capacity of the growing population in different ecosystems. By placing the society–Nature relationship as determined by the social relations of production, ecological Marxism politicizes the environmental question that drives the transformation of the dominant economic rationality and the construction of productive rationality based upon the articulation of natural, technological and social processes.[15]

Eco-Marxism opens critical thinking to environmental complexity (Leff 2000), in a different approach to that of social constructivism, generalized ecology or general systems theory (Morin 1977, 1980). Eco-Marxism moves from structural totality based on the capital–labor relationship to the study of complex socio-environmental and economic rationalities based on production relations and labor processes where cultural values, ecological processes, and technological progress are interlinked to the development of the forces of production. Eco-Marxism goes beyond systemic approaches that incorporate the environment to the economic system through planning practices.[16] In order to approach co-evolutionary processes between techno-economic, bio-thermodynamic and geo-ecological systems. Here, the environment is not subsumed under the capitalization of Nature nor maintained as a system external to the economic sphere but is instead integrated to the process of production, not only as a production condition but also as a productive force. The environment is thus regarded as the articulation of cultural, ecological, technological, and economic processes that integrate spatial-ecological and cultural-temporal complementarities to organize a complex, balanced, and sustained productive system open to a variety of ecological and cultural conditions.[17]

Orthodox Marxist theory was able to analyze the relations between the different social processes (political, juridical) over-determined by the economic process and based on the fundamental opposition between capital

[15] Cf. Leff, Sobre las relaciones sociedad-naturaleza en el Materialismo Histórico, *op. cit.*; E. Leff, Ambiente y articulación de ciencias, in E. Leff, ed., *Los Problemas del Conocimiento y la Perspectiva Ambiental del Desarrollo*, México: Siglo XXI Editores, 1986.

[16] Thus, in a different strategy than that posed by environmental economics to internalize environmental externalities in a system's approach, Ignacy Sachs proposed to assimilate—and therefore to nullify the critical otherness of the environment to economic rationality—the environment as a dimension of a planning process: "As policies are available to the [economic] system, the environment gets narrower. The success of such policies will be based upon [...] the disappearance of the concept of environment, which will end up being assimilated into the system [...] In fact, in the long run, the environment, assimilated as a permanent dimension of planning, is destined to disappear as a concrete field of action" (Sachs 1980).

[17] See Chap. 3.

and labor. From an environmental perspective, the social relations of production become more complex among capital, labor and natural processes. The sustainability of the productive process is no longer established through strategies of economic re-mastering of the environment and its appropriation as "natural capital." Instead, it stems from a dialectic in which the ecological conditions for sustainable productivity of natural resources, global ecological balance, territorial distribution of productive activities and the participatory management of resources, are fundamental for attaining an equitable and just distribution of ecological wealth in the planet for the whole of humanity. Thus, the environment emerges as a complex system that articulates natural, technological, and cultural processes within the social forces of production, and which is embedded in new social relations of production—relations between civil society, the state and Nature, and between communities and their environment, where Nature is more than a means of labor and production for the self-management of the bio-cultural patrimony of natural resources of a social group, but where their production of existence is embedded in the web of life.

Third, the construction of a productive paradigm integrating Nature and culture as productive forces: in an alternative productive paradigm to that of the dominant economic rationality, production would not primarily depend on the development of productive forces and technological progress propelled by the logic of the market and profit maximization but on the production of use values to satisfy socially and culturally defined necessities, based on the socialization of access to and appropriation of Nature, on the decentralization and ecological distribution of productive activities, and on the self-management of people's and communities' environmental resources. This reconstruction of productive rationality stems from the ecologically normative nature of the global economy and especially from the development of human-scale local micro-economies (CEPAUR 1989). In this paradigm, Nature and culture are not just mediating processes but act as social labor and direct productive forces. In this sense, we have proposed an eco-technological paradigm for the integrated, sustainable, and lasting management of natural resources, based upon three articulated levels of productivity: the first is the level of ecological productivity derived from Nature (negentropic potential, photosynthetic process, ecological organization) which generates a "system of natural resources" with a growing sustainable supply of natural use values. The second is the level of cultural productivity, in which the diversity of cultural organizations and ethnic identities embrace and enact social productive forces. Through reinvention of their cultural identities, the Peoples of the Earth renew their traditional practices incorporating new technologies to self-manage their environmental resources sustainably. The third is the level of technological productivity, which is based on a complex and polyvalent technological system capable of enhancing the ecological and cultural productivity levels without destroying the resiliency mechanisms and the bases for ecological sustainability as well as the cultural identities, creativity, and autonomy of the people (Leff 1986b).

The construction of a sustainable world faces the question of articulating these emergent life-territories, new spaces of environmental management (not oriented directly toward the production of value nor subject to market laws) with an expanding capitalist economy. Two different rationalities confront one another. Various "ecological interests" and rationalities of current consumers and producers collide in the field of political ecology, together with alternative strategies for the reappropriation of the bio-cultural patrimony by the Peoples of the Earth—strategies for property, possession, appropriation, transformation, and usufruct of their environmental resources—in a new field of political economic open to negotiations and complementarities, but that is not free from contradictions and socio-political struggles. These are the conditions for the reconstitution of an eco-Marxist theory in the emergent field of political ecology.

9. Political Conditions for Sustainability: The Environmental Movement

The crises of socialism and the environment have opened up new ways of thinking and constructing the destiny of life on Earth in the perspective of historical materialism. The new social movements are grounded on the conditions of life in the planet, on the rights of existence of Nature and peoples, and oriented through principles of democracy, of political difference and cultural diversity toward building a sustainable world.

Socio-environmental movements reject the idea of a historical transcendence led by class struggle, aiming at the collapse of capitalism (peacefully or by revolutionary means) and the transition to socialism. The aim is to construct a new sustainable, lasting, equitable, and diverse world order, challenging economic rationality, power domination, and bureaucratic organization, under political pluralism and with the participation of the whole of society in the management of its productive and vital processes.

The aim of such a transformation is no longer the ownership of the means of production by a socialist State that will ensure a more accelerated development of the forces of production and a better distribution of wealth; the aim of a sustainable world goes beyond the conservation and enhancement of the resource base for a sustained development, but to ensure the conditions of existence of Nature, of biodiversity and the improvement of the quality of life and the environment through a radical critique of human needs (Heller 1976) and the meanings of life. It is a question of preserving the Earth's common goods for the sake of humanity and for life itself, of recovering the environment as a productive potential, as a means of production and the creative evolution of life. New struggles are thus being waged around ways of preserving, appropriating and socializing Nature.

Environmental thought is thus inscribed within a post-Marxist postmodern thought. It defends the specificity of local action (thinking globally, acting locally), the autonomy of social groups, the rights of existence of the Peoples of the Earth, and a political ontology of difference: difference in cultural values, is development styles, and in modes of being-in-the-world. Political ecology presents new ethical values and a new political culture, but at the same time poses the problem of political efficacy and the real political power held by capitalism and by different social-environmental groups. Although the defense of autonomy and local difference can be seen as part of the struggle against totalitarianism (against vertical and corporative power structures in traditional political organizations), it also rules out any universal demand other than one claiming the legitimacy of all local demands, of the rights of being of life and the rights of existence of different cultural beings. The demand for autonomy of the people as a condition for the development of alternative local productive projects is part of the larger movement for another kind of material existence based on the integration of multiple life-territories, development styles, and life-worlds. This demand for cultural diversity goes beyond the democratization of the State through political representation, which in many cases is creating new forms of authoritarian and "progressive" regimes imposing one-dimensional neoliberal rationality on the people.

The environmental movement, driven by the principles of autonomous and participatory management of the environmental commons, the rights of indigenous peoples and peasant communities over their life-territories, and the reappraisal of aesthetic values in Nature, is generating new forms of participation in the social process of production and the socialization of the environmental conditions for a sustainable living. In contrast to the conventional Marxist concept of a class (the proletariat) as the protagonist of social change, the environmental movement is the expression of diverse interests and identities and rejects a unitary class response to capitalism as the dominating mode of production. Nevertheless, this movement must still redefine a social project that can integrate its particular differences into a strategy based on "common social labor," one that can respond to "post-Marxist" challenge to construct a unifying solidarity among their fragmented social identities.

Unified objectives are fundamental in order to build alliances that allow for a robust environmental movement with real effective power to build a sustainable world. The question is whether the unity/ force of the movement should be based on the unity of labor in its dialectical relationship with capital; or whether it involves a more complex dynamic, in the sense that the purpose of environmentalism would be the conjunction of different interests and identities within common aims: beyond the collapse of capitalism for the transition to socialism, the construction of a new *environmental social rationality* to ground diverse life-territories on the potentialities and the conditions of life in the planet.[18]

[18] In this sense, Carboni was right in asserting that "the challenge of specificity is accepted by all the new social actors […] It is the result of the complex network of policies […] implemented by capital and the state in order to integrate people at the same time as production conditions are changed. On the one hand, this specificity (difference) represents a break with collective and class solidarity. On the other hand, it reveals new micro-networks of social solidarity and a universal network of solidarity based on social citizenship" (C. Carboni, communication to J. O'Connor 1988).

As James O'Connor (1988) stated, "questions relative to the conditions of production are questions of class, although they may be more than questions of class." This means that, despite the multi-class nature of the environmental movement, demands for the ownership of land and means of production, for the democratization of the State and decision-making, and an improved quality of life and living conditions for all peoples, are common to indigenous peoples, peasant societies, proletarian workers, and urban middle-class struggles. Environmentalism is not only an anti-capitalist struggle in favor of socialism but rather a moving force of human life against the hegemonic power of the dominant economic rationality; it is a social movement involved in the deconstruction of the dominant economic rationality and the construction of an *environmental rationality* built from the alliance of the peoples of the Earth in their purpose to legitimize their rights of existence and establish the social and material bases for an equitable and sustainable world order, grounded on the forces and the conditions of life, engaged in building a sustainable productive process, including new strategies for democratic participation in the economic and social appropriation and management of environmental resources.

Bibliography

Althusser, L. (1969), *For Marx*, London, Allen Lane.

———. (1970), "Idéologie et appareils idéologiques d'Etat," *La pensée*, 151.

Althusser, L., Balibar, E., Establet, R., Macherey, P. & Rancière, J. (1965/2016), *Reading Capital*, translated by Ben Brewster and David Fernbach, London: Verso.

Bergson, H. (1907/1998), *Creative evolution*, Mineola, New York: Dover Publications.

Boege, E. (2008), *El patrimonio biocultural de los pueblos indígenas de México. Hacia la conservación in situ de la biodiversidad y agrobiodiversidad en los territorios indígenas*, México: INAH/Comisión Nacional para el Desarrollo de los Pueblos Indígenas.

———. (2018), "Hacia una antropología ambiental para la apropiación social del patrimonio biocultural de los pueblos indígenas", in Toledo, V.M. & Alarcón-Cháires, P. (Eds.), *Tópicos bioculturales: reflexiones sobre el concepto de*

bioculturalidad y la defensa del patrimonio biocultural de México, México: UNAM.
Burkett, P. (1999), *Marx and nature: a red and green perspective*, New York, St. Matyin's Press.
CEPAUR (1989), "Human scale development", in *Development dialogue*, Num. 1, Motala: CEPAUR/Dag Hammarskjold Foundation.
Daly, H. (1991), *Steady-state economics*, Washington: Island Press.
Echeverría, B. (1998), *Valor de uso y utopía*, México: Siglo XXI Editores.
Ely, J. (1988), "Lukacs' construction of nature", in *Capitalism, nature, socialism*, No. 1.
Enzensberger, H.M. (1974), "A critique of political ecology", *New left review*, 84 (3–31).
Faber, D. (1988), "Dependent Development, Disarticulated Accumulation and Ecological Crisis in Central America," *CNS*, 1.
Fossaert, R. (1977), *La société, Tome 2, Les structures économiques*, Paris, Seuil.
Foster, J.B. (2000), *Marx's ecology: materialism and nature*, New York: Monthly Review Press.
———. (2018), "Marx, value and nature", *Monthly review*, Vol. 70, No. 3 July-August.
Foucault, M. (1966), *Les mots et les choses*, París, Gallimard.
Georgescu-Roegen, N. (1971), *The entropy law and the economic process*, Cambridge, Mass.: Harvard University Press.
Godelier, M. (1969), *Rationalité et irrationalité en économie*, Paris, Maspero.
———. (1974), *Economía, fetichismo y religión en las sociedades primitivas*, México, Siglo XXI Editores.
Godelier, M.M. (1984), *L'Idéel et le Matériel*, Paris, Fayard.
Granou, A. (1973), "Le Capitalisme Face à la 'Non-Croissance'", *Les temps modernes*, 236, 1973.
Habermas, J. (1987), *The philosophical discourse of modernity*, Translated by Frederick Lawrence, Cambridge: Polity Press/Blackwell Publishers.
Heidegger, M. (1935/1977), "The question concerning technology", in *Heidegger. Basic writings*, trans. D. Farrell Krell, New York: HarperCollins, pp. 283–317.
Heller, A. (1976), *The theory of need in Marx*. London: Allison and Busby.
Leff, E. (1986a), *Ecología y capital: hacia una perspectiva ambiental del desarrollo*, México: UNAM

———. (1986b), "Eco-technological productivity: a conceptual basis for the integrated management of natural resources", *Social science information*, 25 (3): 681–702.

———. (1986c), "Ambiente y articulación de ciencias", in E. Leff, *Los problemas del conocimiento y la perspectiva ambiental del desarrollo*, Mexico City: Siglo XXI Editores.

———. (1992b), "A second contradiction of capitalism? Notes for the environmental transformation of historical materialism", *Capitalism, Nature, Socialism*, 3 (4):109–116, December, 1992.

———. (1993), "La dimensión cultural del manejo integrado sustentable y sostenido de los recursos naturales", en Leff E. & J. Carabias (Eds), *Cultura y manejo sustentable de los recursos naturales*, México: CIIH-UNAM/Miguel Angel Porrua.

———. (1994), *Ecología y capital: racionalidad ambiental, democracia participativa y desarrollo sustentable*, México: Siglo XXI Editores.

———. (1995), *Green production: towards an environmental rationality*, New York: Guilford.

———. (2000), "Pensar la complejidad ambiental", en Leff, E. (Coord.), *La complejidad ambiental*, Colección "Aprender a aprender", México: Siglo XXI Editores (2nd revised edition, 2003).

———. (2004), *Racionalidad ambiental: la apropiación social de la naturaleza*. México: Siglo XXI Editores.

———. (2014), *La apuesta por la vida: imaginarios sociales e imaginación sociológica en los territorios del Sur*, México: Siglo XXI Editores.

———. (2018), *El fuego de la vida: Heidegger ante la cuestión ambiental*, Mexico: Siglo XXI Editores.

———. (2019), "Negentropic production", in Kothari, Ashish, Demaria, F., Acosta, A., Salleh, A. and Escobar, A. Eds., *Pluriverse: a postdevelopment lexicon*. London: Zed Books.

———. (2020), "Heráclito: A Φύσις e o Desocultamento da Vida", *Desenvolvimento e meio ambiente*, 54.

Martínez-Alier, J. (1987), *Ecological economics. Energy, environment and society*, Oxford/Cambridge: Basil Blackwell.

———. (1989), "Ecological economics and eco-socialism," *CNS*, 2.

Marx, K. (1965), *Œuvres, Économie I*, Paris: Gallimard.

Meillassoux, C. (1977a), *Terrains et Théories*, París, Editions Anthropos.

———. (1977b), *Mujeres, graneros y capitales*, México: Siglo XXI Editores.
Morin, E. (1977), *La méthode. La nature de la nature*, Paris, Editions du Seuil.
———. (1980), *La méthode. La vie de la vie*, Paris, Editions du Seuil.
O'Connor, J. (1988), "Capitalism, nature, socialism: a theoretical introduction", *Capitalism, nature, socialism*, 1 (1): 11–38.
———. (1989), "Combined and uneven development and ecological crisis: a theoretical introduction," *Race and class*, 30, 3, 1989.
Passet, R. (1979), *L'économique et le vivant*, Paris: Payot.
Rappaport, R.A. (1971), "The flow of energy in an agricultural society", *Scientific American*, 224 (3):116–132.
Sachs, I. (1980), *Ecodesarrollo: desarrollo sin destrucción*, México: El Colegio de México.
Schmidt, A. (1971), *The concept of nature in Marx*, London: New Left Books.
Schmidt-Bleek, F. (2008), "Future. Beyond climate change", Position paper 08/01, Factor 10 Institute, Provence, France.
Schumpeter, J. (2014/1942), *Capitalism, socialism and democracy* (2nd ed.) Floyd, Virginia: Impact Books.
Toledo, V.M. (1980), "Ecología del modo campesino de producción", *Antropología y marxismo* No. 3, pp. 35–55.
WCED (1987), *Our common future*, Oxford: Oxford University Press.
Weiszsäcker, E.U. von, Lovins, A.B. & L.H. Lovins (1997), *Factor four. Doubling wealth halving resource use*, a report to the Club of Rome, London: Earthscan.

8

De-growth or Deconstruction of the Economy: Towards a Sustainable World

1. The De-growth Wager

The 1960s decade was a turbulent period in the modern world that announced the rise of a new age. While emancipatory and countercultural movements (labor, youth, students, gender) irrupted, the "population bomb" exploded (Erlich 1968), seemingly updating the Malthus's claim that rapid demographic growth was the primary cause of land erosion and fertility loss, which would trigger a global ecological crisis. The publication of *The Limits to Growth* (Meadows et al. 1972), a book that was disseminated worldwide as a diagnosis and response to this environmental crisis came as a shock to the ontological security of modernity. For the first time since the treadmill of production and market mechanisms

---

This text was prepared as a conference for the V Colloquium, The Energy Transition in Mexico to a Post-oil era, organized by Ecomunidades, Mexico City, July 24, 2008. The original Spanish version was published as Leff, E. (2008), Decrecimiento o deconstrucción de la economía: Hacia un mundo sustentable, *Polis*, No. 21, pp. 81–90. (Universidad Bolivariana, Santiago de Chile). The English version was published as Leff, E. (2009), De-growth or deconstruction of the economy: Towards a sustainable world, in U. Brand, N. Bullard, E. Lander, and T. Mueller (2009). *Contours of Climate Justice: Ideas for Shaping New Climate and Energy Politics* (Critical Currents, Dag Hammarsjöld Foundation, Occasional Paper Series, No. 6), pp. 101–107, Uppsala.

E. Leff, *Political Ecology*, https://doi.org/10.1007/978-3-030-63325-7_8

had been turned on by the nascent capitalism of the Renaissance, the first time since the West had historically entered modernity, guided by ideals of freedom and enlightened reason, one of the pillars of Western civilization cracked. The myth of progress, impelled by the power of science and technology, had become the most servile—and serviceable—tools of capital accumulation, and of unlimited economic growth.

Environmental crises came to question some of the most ingrained beliefs of modern society: not only was the supremacy of man over all other creatures of the planet and the right to dominate and exploit Nature for his own profit interrogated, but also the very meaning of human existence that has been grounded in economic growth and technological progress. This ontological regime had been forged in economic rationality and shaped by the tools of classical science, had set up the social structure and the imaginary model that established the idea of progress. Human destiny was no longer based on the co-evolution of cultures with their environments, but on economic development based on a mode of production that carried in its genetic code the imperative of limitless growth.

Limits to Growth gave way to the proposal for "zero growth," and for a "steady-state economy" and "Blueprint for survival." The precursors of new ecological economics unveiled the ineluctable relation between the economic process and the degradation of Nature. This, in turn, led to acknowledging the need to internalize ecological costs and to deploy distributive counter-measures to the ecological unbalance generated by market mechanisms. In his book *The Entropy Law and the Economic Process*, Nicholas Georgescu-Roegen (1971) disclosed the fundamental connection between economic growth and the natural limits set by the second law of thermodynamics. The production process generated by the economic rationality that inhabits the machinery of the industrial revolution is defined by an impulse to grow or die, unlike living beings, who are born, develop and die, and human populations, which can establish policies to stabilize their growth. Economic growth, industrial metabolism, and exosomatic consumption imply a permanently growing consumption of natural resources (matter and energy), which not only runs up against the limits of the planet's provision, productivity, and renewal of resources, but becomes degraded in the process of production and consumption, following the entropy law.

Four decades after the eye-opening book *Silent Spring* by Rachel Carlson (1962) that divulged the effects of DDT, ecological destruction has increased dramatically, accentuating global warming caused by greenhouse gases, and by the inescapable laws of thermodynamics, which have set in motion the planet's entropic degradation. The remedies generated by scientific and technological innovation are difficult to integrate into a sustainable economic system. The paradigms and policies that since then have been generated within the emergent geopolitics of "sustainable development" are showing to be short-lived, because they are not sustainable (Park et al. 2008); they are not rooted in the thermodynamic, ecological, symbolic, and cultural conditions of life.

In its globalizing drive, the economic world order that is supported by the discourse of "sustainable development" has obscured the fundamental problem of the environmental crisis. Rather than internalizing the ecological conditions for sustainability, the geopolitics of "sustainable development" ended up commoditizing Nature and over-economizing the world. "Mechanisms for clean development" were put in place, alongside economic instruments for environmental management that established private property rights over the monetary value of environmental goods and services (Brand/Görg 2008). Natural commons (water, air, the sea, and the atmosphere) have been progressively privatized, while new market devices have been created for trading pollution rights (carbon bonuses) and giving a price to the ecological balance of the planet (carbon offsetting).

Today, faced with the failure of all efforts to mitigate global warming (such as the Kyoto Protocol that had established the need to reduce greenhouse gas emissions to the baseline of 1990), awareness of the limits to growth returns, and with it, a clamor for de-growth. The de-growth wager is not merely a critical and reactive moral position: resistance to an oppressive, destructive, unequal and unfair power structure; a manifestation of alternative beliefs, tastes, and lifestyles. De-growth is more than a pure loss of faith in the benefits of the economic process: it is the active conscience of a force right at the heart of the civilizing process which puts the quality of human life and life on the planet as a whole at risk. The call for de-growth is more than just a rhetorical recourse in the critique of the unsustainability of the present world order. However, it must be sustained on solid theoretical grounds and a consistent political strategy.

The call for de-growth is not a mere ideological slogan against the myth of unlimited growth, a *mot d'ordre* to mobilize society against the evils of growth, or its deadly trends. It is not a counter-order to de-link from growth, in the way the hippies could flee from the dominant culture, nor the apology of communities marginalized by "development." Today, not even the most isolated indigenous cultures are safe from or can detach themselves from the effects of globalization driven forward by the engine of economic growth. How to defuse growth in a process that has inscribed in its original structure and genetic code the force that impels it to grow or die? How to de-grow without generating an economic recession with social and environmental consequences on a global, planetary scale? For if the economy itself, through its internal crises, cannot arrive at the level of growth desired by heads of state, corporations, and entrepreneurs, then to deliberately brake growth would amount to willingly kicking off a crisis with incalculable effects. This is the reason why we cannot only think in terms of de-growth, but also in terms of a transition towards a sustainable economy. The latter could not be a mere ecological grounding of the existing economic rationality, but rather another economy, grounded in other productive, social, and ethical principles. De-growth thus implies a *deconstruction of the economy*, together with the construction of new productive rationality built on an ontology of life (Leff 1995, 2004, 2014, 2018).

De-growth implies not only downshifting or unlinking from the economy. It is not synonymous with dematerializing production, since that would not prevent a growing economy to go on consuming and transforming Nature until it reaches the very limits of the planet's sustainability. Abstinence and frugality on the part of some "responsible consumers" do not defuse the *growthmania* at the core of economic rationality, with the impulse towards capital accumulation, economies of scale, urban agglomeration, globalization of the market, and concentration of wealth. To jump from a moving train is not to change tracks and to envision other horizons. De-growth does not entail moving down in the economy's wheel of fortune; it is not enough to wish to make it smaller or to stop it. Beyond the refusal of the commodification of Nature, it is necessary to deconstruct the principles on which the economy was built and to build another economy on the sustainable conditions of life on the planet.

2. From De-growth to the Deconstruction of Economic Rationality

The economistic strategy that purports to contain the overflowing of Nature's downfall degradation by constraining it to the cage of modern rationality has failed. This is the strategy that while restraining Nature within economic instruments and market mechanisms, submitted it to dominant forms of calculation and valuation. The restlessness that demands de-growth is born from anxiety in the face of ecological disaster and disbelief in the efficacy and morality of the market. However, the solution to the problem of growth is not de-growth, but the deconstruction of the economy and the transition towards new rationality that can guide the construction of sustainable worlds.

Deconstructing the economy means more than a mental exercise in order to unravel and identify ideas and social forces that came together and gave birth to the modern economy, a child of the Enlightenment and of a nascent capitalism; it entails a much more complex philosophical, theoretical, political, and social exercise. The economy exists not only as a theory, as the science of production and distribution of wealth. The world economy has been constructed within an ontological regime, as an order of rationality—a form of world construction, of interpreting and acting in the world—that has become institutionalized and incorporated into our subjectivity. Rational choice, the drive for "having," "controlling," "accumulating," "selecting," is in itself a reflection of the configuration of subjectivity constituted within the economic structure and the rationality of modernity.

Deconstructing the unsustainable economy implies questioning the ontological regime where thought, knowledge, science, and technology are inscribed, where the techno-economic process that drives the historical destiny of humanity has been institutionalized; it has created the "cage of the rationality of modernity" (Weber 1922/2014). Economic rationality is not merely a superstructure to be investigated and deconstructed in thought: it is a mode of production of knowledge and commodities; the

Nature-swallowing monster exhaling Faustian fumes, polluting the environment and warming the planet.

A growing economy that feeds on finite Nature cannot be sustained. It is especially impossible for an economy based on oil and coal, which by the metabolism of industry, transport, and human consumption is transformed into carbon dioxide, the main cause of the greenhouse effect and planet-threatening global warming. The fundamental problem with the oil economy is not that of its management as a public or private good. Neither is it bringing fuel costs down by increasing its supply, exploiting protected reserves and submarine fields, building ocean platforms and penetrating the heart of the planet with geological *fracking* technologies. The end of the oil era does not result from its growing scarcity, but rather from its transmutation into carbon dioxide and the abundance of greenhouse gases in relation to Nature's capacity of absorption and dilution; that is from the transformation of the atmospheric equilibrium reached after eons of evolution of life in the planet and the risks of hydrometeorological phenomena and climate change. The search for economic balance by way of the overproduction of hydrocarbons in order to continue feeding the machinery of industry (and the production of agrofuels) puts at risk not only the sustainability of the planet but that of the economy itself. It is imperative to free the economy from its dependence on oil in light of the catastrophic risks of climate change that lie beyond the threshold of 450–550 parts per million of greenhouse gases, as reported in the Stern Report (2008) and the rise of the mid-temperature of the Earth by over 1.5 degrees, as warned by the Intergovernmental Panel on Climate Change.

De-growth of the economy implies not only the theoretical deconstruction of its scientific paradigms but also of its social institutionalization and the internalization of the principles that try to legitimate economic rationality as the ultimate, inevitable mode of being-in-the-world. Nevertheless, the various reasons for deconstructing economic rationality do not directly translate into strategic actions that can defuse the capitalist will to power. It is not merely a matter of "greening" the economy, moderating consumption or enhancing alternative and renewable sources of energy within the niches of opportunity that appear profitable in the context of the increase in energy costs. These principles, even if converted into social movements, do not in and of themselves affect a defusing of production. Instead, they constitute a mere normativity of the system, a counter-current that fails to arrest the overflowing torrent of entropic commodification of the machinery of growth. This is why we need to deconstruct economic reasons by understanding and legitimizing other principles, values, and non-economic potentials; we must forge a new production paradigm and a strategic political program to deconstruct economic rationality while at the same time constructing environmental rationality.

Beyond the purpose of dismantling the dominant economic model, it is a matter of unravelling economic rationality while weaving new matrix of rationality to fertilize new life-territories. This leads to a strategy of deconstruction and reconstruction; not making the system crumble as its first intention, but reorganizing production and consumption based on the principles of environmental rationality, unbinding from the cogs of market mechanisms and economic valuation of environmental goods and services as the dominating principle that organizes the global economy, and incorporating negentropic principles while enhancing the ecological productivity of the Earth's ecosystems. This reconstruction, however, is not only guided by principles of "ecological rationality," but by cultural imaginaries and processes of re-signification of Nature in order to embody the potentialities of Nature in sustainable cultural production practices and to ground them in life-territories. In this sense, the construction of environmental rationality capable of deconstructing economic rationality implies processes of reappropriation of Nature and re-territorialization of cultures.

Economic growth carries with it the problem of measuring. The omnipresent measure of GDP, by which national economies are evaluated in their success or failure, does not measure negative externalities. However, the fundamental problem cannot be solved with a multiple scale or multicriteria methods, or with "green accounting," the calculation of the hidden costs of growth, a "human development index" or an "indicator of genuine progress." The issue is to defuse the economy's internal device (the genetic code) and to do it without causing a recession of such magnitude that it would bring about yet more poverty and environmental destruction.

The decolonization of the social imaginary sustaining the dominant economy will not emerge from responsible consumption or pedagogy of socio-environmental catastrophes, as Serge Latouche (2009) suggested when focusing on the de-growth wager. Economic rationality has become institutionalized and integrated into our way of being-in-the-world: *Homo economicus*. What is needed is the decolonization of our minds and a change of skin. The existing economy cannot be deconstructed by ideological reactions, emancipatory discourses or revolutionary social movements only. It is not enough to moderate its growth by incorporating ecological values and social norms. Unless we want to remain forever at the purely theoretical level, deconstruction entails practical measures to build other modes of production over ecological potentialities and an alternative productive rationality. But praxis without a sound theory can leave us striking blindly in the dark with our desires of a better and more sustainable world but without an understanding of the conditions that guide our purpose and underlie our actions. It is, therefore, necessary to forge *another economy*, grounded on the forces of Nature and the creativity of cultures; on the potentials and possibilities arising from the Real forces of Life and under the conditions posed by the Symbolic orders of Life: under the principles and values of an environmental rationality and on the thermodynamic, ecological and cultural conditions of life in the planet.

3. Limits to Growth, Resignification of Production and Construction of a Sustainable Future

The limit of the techno-economic rationality that dominates the world is the end-point from which life is re-emerging to build a sustainable world. It is from the death drive that we humans derive the meaning of our existence. The limit-law of entropy gave new foundations to the natural sciences analogous to the human world is sustained by the recognition of its cultural and genetic condition in the prohibition of incest. In the face of this cultural landscape and knowledge of the world, we must ask by which strange design the economy has managed to bypass the question of its limits and its dependence on the conditions of life, as it attempts to rule the world as a system of mechanical equilibrium among factors of production and circulation of value and market prices. The limit to this unbridled process of accumulation has not been the "law of value," or the cyclical crisis of overproduction or under-consumption of capital.

The limit is marked by the law of entropy, which, as indicated by Georgescu-Roegen, functions as the limit-law of production. The law of entropy imposes to every economic process, as a productive process, the condition of an ineluctable process of degradation that drives the metabolism of the biosphere towards the *entropic death of the planet*. This entropic condition of production implies that every productive process, like the metabolic process in all living organisms, feeds on matter and low-entropy energy; in this transformation process, it produces consumer goods with a residue of degraded energy, which finally expresses itself as heat. This process is irreversible; notwithstanding the advance of recycling technologies, pollutant residues are only partially reconverted into useful matter and energy. This is what manifests itself as the limit to the accumulation of capital and economic growth: the de-structuring of productive ecosystems, and their saturation with regards to their capacity of dilution of contaminants in common environments (seas, lakes, air and soils), which ultimately appears as a process of global warming, and the possibility of an ecological collapse that crosses the thresholds of the planet's ecological equilibrium.

While the "bio-economy" of Georgescu-Roegen takes the material conditions of Nature as the root of production, the "economy" searches for a way out through the dematerialization of production. The economy flees towards the fictitious and the speculation of financial capital. Nonetheless, for as long as the economic process must produce material goods (houses, means of transportation, clothes, food) it cannot escape the law of entropy. This is the ultimate limit of economic growth. The only antidote to this inevitable trend towards entropic death is the process of *negentropic production* of living matter, which translates into renewable natural resources.

The transition towards *this* bio-economy would mean a decrease in the rate of economic growth as it is measured today, and a negative rate in time, while indicators for a sustainable, negentropic eco-technological production are developed. In this sense, the new economy is based on ecological potentials, technological innovation and cultural creativity, a new paradigm of negentropic sustainable production (Leff 1986, 1995, 2019). Thus, a post-growth society could be forged bases on an economy in balance with the planet's conditions of sustainability. From environmental rationality emerges a new mode of production and renewed ways of being-in-the-world: new processes of signification of Nature and new existential meanings that open new paths towards a sustainable future through the confrontation of different life-worlds in the horizon of a sustainable world order. These historical processes are fertilizing the emergent field of political ecology.

Bibliography

Brand, Ulrich/Görg, Christoph (2008), Sustainability and Globalisation: A Theoretical Perspective, in Park, Jacob/Conca, Ken/Finger, Mathias (eds.), *The Crisis of Global Environmental Governance. Towards a new political economy of sustainability*. London and New York: Routledge, 13–33.

Carlson, R. (1962), *The silent Spring*, Boston: Houghton Mifflin.

Erlich, P. (1968), *The population bomb*, New York: Ballantine Books.

Georgescu-Roegen, N. (1971), *The entropy law and the economic process*, Cambridge, Mass.: Harvard University Press.

Latouche, S. (2009), *La apuesta por el decrecimiento: ¿cómo salir del imaginario dominante?*, Barcelona: Icaria.

Leff, E. (1986), "Eco-technological productivity: a conceptual basis for the integrated management of natural resources", *Social science information*, 25 (3): 681–702.

———. (1995), *Green production: towards an environmental rationality*, New York: Guilford.

———. (2004), *Racionalidad ambiental: la apropiación social de la naturaleza*. México: Siglo XXI Editores.

———. (2014), *La apuesta por la vida: imaginarios sociales e imaginación sociológica en los territorios del Sur*, México: Siglo XXI Editores.

———. (2018), *El fuego de la vida: Heidegger ante la cuestión ambiental*, Mexico: Siglo XXI Editores.

———. (2019), "Negentropic production", in Kothari, Ashish, Demaria, F., Acosta, A., Salleh, A. and Escobar, A. Eds., *Pluriverse: a postdevelopment lexicon*. London: Zed Books.

Meadows, D.H., D.L. Meadows, J. Randers & W.W. Behrens III (1972), *The limits to growth*, New York: Universe Books.

Park, Jacob/ Finger, Matthias/Conca, Ken (2008), The Death of Rio Environmentalism, in Conca, Ken/Finger, Matthias/ Park, Jacob (eds.), *The Crisis of Global Environmental Governance: Towards a New Political Economy of Sustainability*. London: Routledge.

Weber, M. (1922/2014), *Economía y sociedad: esbozo de sociología comprensiva*, México: Fondo de Cultura Económica.

9

Bioeconomics, Negentropic Productivity and Eco-social Sustainability

Introduction

The environmental crisis has imposed an urgency on revising the role of Nature in the economic process. New approaches to the relations between Nature and society have emerged, imposing the need to internalize ecological and cultural conditions for an equitable, diverse, and sustainable development. Economic rationality has impinged on the self-organizing mechanisms of biological and ecological systems, which are responsible for global ecological balance and regeneration of natural resources. It destroys the ecological and social conditions for sustainable production (Naredo 2006). Market mechanisms do not assign just value either to the

---

A first version of this chapter was presented at the Second International Conference Implications and Applications of Bioeconomics, European Association for Bioeconomic Studies. Dragan European Foundation, Palma de Mallorca, March 11–13, 1994. A revised version was presented in the Conference Getting Down to Earth: Practical Applications of Ecological Economics, International Society for Ecological Economics, San José, Costa Rica, 24–28 October, 1994 and published in Leff, E. (1996). From ecological economics to productive ecology: perspectives on sustainable development from the South," in Costanza, R., Segura, O. and Martínez-Alier, J., *Getting Down to Earth: Practical Applications of Ecological Economics*, International Society for Ecological Economics, Island Press, Washington, DC, pp. 77–89. This is a new version based on those previous papers.

E. Leff, *Political Ecology*, https://doi.org/10.1007/978-3-030-63325-7_9

long-term synergistic productivity of socio-environmental processes that support sustainable development, nor to social equity and cultural diversity. Economics cannot be subsumed in biological laws nor incorporate the specificity of social rights, interests, and institutions that define the conditions for the participatory and democratic management of natural resources. The emergent paradigm of *bioeconomics* has proposed a thermodynamic approach to analyze the unsustainability of the economic process grounding it on the concept of entropy. Yet, this critical approach is a limited one, since it disregards the fundamental process on which life depends: negentropy. The negentropic potential that mobilizes the ecological organization of life on the planet is the basis for an alternative productive paradigm, built on the integration of ecological, technological, and cultural processes that contribute to sustainable production of socially defined wealth. This eco-technological paradigm for production enhances the capacities of the biosphere to produce natural use values from the negentropic production of biomass generated by photosynthesis and symbiogenesis, through a technological system designed to couple and expand this potential, harnessing and limiting entropic decay, rooting sustainable livelihoods in the productive character of Nature and the conditions of life in the planet (Leff 1986b, 2004, 2019).

Sadi Carnot introduced the concept of entropy, as the condition—the ineluctable degradation and loss of usable energy—for every matter and energy transformation processes, in 1924. A century later, Alfred Lotka (1922) extended its meaning as a key concept in the thermodynamics of the evolution of life. It took 50 years more to introduce the entropy law to the critique of standard economics (Georgescu-Roegen 1971). The emergent bioeconomic paradigm analyzes the unsustainability of economic rationality from the standpoint of the inexorable increase of entropy of current processes of production and consumption fueled by economic growth and driven by free (but blind) market forces dominating techno-economic ontological order.

Environmental crisis confronts us with the need to review the scientific and moral foundations of modern civilization, and how science and knowledge are conceived. New scientific approaches that integrate life, cultural, economic, symbolic, and power structures are emerging and guiding socio-environmental processes. Systems theory, bio-thermodynamics and dissipative complex systems have become new methods, challenging the paradigms of normative science (Prigogine 1955, 1981; Prigogine and Nicolis 1977; Prigogine and Stengers 1984; Urry 2003, 2005; Mol and Spaargaren 2006; Spaargaren et al. 2006). Notwithstanding the importance of these new approaches to harness the economic process by recognizing the flows of entropy and the workings of emergent complex systems in the biosphere, they have neglected the material source for constructing an alternative productive paradigm of bio-economics and sustainability: the potential of biomass formation through photosynthesis. Negentropy is the fundamental source from life that balances a sustainable economic process that entails entropic decay during the technological transformation of matter and energy in the process of production.[1]

This alternative productive paradigm integrates ecological, technological, and cultural levels of productivity to increase the sustainable production of socially and culturally defined wealth. As a strategy it seeks to enhance the negentropic capacities of the planet to produce natural use values and develop a technological system to reduce the entropic trends of technological transformation processes. Here Nature and culture cease to be considered resources to be appropriated by a techno-economic process, and are no longer a set of conditions for production, or a subsystem interrelated with the economic process. Nature and culture emerge as *productive potentials*, as contributing processes in the development of productive forces, subject to thermodynamic, ecological, symbolic and cultural conditions of life in the planet (Leff 1993).

[1] Originally, I conceived this alternative productive rationality as the articulation of two levels of productivity: the primary productivity of ecosystems—the production of biomass through photosynthesis—and technological productivity transforming natural resources into use values and commodities for human consumption (Leff 1975). I intended to root my environmental critique of standard economics and economic growth on the possibility of conceiving production as integrated levels of ecological, technological and cultural productivity. Thus, the concept of eco-technological productivity was forged to conceive an alternative productive rationality as the transit towards a *negentropic society*. With this conceptual strategy I intended to view sustainable production as the synergistic potential arising synchronically from the articulation of ecosystem's resources, an ecological–technological system and cultural values, and diachronically from the co-evolution of ecological succession, scientific–technological innovation, cultural creativity and social change (Leff 1975, 1986a, c).

The sustainability of life on the planet depends on a *new bioeconomic paradigm of production* based on the negentropic process of matter formation. This free source of organizing matter and energy is based on the capacity to draw a sustainable production of use values to satisfy culturally defined human needs from the photosynthetic capacity of the biosphere. This fact has been overlooked both by ecologists concerned with measuring primary productivity, deforestation and greenhouse gas emissions, and by ecological economists concerned in the critique of standard economics from the vantage point of entropy, for the positivistic views of economic science as pursuing unlimited production. The applications of the heuristic meanings of the entropy law to the purposes and goals of an equitable, democratic, and sustainable development, viewed as an emergent complex socio-environmental system, requires the deconstruction of fixed scientific concepts from thermodynamics and biology.

Standard economic theory has legitimized a ruling social order that destroys the organization of the biosphere as a life-support-system and its conditions of ecological balance. A new economic rationality cannot be subsumed under a broader ecological order without grounding it in thermodynamic and ecological conditions of Nature and of cultural and social organization (symbolic processes, political power, social interests, and institutional structures) that determine the sustainable and equitable appropriation of natural resources. Economics emerged in modernity as the science of alternative allocation of scarce resources. In post-modern times, economic rationality no longer faces relative resources' scarcities, but global scarcity: the limit state of entropic degradation as a condition for life. The extension of the laws of entropy in closed, near to equilibrium systems, as foundations of a new economics, leads to an ineluctable dead-end of the process of production. The interpretation and extension of the concept of entropy to economic processes and living systems in view of building sustainable human societies, has strong moral and practical consequences that need to be critically addressed.

The *fatal strategies* of capital accumulation have come to surpass the carrying and dilution capacities of the planet, leading to entropic decay, environmental degradation, biological extinction, and human alienation.

Today, not only economic rationality is being contested, but so is the human project built on the concept of production (Baudrillard 1974, 1976, 1993). Thus, eco-technological-cultural productivity as a concept of a sustainable source of wealth demands an alternative productive and social rationality, one capable of incorporating the values and potentials of ecological processes, cultural diversity, and participatory democracy.

The need for new paradigms to apprehend the complexity of socio-environmental systems and to democratize knowledge as a condition for the self-reliant management of production, has led to the revision of preconceived notions of reality and of science, shifting from exact quantitative, unified, and mathematized knowledge, to more comprehensive, though less accurate and less measurable, heuristic paradigms. While hard science (quantifiable, measurable, predictable) has yielded efficient technology, a soft post-normal science could enable us to assess the systemic effects of techno-economic and scientific rationality on the degradation of life and human existence, and foster a participative decision-making process with people's involvement and responsibility in the destiny of life in the planet (Funtowicz and Ravetz 1993).

The concept of entropy as a measure of order/ disorder and organizational change has opened a new theoretical breakthrough in the fields of bio-thermodynamics and bio-economics. Yet, the concept of entropy has not been applied in a more comprehensive way to assess sustainable productivity. A new perspective of negentropic productivity sheds new light and provides a new basis to address the issue of sustainability. From this viewpoint, biomass formation through photosynthesis appears as a negentropic productive process that balances the ineluctable tendency to maximize entropy production in every metabolic and technological process involved with the transformation of mass and energy for the reproduction of capital, and in the sustainability of life on the planet. Entropy as a measure of order is meaningful for this purpose inasmuch as the ecosystem organization—its complexity, diversity, and stability—determine its ecological productivity. Thus, an entropic/negentropic balance will set the limits and open new perspectives for sustainable production and livelihood.

Flows and Flaws in the Thermodynamics of Economics

Environmental crisis has imposed the need to internalize the ecological conditions of a sustainable economic process. The real world appears now as a complex order, overriding the capacity of natural and social sciences to apprehend its emergent and unpredictable dynamics. Thus the environment has imposed the reconstruction of knowledge through new systemic and interdisciplinary approaches, beyond the articulation of established segmented disciplines on fragmented spaces and ontological orders of the Real (Leff 1986c; Funtowicz and Ravetz 1994).

While in the 19th and 20th centuries the physical sciences developed challenging their Newtonian origins, economics reinforced its mechanistic foundations as an ideal system, close to equilibrium, based on two basic productive factors: capital and labor. Thus, natural processes were valued only for their contribution to the productivity of capital, labor force, and technology. Environmental services were considered an eternal gift or an external system from which the economic activity could draw unlimited sources of energy raw materials. However, Nature did not contribute to value formation.[2] Thus, the destruction of the ecological basis and conditions of production was obscured by the prevalent paradigms of economic rationality. Environmental services and ecological processes that support ecological balance and natural productivity have not been properly valued. Externalized and rejected, Nature is taking its revenge on humanity, as the irruption of an ontological truth that has been negated by that scientific knowledge that rules the destiny of life in the planet.

Several authors have criticized the limits of economics and tried to view it as a subsystem of the wider process involving the biogeochemical, ecological, and thermodynamic conditions of production. Thus, economic behavior must develop as the extension of life systems, reconstructing economic rationality with principles of ecology and thermodynamics (Georgescu-Roegen 1971; Passet 1979; Martínez-Alier 1987; Grinevald

[2] See Chaps. 5, 6, and 7.

1993). The mainstream in the emergent paradigm of bio-economics is founded in the concept of entropy and a systemic view of the interrelations of economic processes with its biogeochemical environment. The entropic condition of production arising from the laws of thermodynamics has helped understand the increasing flow of degraded energy leading to the heat death of the economic process, as well as climate change and the extinction of biodiversity, exacerbating conflicts arising from struggles of human society for survival and access to scarce resources.

From an ecological and systemic approach, linking economics to the sources of life, the claim has been made to subsume economics under the ecological conditions of its surrounding environmental system to achieve a sustainable productive process (Passet 1979). Thus, economic processes should be viewed as thermodynamic and informational processes. However, this cannot be done by simply adding to standard economic calculation an ecological evaluation and an energetic measure of the inefficiency of the externalities of the economic processes—diminishing energetic yields, matter, and energy dissipations, deforestation and loss of soil fertility, diseconomies of growth and environmental degradation—as these are incommensurable processes and cannot be properly valued through market prices (Passet 1985, p. 241). Rene Passet has thus stressed the need to conceive the productive sphere in close interdependency with the environment, without reducing the ecological processes to market logic and without diluting economic specificity in any sort of biologism or ecologism. His view for "regulating mechanisms by which the natural environment and societies ensure their reproduction", is conceived as a set of norms that constrain the economic process (*gestion normative sous contrainte*) (ibid., p. 841).

However, viewing the environment as a constraint, as a cost and as a limit set by the laws of entropy and ecology, is insufficient to reverse current trends. What is at stake is the possibility of internalizing the contribution of ecological productivity to a sustainable production process. *The environment must be conceived as a productive potential for an alternative productive rationality* (Leff 1995). Economic rationality has hampered the self-organization of biological and ecological systems that sustains the negentropic production of biomass and regeneration of ecosystems.

Capital accumulation has undermined the ecological conditions of production. Social equity and ecological distribution have been sidelined relative to the primary objective of economic growth. Thus, the economic process generates increasing entropy, not only as degraded heat and unavailable matter and energy, but also in terms of the destruction of the self-organizing and productive ecosystems, capable of generating and regenerating natural resources.

Biological processes that contribute to the stock of raw materials have been considered a free asset, not as a potential for the process of production that depends on physical limits and resource scarcities driven by entropic degradation fueled by the economic process, as well as on social strategies for the sustainable management of negentropic and ecological potentials. Living systems do not merely establish a set of conditions to be respected by the economy (ecological resiliency and carrying capacity fixes). Nature must be conceived as the precondition of life-support systems for sustainable production of use values, goods and services.

Ecological conditions for the production of biological resources overlooked by mainstream bioeconomics should be viewed as a potential for alternative productive processes. The negentropic potential fueled by solar energy as the primary source of life in the biosphere is the source of self-organizing emergent complex systems that balances entropic degradation. This approach challenges the mechanistic model of the economy from the perspective of its energetic inefficiency and increasing entropy of throughput in the productive process that destroys the biological and ecological basis of production. The workings of the entropy law and dissipative open systems have demystified the false foundations of the *physics of economics*, as Nicholas Georgescu Roegen named it. However, the process of biomass formation from photosynthesis and its contribution to economic production has been underestimated. Therefore *sustainable development* has been shaped as an ideological proposal for launching a new era of economic growth, while alternative paradigms based on the productivity of Nature—"an economy fueled by solar energy harnessed by green plants" (Georgescu-Roegen 1993)—have been disregarded as non-economic conservation strategies.

The concept of entropy has been often transposed to ecology as a measurement of order, complexity, and productivity of ecosystems. Ecologists like Ramon Margalef have related the higher natural productivity of the more diverse tropical ecosystems to their higher complexity and diversity, in spite of the fact that strict quantitative correlations have not been clearly established between species diversity and the complexity and productivity of ecosystems.[3] Following Lotka's *maximum power principle* (Lotka 1922), other authors have claimed that the more complex and ordered an ecosystem the more it changes toward maximum entropy production, which is expelled to its environment.[4] Thus, the entropy law is conceived as the inescapable condition that drives the economic process, as well as the organization of life in the biosphere, towards an entropic increase for the sake of the ever-increasing complexity of the system. However, this has remained an abstract scientific hypothesis of an idealistic vision of life's creative evolution. In fact, the life-support system has been continually and increasingly degraded by the capitalistic appropriation of the biosphere, triggering ecological decay and destining life towards the entropic death of the planet for the sake of supporting an unsustainable economic process. Lotka's principle of "maximal level of entropy production" to organic living systems, as well as its meaning for socio-environmental systems is ambiguous and questionable. If Lotka's principle were to be true as a general law of Nature, and extendable to

[3] Thus, Margalef indicated that: "The formal analogy of the expressions used to compute an index of diversity from the proportions of individuals falling in different species, with expressions on entropy, does not in itself justify basing thermodynamic properties of the ecosystems on values of diversity indices." However, "It is appropriate to speak of the entropy produced in sustaining a unit of biomass in the ecosystem; this entropy is roughly proportional to the total flow of energy. If the system has many trophic levels, the energy flow per unit biomass is lower because a fraction of energy passes through different levels. In a system subjected to frequent changes in which a high proportion of the primary producer's substance is decomposed by bacteria, energy is inefficiently used and relatively more entropy per unit time and unit biomass is produced than in a more diverse and more efficient ecosystem" (Margalef 1968, pp. 19, 20–21).

[4] "The ordering process in this autopoietic system is associated with the succession of the ecosystems towards a maximized capacity to catch solar energy and produce entropy (in this case: thermal heath that is exported from the system). This could be a reflection of a change towards maximum entropy production of the system [...] This view of man's autopoietic organization of interactions with ecosystems driven by solar energy could throw some light on the present imbalance of man's interaction with nature" (Günther 1993, pp. 268, 271). Entropy production is here distinguished from entropy content, "since the (living) system increases in organization and therefore decreases its internal entropy by the increasing cycling and feed-back mechanisms that evolve as the system moves away from equilibrium. Thus, the maximum entropy production could be expected to increase the complexity of the system..." (ibid., p. 265).

social systems, then all efforts to escape this ineluctable trend towards entropic death would be futile. Yet, entropy and negentropy can be valuable heuristic concepts for phenomenological descriptions to assess trends towards sustainable production without the pretension to provide a concrete unifying measure of order or degradation.[5]

The concept of entropy in ecology appears as a measure of order, complexity, and ecological balance, assuming that increasing organization leads to maximum entropy exported from the ecosystem. However, this interpretation of entropy in ecological systems opens up the following critical questions: What would be material expression and the ecological effects of entropy generated by the self-organizing mechanisms of ecosystems in climax equilibrium or in ecological succession? Towards which external system—a larger terrestrial ecosystem, or the atmosphere and stratosphere?—is entropy expelled from the complex ecosystem in order to maintain its order and stability? What is the production of maximum entropy in the process of photosynthesis or ecological succession and how can this production of heat be measured? Does evolution of living organisms have the effect of breaking down and degrading ecosystem complexity that supports the life of individual organisms? There seems to be conceptual inconsistency, an aporia, in the logic of ecological scientific discourse.

Jacques Grinevald observed that living organisms perform self-organizing and productive functions through complex interrelations established by biological communities with the biogeochemical environment (Grinevald 1993); through energy and matter cycles and feedbacks that drive biological evolution and ecological successions, losses of available energy are constantly replaced by solar energy. Only when these complex mechanisms are altered by man's intervention, as in highly artificialized energy intensive agro-ecological systems, entropic degradation grows because of increasing energetic resources applied to the system,

[5] Przybylski Tadeusz (1993) has stressed the value of entropy in relation to the ecological balance of two of the planet's atmospheric gases (O_2 and CO_2). Forest destruction diminishes the production of biomass and as a result reduces the assimilation rate of CO_2 in the atmosphere. Thus, the concept of entropy is related to ecological balance. Reduced biomass increases entropy by degrading ecological order and productivity. By destroying the forest ecosystems global warming is being produced through erosion and desertification, and by the greenhouse effect that alters the O_2–CO_2 balance in the atmosphere.

while natural productivity in terms of input–output energy yields diminishes. Mario Giampietro reformulated the maximum power principle asserting that "self-organizing systems show a natural tendency to evolve toward designs—energy transformations, hierarchical patterns, feedback controls, amplifier actions—that enable an increase in the quantity of available energy, and the efficiency of using it, for sustaining their structure/functions" Giampietro (1993, p. 206). However, no method has been developed to measure the order, the internal entropy and the entropy discharges to its external environment to maintain the evolving and ever-more complex structure of these open, dissipative, becoming systems (Prigogine). The applicability of this principle is restricted by "the difficulty in defining clear boundaries in space and time for interacting components of a multilevel hierarchical system" (ibid., pp. 207–208).

In the progressive effort to construct a new paradigm of bioeconomics deriving from the laws of entropy, the non-critical extension of the concept has generated "many conceptual and terminological confusions [...] about entropy, life and economic activity" (Grinevald 1993, p. 251). New controversies have arisen concerning the use and meaning of entropy, from its strict formulation in the second law of thermodynamics to new heuristic and often metaphoric uses. Georgescu-Roegen applied the concept to his "fourth law of entropy", to account for the degradation of available matter, while conceiving the biosphere and economics as closed systems. However, Kozo Mayumi (1993) has exposed good reasons why the "fourth law" cannot be justly considered a quantitative and general scientific law, such as in Clausius or Boltzmann's formulations of the laws of entropy. Mayumi argues: "The concept of entropy is, in essence, tantamount to entropy of energy diffusion. Therefore, the degradation of matter in bulk at the level of our senses cannot be treated in terms of entropy in thermodynamics" (Mayumi 1993, p. 403). Nevertheless, it still holds that there is a loss of matter in every manipulation and transformation of material objects, so the heuristic use of the concept conveys a practical meaning. However, the loss of matter in production and consumption and the degradation of available energy at larger scales result from planned obsolescence and the drive for economic growth, rather than by the normal wearing-out of use values.

Gabriel Lozada has gone one step further to criticize the validity of Boltzmann's concept of entropy in statistical mechanics, stating, with Prigogine, the impossibility of unifying the fields of entropy at the micro and macro levels, from thermic machines to economic processes and to the cosmic scale of the universe. Thus Gal-Or argues: "the explanation of time irreversibility is not in the microscopic scale, but in the cosmic scale; and in particular in the fact that the universe is expanding" (Lozada 1993, p. 396). The biosphere is an open system deriving its sources of life from the universe while the economic process undermines sustainability and productivity of life. Two different time-space scales are at work in this encounter of differentiated negentropic-entropic processes. Negentropic productivity is as important for bioeconomics as the irreversible entropic decay in technological and economic processes. This does not imply that life processes can reverse the laws of entropy in the universe; it only stresses a conception of life and the structural functioning of ecosystems as dissipative open systems (Prigogine 1955, 1981; Prigogine and Nicolis 1977; Prigogine and Stengers 1984) that do not follow the entropic laws of thermodynamic systems in dynamic equilibrium.

Erwin Schrödinger (1944) conceived life on earth as being formed by extracting *negative entropy* from the universe. This source of life—capturing negative entropy from the Sun to fuel the life self-organizing process—becomes a productive potential, a resource from Nature for the economic process. While self-organizing and metabolic processes of living systems in the biosphere maintain their order by externalizing degraded energy and pollution to the environment, biomass production from photosynthesis does not alter the entropy of the universe. Clearly we are dealing here with different spatial and time scales, from the photosynthetic and metabolic processes in the biosphere to the cosmic space of the universe from which life draws its sources. The entropic drive of any individual living organism on earth responds to different spatial-time scales of the life processes from the negentropic formation of biomass through photosynthesis and symbiogenesis[6] (Margulis and Sagan 1995) in the creative evolution of life.

[6] Lynn Margulis and Dorian Sagan define symbiogenesis as "an evolutionary term referring to the formation of new life forms, new organs, or new cell organelles by permanent association of older, pre-established life forms" (Margulis and Sagan 1995, p. 263).

Thus we can conceive of two differentiated biological phenomena where entropy flows have different space–time scales and meanings. The negentropic process implies that the economy is an open system, in which biomass is created by extracting "negative entropy" from the Sun. Entropy discharged to outer space from this self-organizing process will not alter the timing for the extinction of the Sun and its negentropic input to the Earth's biosphere. Nothing that happens to man and the earth will affect the pace of expansion of the universe, nor its increasing entropy, except for any unpredicted effects of the human conquest of our solar system. It isn't clear either how photosynthesis expels entropy to its near environment, as primary productivity is not an isolated process, but depends on the ordered structure of the ecosystem itself.

In order to construct a new paradigm of production based on an entropic–negentropic approach, we must clearly understand the different meanings of the concept of entropy in different space–time scales and ontological regimes affecting the courses of Nature on Earth. The laws of entropy in closed and near-equilibrium systems do not apply the same in open, dissipative, far-from-equilibrium systems, which is the case for living organisms, technological processes, and social organization. We must distinguish between the negentropic process by which vegetal matter (biomass) is formed through photosynthesis, from self-organizing processes of ecological succession, biological evolution, and biosphere metabolism; and differentiate the economic, technological, industrial, and urban processes that degrade available energy and generate high entropy as heat, waste, and pollution. Thus, in ecological systems, higher-order, complexity and stability are associated with biological negentropic productivity, while technological systems appear as productive of entropy in spite of their complex cybernetic feedbacks, because they are incapable of producing new matter and available energy. Thermodynamic equilibrium in technological systems appears as its highest entropic state, while the dissipative structure of ecological systems is a state of productive order.

The conception of the economic process as a closed system has limited the perspectives of bioeconomics to a negative and fatalistic perception of production that inscribed in the capitalist mode of production geared towards unlimited growth, is the economic reason and ineluctable cause of entropic degradation in the planet. However, the primary productivity of ecosystems and biological evolution are only possible because the Earth is an open system, fueling solar energy into the self-organizing of life. Grinevald has rightly brought this important issue into the discussion:

> Vernadsky adopted the idea, shared by Bergson, Auerbach and many other pre-Schrödinger thinkers, that life is a process reversing the increase of entropy, and not, as emphasized by Georgescu Roegen, accelerating the increase of entropy flow. The matter is a question of difference between closed and open systems, total and boundary systems. Living structures are open and boundary dissipative systems, always coupled with a global system, the environment. Contrary to the view that life is an order opposed to the degradation of energy and increase in entropy, or at least retarding it, the sound conclusion of Georgescu Roegen is that life activity accelerates in fact the entropy flow connecting the living organism and the total environment. (Grinevald 1993, p. 247)[7]

Thus, bioeconomics was founded on the interpretation of the law of entropy as the ineluctable death-drive of all economic process and of living activity, instead of constructing a sustainable economy on the production of ecological order and human creativity, of order generating a sustainable productivity of goods by enhancing the vital negentropic potentials of biological processes. However, in building sustainable societies we must not confuse the entropy of the universe—which is not affected by our living planet—and the validity of entropy laws in closed, near-equilibrium thermodynamic systems, with the time and space scales of biological and ecological dissipative systems and economic processes involved in negentropic production.[8]

---

[7] Georgescu Roegen stated that, "life is characterized by the struggle against the entropic degradation of mere matter. But it would be a gross mistake to interpret this statement in the sense that life can prevent the degradation of the entire system, including the environment. The entropy of the whole system must increase, life or not life" (Grinevald 1993, p. 252).

[8] Thus, Günther stated: "Living systems exist only far from thermodynamic equilibrium. This means that one has to be very careful when making analogies between reaction assemblages that are close to equilibrium" (Günther 1993, p. 264).

The problem here is the definition of the different space–time scales and the differentiated processes taking place in negentropic–entropic ecological productivity, and of the boundaries of what is considered the environment for the bioeconomic system. The basic question is the balance between the negentropic formation of biomass and the degradation of mass and energy through food chains and matter–energy flows in the metabolism of the biosphere, and in technological transformations of rural-ecological and industrial-urban systems. Ecological engineering can control and retard entropic degradation, but will not avoid it nor achieve complete recycling of mass and energy in the ecosystem. Even life within a specific organism is irreversible and follows the entropic path. However, living systems do not feed only on negative entropy from their close biosphere environment, to which they expel entropy from their self-organizing and metabolic processes; they feed on negative entropy mainly from the Sun, to create living matter through photosynthesis and ecological organization.

The flaws of Georgescu-Roegen's "fourth law" arise from his conception of the economy as a closed system, while in fact the biosphere is an open dissipative process of evolution, by transforming energy from the Sun. Thus, the bioeconomic approach to sustainability of life should not be limited to the Earth's ecosystemic conditions for sustaining economic growth activities—the resiliency and carrying capacity of ecosystems and the biosphere to metabolize and support increasing production, population growth, and waste disposal, renewing the availability of scarce resources; rather, bioeconomics should illuminate an alternative economic paradigm, conceived as an open negentropic productive process of biomass formation through photosynthesis supported by self-organizing ecosystems. Thus, the perspectives for the sustainable management of resources should not be limited to ecological constraints to control the unavoidable entropic tendencies of current production and consumption patterns, but should open up new possibilities for eco-technological productive processes based on the potentialities of life processes in ecological systems and cultural organizations.

The Negentropic Potential: Photosynthesis and Primary Productivity of Biological Resources

In viewing the possibilities of putting into practice this bioeconomic approach based on the concept of eco-technological productivity, it is important to assess the actual trade-off of negentropic production of biomass countering entropic decay in today's global economy and to forecast prospectively the construction of knowledge and institutional changes to make a transition towards a "coevolutive economy" (Norgaard 1984, 1994) to balance the negentropic productivity of natural resources and entropic degradation of biotechnological transformation processes. This balance cannot be established in abstract terms by the theory of dissipative open systems, self-organization of living systems and entropy laws (O'Connor 1991, 1993a). The actual capacity for biomass formation in the biosphere through photosynthesis must be evaluated and policies and strategies designed to promote the sustainable productivity of natural resources, increasing ecological productivity through high-efficiency photosynthesis, and sustainable ecological productive practices, harnessing technology to reduce entropic decay in productive processes.

This paradigm of eco-technological productivity conceives sustainability as a balance of entropic and negentropic paths in ecological cycles and technological processes, and relates matter and energy flows to actual productivity of goods and services. At the ecological level of production, the reordering of ecosystems enhances negentropic formation of biomass but cannot evade entropic degradation in the overall metabolism of the ecosystem. At the technological level, although all mechanical and thermodynamic processes follow the trend toward increasing entropy, biothermodynamic innovations can be oriented to reduce the degradation of usable energy. Moreover, biotechnology can improve the efficiency of photosynthetic processes and ecological succession can be engineered to maximize ecological productivity.[9]

[9] A formal structure of a joint-production approach is discussed by Martin O'Connor (1993a, 1994).

The practical application of this eco-technological paradigm is shaped by cultural values. Culture mediates and is integrated into sustainable negentropic productivity. This leads to the difficult task of assessing in terms of entropy–negentropy cycles, processes, and balance, the autopoietic features of cultural organization and the effects of culturally embedded values, knowledge and practices that result in *cultural productivity*. Linking cultural enactment of entropic–negentropic processes in the biosphere's metabolism and in the sustainable management of natural resources questions anthropological approaches that view cultural evolution as the constitution of increasingly complex and hierarchical power structures associated with an inexorable rise in the use of energy sources (White 1949; Adams 1975).[10]

The energetic approach in anthropology has led to the envisioning of cultural evolution and economic development as a progressive hierarchization of social and power structures involving increasing organization and greater order, which would imply an increase in entropy in the socioproductive system exported to its environment. This extension and generalization of the concept of entropy becomes contradictory and paradoxical when applied to cultural organization. If lower entropy is related with increasing order of the socioeconomic system, then we have to face the paradox that "development" associated to increasing order in the evolution of civilization has brought about an exponential increase in entropy dissipation as environmental disorder, evident in ecological destruction, loss of biological and cultural diversity, and in global warming of the planet. This paradox is the result of Cartesian ontological dualism generated the epistemological paradigms of science that separates culture and Nature: cultural organization from the territory and the environment where culture is embedded. The inconsistency of this paradox when stating that the organization of the system is achieved at the expense of dissipating entropy to its environment becomes evident when the environment's resiliency and productivity becomes a condition for

[10] Thus, Leslie White (1949) postulated a *basic law of evolution*, as an axiom establishing a direct relation in the evolution of cultural organizations with the increasing control and energy use. Claude Lévi-Strauss also believed that cultural organization could resist the ultimate principle of the law of entropy and that anthropology would eventually become an *entropology* (Lévi-Strauss 1955).

sustaining the socioeconomic practices of any cultural organization. Yet this explanation can only arise from a view that dissociates the economic, social, and cultural systems from the environment. This approach to economic development, cultural evolution, and social change disregards the fact that socio-environmental processes are integrated open systems. This has been the dominant "scientific" view that has legitimized an entropic unsustainable society by destroying its material basis for sustainable development. This leads to an unsolvable question about forms of social organization that could be considered more or less "ordered." Clearly a more hierarchical social order based upon the concentration of power and production has generated ecological disorganization, leading, as an exemplary case in history, to the collapse of the Maya civilization. However, this emblematic case does not provide a historical rule. Likewise, the thermodynamic analysis of social structure and production as free energy potentials does not provide scientific basis to orient social organization and production towards sustainable development.

Socio-environmental systems are not self-contained entities; they cannot be analyzed as physical and biological systems. They are open systems guided by cultural meanings and social interests that are beyond the self-organization of physical processes and the purposive and teleological development of living systems. Meaning emerging from symbolic and ideological order cannot be contained in the concept of information as a notion of order. The organization of a social system is conditioned by its capacity to survive in balance with its environment (a capacity to harness energy sources and to control destruction of its underlying environment. Social decay depends on the entropic decay of the overall socio-environmental system. Sustainability cannot be assessed as entropy in information (Shannon), as a measure of entropy in social organization. Meaning is a symbolic process that cannot be reduced to any physical, thermodynamic, or biological laws of entropy. Negentropic eco-technological productivity is fueled by different thermodynamic potentials organized in cosmic forces, ecological processes, and cultural organizations. The entropic–negentropic flows, cycles, and balances in the interaction of these different orders is not provided by homogeneous unitary measure of bio-thermodynamic processes that occur at different time–space scales in Nature and culture. They are rather diverse co-evolutionary potentials arising from multiple physical potentials, cultural meanings, and social interests that organize society for the appropriation and management of natural resources.

Martin O'Connor conceives the integration of economics and the environment as "coupled and concatenated open—far from equilibrium—systems, simultaneously dissipative and self-organizing" (O'Connor 1991, p. 116). The question, in the environmental rationality perspective, goes beyond the coupling of existing economic system with ecology as a conditioning system, to envision the social reorganization of production under a new paradigm, based upon the negentropic potentials of Nature, the entropic tendencies of technology and the organizing traits of culture. What is at stake for the applicability of the paradigm of eco-technological productivity at the macro level is the possibility of rebuilding the world economic order based on productive principles other than capital and labor as productive factors. A truly bioeconomic paradigm for sustainable production must be based on ecological and cultural conditions that can balance (rather than reverse or retard) the inevitable increase of entropy in the process of production with the potential for biomass formation through photosynthesis. This brings into question the actual potential to sustain a human population that might well reach 10 billion people in the course of this 21st century, without accelerating the trends toward exhausting non-renewable resources, increasing environmental pollution and triggering climate change.

Ecotechnological productivity as a production process, limited by an entropy/negentropy balance, sets the whole idea of sustainability in new light. Steady-state economics cannot be based on Herman Daly's assumption of maintaining a constant capital and minimizing *throughput* by technological progress (Daly 1991). Even though one can conceive a steady-state production based on a social control of population growth, moderation of people's wants, and capital obsolescence, relying on technological progress for dematerialization of production, diminishing resource scarcities, and entropy decay, the question arises: Is it possible to constrain the economy to such steady-state based on social control and restraints, instead of liberating new sustainable productivity potentials subject to the conditions of life in the planet?

Yet, not all open, far-from-equilibrium processes, by liberating their self-organizing potentials, result in equal negentropic/ productive yields. The same can be said of entropic processes in the transformation of matter and energy for the production of commodities, as there is no unitary energetic process to measure this balance of useful sustainable productivity, and their impacts and effects on the quality of life of different human groups in different ecological contexts. If entropy/ negentropy balances cannot provide a definitive and unambiguous measure of order and disorder in social organization (Prigogine and Stengers 1984), it contributes to evaluate general trends of ecological balance and sustainable productivity. Because of this ambiguity in relating entropy with orderliness, Norgaard's assertion that "evolution has been a negentropic process" remains a heuristic hypothesis on co-evolution towards diversity, but that cannot give any measure of socio-environmental orderliness in the evolution of modern civilization. To expand the concept of negentropy to social order is not without theoretical controversy: the assertion that technological progress and economic growth evolve towards higher order and less entropy is questionable when cultural identities are disintegrated and ecological balances disrupted as a result.

Thus, if eco-technological productivity can be evaluated in terms of entropic/ negentropic balance, its function within an environmental rationality is not a new energetic approach to sustainable development. Insofar as eco-technological productivity is conceived as the integration of ecological, technological, and cultural systems—with culture providing meaning and purpose to alternative sustainable societies—this process cannot be understood as a thermodynamic open system nor can its potentialities and creativity be reduced to potential energy differentials. Its purpose is supported by and depends on laws of physical and living matter, but is not reduced to them.

The potential for drawing production from biomass formation has been regarded as insufficient to respond to the needs of present human population.[11] However, the annual rate of biomass formation in the

[11] "solar technology has been here since the emergence of the chlorophyll plants. We have been using it for millennia [...] but not in the measure that could sustain a development of the vital sectors of an exosomatic living to which we are addicted..." Georgescu-Roegen 1993, p. 14).

biosphere was estimated to be in the order of 2.4×10^{12} tonnes, having a net primary productivity of 1.7×10^{11}, equivalent to 10^{19} kcal (Rodin et al. 1975).[12] Energy expenditures in the world economy were estimated to be of the order of 340,000 petajoule, equivalent to 8.1×10^{16} kcal back in 1990 (WRI 1990). However, primary productivity has been declining for the last half-century because of deforestation trends; every increase in the extraction and consumption of non-biological sources of energy (oil, natural gas, coal, hydropower) increases total entropy, not only through the production of heat, but by altering ecosystems' structures (deforestation, soil erosion) and the self-organizing biological processes that determine biomass formation, affecting the geophysical balance between oxygen and greenhouse gases in the atmosphere, and consequently triggering the planet's global warming.

The potential of solar energy for building a natural economy on biological resources has been underestimated. It is claimed that the amount of solar energy that can be captured and transformed by the biosphere is

[12] These are theoretical estimates, as the equivalent between biomass in weight and its caloric value depends on the state of oxidation of carbon molecules being produced, which varies between 3 and 10×10^3 kcal kg^{-1}. Total net primary production of the biosphere, the community of all organisms of the earth's surface, was estimated 50 years ago by Whittaker to be about 170×10^9 tons of dry organic matter per year. Land communities are, on average, more productive than those of the sea, and about two-thirds of global productivity occurs on land. Because of the accumulation of woody biomass on land, the biomass disparity is even greater; biomass on land is about 1800×10^9 tons, which is more than one thousand times the plant biomass of the marine plankton. The global efficiency of primary production is about 0.27% for net, and 0.6% for gross primary production relative to the energy of sunlight in the visible region at the earth's surface. Man harvests every year about 1200×10^6 tons of plant food, and about 90×10^6 tons of animal food from the biosphere. These harvests, and man's release of industrial energy, are still small compared with the biosphere as a whole, but man's pressures on the biosphere are increasing exponentially (Whittaker 1975). Other authors have estimated yearly total biomass production from photosynthesis to be around 220,000 million tons of dry matter (Hall and Rosillo-Calle 1999, pp. 101–102, 109, 118). This information contrasts with Scheer's findings of: "an annual energy extraction of about 3500 million tons from oil, 2000 million Toe (tonnes of oil equivalent) of natural gas, and 2400 million Toe of carbon; that is, 8000 million Toe of scarce fossil resources used to cover the current demand of electricity, fuel, heating, and basic material for the chemical industry" [The tonne of oil equivalent (toe) is a unit of energy defined as the amount of energy released by burning one tonne of crude oil. It is approximately 42 gigajoules or 11.630 megawatt-hours]. In order to "substitute the oil, natural gas and carbon as fossil energies to cover the world's energy needs (in the hypothesis that all fossil energy were replaced by biomass only, and it were cultivated exclusively for direct combustion, without using residual substances from edible products crops, nor the gas potential from organic waste), a growing area or woodland of 12 million km^2 would be needed for an average production of some 15 tonnes of dry matter per hectare (Scheer 2000, pp. 81–82).

only 1–2% of the total amount that reaches the planet. However the main point is not how small is that percentage, but *how much is enough* to support a sustainable economy for 10 billion human beings. The energy equivalent of the net primary productivity of ecosystems still exceeds the amount of energy being produced and consumed by the economic process. These findings are important to view the actual potentialities of natural processes to shift from a non-renewable and polluting economy, to an economy rooted in the sources of living systems and biologically renewable resources. The question at stake is not to compare the energy of proven reserves of non-renewable resources with the stock of biomass in the biosphere or to hope for a technological breakthrough in atomic fusion that will realize the illusions of an endless source of energy and permanent movement. What matters, is the amount of solar *exergy* (available energy) that is captured and transformed into biomass and other usable forms of matter and energy resources.

The biological potential for biomass formation has been estimated as an average annual rate of about 8%. Obviously, this differentiated biomass must undergo ecological and technological processes in order to satisfy human needs. However, this net primary productivity can be enhanced through high-efficiency photosynthesis and biotechnology processes without compromising sustainable ecological yields by impinging on the very sources of ecological productivity, as in high-intensity agrochemical-based agricultural systems. Yet, this annual yield of eco-technological productivity cannot fuel a sustainable growth process. There will certainly be limits in the amount of biomass in the biosphere to keep global geophysical balances. However, it is possible to foresee a scenario where human population will reach a "steady-state" in the course of this century while a transition is operated from present unsustainable economic growth to a dynamic and complex bio-economic productive system.

The main source of the bio-economic process will be the Earth's potential for biomass formation. However, biomass formation does not appear as direct production of use values or utilities. In the management of biological resources, not all matter is available matter, and it holds true that all matter and energy will dissipate irrevocably after this starting point. However, scientific and technological development can be oriented to increase and transform primary productivity into low-entropy use values. Primary productivity can be transformed through selective regeneration to generate sustainable high yields of those species of utmost interest for human consumption, without disrupting the productive potential of ecosystems. Biotechnology can increase ecological productivity while preserving the complex productive capacity of ecosystems, yielding a productive and sustainable management of natural resources (Leff 1986a).

Internalizing entropic processes and ecological conditions of production are insufficient to reverse environmental degradation generated by economic rationality. Social values not reducible to any physical laws—autonomy, self-reliance, existence rights—are leading society to develop strategies more akin to sustainable patterns of resources use, cultural co-evolution, social equity, and environmental justice. Biomass sources of energy, management of biodiversity, and integrated productive systems are more likely to work effectively at community scale (Leff 1992b). The social organization that can more easily internalize the ecological and thermodynamic conditions for sustainable development are decentralized, autonomous, and self-reliant human communities, co-evolving with their ecological environments. These are the living cells of negentropic societies, while the market economy develops a homogenizing tendency in society that dries out the sources and co-evolutionary potential that arise from biological and cultural diversity.

Social change and productive transformations towards sustainability will not be guided by the laws of thermodynamics and by preconceived principles of cultural evolution that are ambiguous when applied to socio-environmental complex systems.[13] Order and complexity of ecological and productive systems are the basis for sustainable productivity through the selective regeneration of natural use values. The transition to

[13] Martin O'Connor has discussed at length the inconsistencies and difficulties to abide to a general theory relating entropy changes with structural organization. He has rightly argued that entropy cannot offer a unifying concept to explain the diverse and heterogeneous processes involved in organizational changes as well as environmental degradation, including socio-environmental systems characterized as open, far-from-equilibrium, systems (M. O'Connor 1991). Mario Giampietro, has also asserted that "When the system under analysis is a complex, multilevel, dynamic system, the notion of 'ordered' and 'disordered' behavior becomes very elusive; this means that the frequent association of increase in entropy with increase in 'disorder' can be explained only by 'anthropomorphic' assessment rather than by entropy changes" (Giampietro 1993, p. 219).

equitable and sustainable negentropic societies will be conducted through strategies that include cultural values, human desires and political power, beyond the compulsive application of any ecological engineering and social energetics based on scientific concepts of entropy.

Thus a *positive paradigm of negentropic production* must be constructed beyond the critical approach to production built on the concept of entropy and its extension to biological and economic processes. To respond to this challenge, an alternative productive rationality must be constructed; one where Nature does not only set conditions to the prevailing economic rationality, but where Nature becomes an active source and potential for sustainable life. This new paradigm cannot be constructed based only on human rights and social values. People have to empropriate new productive principles and innovative political strategies and embody them in their social imaginaries to institute sustainability principles in human nature. Nature and culture emerge in this perspective as productive processes, challenging the mechanistic economic rationality of capital imposed on objectivized Nature and human labor, giving support to a new economic paradigm built on principles of eco-technological productivity and environmental rationality.

Bibliography

Adams, R.N. (1975), *Energy and structure: a theory of social power*, Texas University Press, Austin.

Baudrillard, J. (1974), *Crítica de la economía política del signo*, Mexico, Siglo XXI Editores.

———. (1976), *L'échange symbolique et la mort*, Paris: Gallimard.

———. (1993), *The transparency of evil*, London, Verso.

Daly, H. E. (1991), *Steady-state economics*, Washington, D. C., Island.

Funtowicz, S. & J. Ravetz (1993), *Epistemologia política. Ciencia con la gente*, Buenos Aires, Centro Editor de América Latina.

———. (1994), "Emergent complex systems", *Futures*, 26 (6): 568–582.

Georgescu-Roegen, N. (1971), *The entropy law and the economic process*, Cambridge, Mass.: Harvard University Press.

———. (1993), "Looking back", in J.C. Dragan, E.K. Seifert, M.C. Demetrescu Eds. *Entropy and bioeconomics*, EABS, Nagard, Milano, pp. 11–21.

Giampietro, M. (1993), "Escaping the Georgescu-Roegen Paradox on Development: Equilibrium and Non-Equilibrium Thermodynamics to Describe Technological Evolution", in *Entropy and Bioeconomics*, Op. cit., pp. 202–229.

Grinevald, J. (1993), "The biosphere and the noosphere revisited: biogeochemistry and bioeconomics", in *Entropy and bioeconomics*, op. cit., pp. 241–258.

Günther, F. (1993), "Man in living systems", in *Entropy and bioeconomics*, op. cit., pp. 259–275.

Leff, E. (1975), "Hacia un proyecto de ecodesarrollo", *Comercio Exterior*, Vol. XXV, No. 1, México, pp. 88–94.

———. (1986a), *Ecología y capital: hacia una perspectiva ambiental del desarrollo*, México: UNAM

———. (1986b), "Eco-technological productivity: a conceptual basis for the integrated management of natural resources", *Social science information*, 25 (3): 681–702.

———. (1986c), "Ambiente y articulación de ciencias", in E. Leff, *Los problemas del conocimiento y la perspectiva ambiental del desarrollo*, Mexico City: Siglo XXI Editores.

———. (1992b), "A second contradiction of capitalism? Notes for the environmental transformation of historical materialism", *Capitalism, Nature, Socialism*, 3 (4):109–116, December, 1992.

———. (1993), "Marxism and the environmental question: from critical theory of production to an environmental rationality for sustainable development", *Capitalism, nature, socialism*, Vol. 4 (1): 44–66, Santa Cruz, California.

———. (1995), *Green production: towards an environmental rationality*, New York: Guilford.

———. (1996), "Los nuevos actores sociales del ambientalismo en el medio rural", en Carton de Grammont H. & H. Tejera, *La sociedad rural frente al nuevo milenio*, Vol. 4 "Los nuevos actores sociales y los procesos políticos en el campo", México: UNAM/INAH/UAM/Plaza y Valdez Editores.

———. (2004), *Racionalidad ambiental: la apropiación social de la naturaleza*. México: Siglo XXI Editores.

———. (2019), "Negentropic production", in Kothari, Ashish, Demaria, F., Acosta, A., Salleh, A. and Escobar, A. Eds., *Pluriverse: a postdevelopment lexicon*. London: Zed Books.

Lévi-Strauss, C. (1955), *Tristes tropiques*, Paris: Librairie Plon.

Lotka, A. J. (1922), "Contribution to the energetics of evolution" *Proc. Natl. Acad. Sci. USA*, 8: pp. 147–51.

Lozada, G.A. (1993), "Georgescu-Roegen's critique of statistical mechanics revisited", in *Entropy and bioeconomics*, op. cit., pp. 389–398.
Margulis, L. & Sagan, D. (1995), *What is life?*, Berkeley and Los Angeles: University of California Press.
Margalef, R. (1968), *Perspectives in ecological theory*, The University of Chicago Press, Chicago.
Martínez-Alier, J. (1987), *Ecological economics. Energy, environment and society*, Oxford/Cambridge: Basil Blackwell.
Mayumi, K. (1993), "Georgescu Roegen's 'fourth law of thermodynamics' and the flow-fund model", in *Entropy and bioeconomics*, op. cit., pp. 399–413.
Mol, A. & G. Spaargaren (2006), "Toward a sociology of environmental flows: a new agenda for twenty-first Century environmental sociology", in Spaargaren, G., Mol, A.P.J. y F.H. Buttel (2006), pp. 39–82.
Naredo, J.M. (2006), *Raíces económicas del deterioro ecológico y social*, Madrid: Siglo XXI Editores.
Norgaard, R. (1984), "Coevolutionary development potential", *Land economics*, 60:160–173.
———. (1994), *Development betrayed*, Routledge: London.
O'Connor, M. (1991), "Entropy, structure and organizational change", *Ecological economics* 3:95–122.
———. Ed. (1993a), "On Steady State: A Valediction", in *Entropy and bioeconomics*, op. cit., pp. 414–457.
———. Ed. (1994), *Is capitalism sustainable?*, New York: Guilford.
Passet, R. (1979), *L'économique et le vivant*, Paris: Payot.
———. (1985), "L'Économie: des choses mortes au vivant", *Encyclopaedia Universalis "Symposium"*, pp. 831–841.
Prigogine, I. (1955), *Thermodynamics of irreversible processes*, New York: John Wiley & Sons.
———. (1981), *From being to becoming: time and complexity in the physical sciences*, New York: W.H. Freeman & Co.
Prigogine, I. & Nicolis, G. (1977), *Self-organization in non-equilibrium systems*, New York: John Wiley & Sons.
Prigogine, I. & I. Stengers (1984), *Order out of chaos*, New York: Bentam Books.
Przybylski, T. (1993), "Entropy as a measure of forest damages", in *Entropy and bioeconomics*, op. cit., pp. 476–481.
Rodin, L.E., N.I. Bazilevich e N.N. Rozov (1975), "Primary productivity of the main world ecosystem", *Ecology First International Congress*, The Hague, September 1974, in W.H. van Dobben e Lowe-McConell, *Unifying concepts in ecology*, The Hague, W. Jung B. V. Publishers and Wageningen: Centre for Agricultural Publishing and Documentation.

Scheer, H. (2000), *Economía solar global. Estrategias para la modernidad ecológica*, Barcelona, Galaxia Gutenberg.
Schrödinger, E. (1944/1969) *¿What is life? The physical aspect of the living cell*, London/New York: Cambridge University Press.
Spaargaren, G., Mol, A.P.J. y F.H. Buttel (2006), Governing environmental flows. Global challenges to social theory, Cambridge, Mass: The MIT Press.
Urry, J. (2003), *Global complexity*, Cambridge: Polity Press.
———. (2005), "The complexity turn", in *Theory, culture & society*, Vol. 22 (5): 1–14.
White, L. (1949), *The science of culture*. Nueva York: Grove Press.
Whittaker, R.H. (1975), *Communities and ecosystems*, Macmillan, New York/London.
WRI/UNEP (1990), *World resources 1990–1991*, Oxford University Press, Oxford/New York.

[illegible]

10

Political Ecology: A Latin American Perspective

The Emergence of Political Ecology

Allegedly, the term "political ecology" appeared for the first time in academic literature in a 1935 article by Frank Throne (Throne 1935). However, if political ecology refers to power relations related to human–environmental interactions, in hierarchical and class structures as well as in the processes of production and the social appropriation of Nature, we should trace its precursors back to the historical dialectical materialism of Karl Marx and Friedrich Engels—though concealed under the primary contradiction between capital and labor– and to the social cooperative anarchism of Peter Kropotkin and his emphasis—against social Darwinism—on mutual aid in evolution and survival (Kropotkin 1902/2005; Robbins 2012). Political ecology was forged at the crossroads of human geography, cultural ecology, and ethnobiology to refer to the power relations regarding human intervention in the environment. It was

Published in Leff, E. (2015), Political ecology: a Latin American perspective, in Leff, E., Floriani, D., and de Oliveira Cunha, L. H. (Eds.), dossier temático, Pensamento Ambiental Latino-americano: movimentos sociais e territórios de vida, *Revista Desenvolvimento e Meio Ambiente* No. 35, Programa de Pós-Graduação de Meio Ambiente e Desenvolvimento, Universidade Federal do Paraná (Curitiba-Paraná-Brasil).

E. Leff, *Political Ecology*, https://doi.org/10.1007/978-3-030-63325-7_10

established as a specific discipline and a new field of inquiry and social conflict in the early 1960s and 1970s, triggered by the irruption of the environmental crisis, with the pioneering writings of authors like Murray Bookchin, Eric Wolf, Hans Magnus Enzensberger and André Gorz.

In 1962, at the time of Rachel Carson's *Silent Spring*, Murray Bookchin published *Our Synthetic Environment*. A decade later, in his article "Ownership and political ecology," Eric Wolf discussed how local rules of ownership and heritage "mediate between the pressures emanating from the larger society and the exigencies of the local ecosystem" (Wolf 1972, p. 202). Hans Magnus Enzensberger also published an influential article titled "A critique of political ecology" in 1974, while André Gorz published his early writings in the ecologist monthly journal *Le Sauvage*, founded by Alain Herve, creator of the French section of Friends of the Earth. *Écologie et politique* was published in 1975, followed by *Écologie et liberté* in 1977 and *Ecologica* in 2008.

As a new discipline—a new field of theoretical inquiry, scientific research and political action—political ecology emerged primarily from a neo-Marxist approach to evolving issues that when linked to the irruption of the environmental crisis and the configuration of an *ecological episteme* were meant to fulfil the oblivion of the conditions set by Nature to orthodox Marxism. Bookchin, Enzensberger, and Gorz inaugurated the field of political ecology as neo-Marxian inquiry on the conditions of Nature for historical transcendence. Enzensberger conceived political ecology as the practice of unmasking the ideology—the class interests and capitalistic appropriation of ecological concerns– behind the new ecological discourses on issues such as the limits of growth, population growth, and human ecology. Notwithstanding this critique, Enzensberger acknowledged the environmental crisis as being produced by the capitalistic mode of production. His critique of the "critique of ideology as ideology" led to a review of Marxist established views on the development of productive forces in the *abolition of want*. Following Marcuse, Enzensberger states that "productive forces reveal themselves to be destructive forces [...that] threaten all the natural basis of human life [...] The industrial process, insofar as it depends on these deformed productive forces, threatens its very existence and the existence of human society". He viewed the *society of superabundance* as "the result of a wave of plunder and pillage unparalleled in history; its victims are, on the one hand, the peoples of the third world and, on the other, the men and women of the future. It is, therefore, a kind of wealth that produces unimaginable want" (Enzensberger 1974, p. 23).

Andre Gorz argued that political ecology springs from the critique of economic thought:

> Starting from the critique of capitalism, we arrive to political ecology that, with its indispensable critical theory of needs, leads to deepen and radicalize even more the critique of capitalism [...] Ecology only acquires all its critical and ethical load if the devastations of the Earth, the destruction of the natural basis of life are understood as the consequence of a mode of production; and that this mode of production demands the maximization of profits and uses techniques that violent biologic equilibriums. (Gorz 2006, p. 15)

Following Karl Polanyi (1944), Andre Gorz underlined the market's tendency to appropriate domains of social and human life that respond to ontological orders and meanings other than economic logic. For Gorz, and counter to orthodox Marxist doctrine, the question of alienation and separation of the worker from the means of production was not merely the result of the social division of labor. This would ignore its metaphysical causes, and the ontological difference inscribed already in economic rationality and stamped in the world order that organizes and determines human life in the metaphysical age that established the technical way of understanding and objectifying the world (Heidegger 1935/1977). Gorz derived his *techno-critique* from the deconstruction of economic reason and reconstruction of the subject, opening new spaces for self-autonomy of community life and emancipation from the technological–bureaucratic machine driven by the economic process (Gorz 1989).

The critique of technology was the focus of reflection of many precursors of political ecology: from the questioning of *technology* (Marcuse 1968) and the *mega-machine* (Mumford 1970), an ample debate was initiated around the adaptation and appropriation of small and intermediate, soft and sweet technologies (Schumacher 1973), calling for a *social harnessing of technology* (Hetman 1973). Ivan Illich distinguished *convivial technologies* that propitiate autonomy and self-management, from *heteronomous technologies* that restrain them (Illich 1973). Gorz further differentiated *open technologies*—that favor communication, cooperation, and interaction—from *bolt technologies* (Gorz 2008, p. 16).

Prior to these critical views on technology, Walter Benjamin had contested the technocratic and positivistic conception of history driven by the development of productive forces. He criticized the *decay of the aura* of historical objects and Nature (Benjamin 1936/1968), and envisioned a kind of labor which "far from exploiting nature, is capable of delivering her of the creations which lie dormant in her womb as potentials" (Benjamin 1940/1968). Other thinkers saw in technology the core and roots of a crisis of humanity in modernity that would manifest later as the environmental crisis: Weber's *iron cage*; Heidegger's age of *Gestell*. These authors are forerunners of political ecology by having pointed out the limits of a civilization process from which the environmental crisis emerged and the power struggles involved in the social reappraisal and reappropriation of Nature.

Among the precursors of political ecology, Murray Bookchin was the more comprehensive, radical and polemic thinker. He was one of the first to anticipate climate change back in the early sixties:

> Since the Industrial Revolution, the overall atmospheric mass of carbon dioxide has increased by 13 percent over earlier, more stable, levels. It could be argued on very sound theoretical grounds that this growing blanket of carbon dioxide, by intercepting heat radiated from the earth into outer space, will lead to rising atmospheric temperatures, to a more violent circulation of air, to more destructive storm patterns, and eventually to a melting of the polar ice caps [...] rising sea levels, and the inundation of vast land areas. Far removed as such a deluge may be, the changing proportion of carbon dioxide to other atmospheric gases is a warning of the impact man is having on the balance of nature. (Bookchin 1964, p. 4)

Bookchin was the founder of the social ecology movement framed within an anarchist, libertarian socialist and ecological thought that led to *communalism* and *libertarian municipalism*, conceived as the decentralization of society according to ecological and democratic principles. His essay "Ecology and revolutionary thought" (Bookchin 1964) introduced *ecology* into radical politics, that evolved into *The Ecology of Freedom* (Bookchin 1982/1991) and to his *Philosophy of Social Ecology: Essays on Dialectical Naturalism* (Bookchin 1990).[1] Postulating hierarchy and domination as essential foundations of historic power relations—more extensive in scope than Marxist class struggles—he proclaimed ecology as critical and political in nature and as the organizing power that guides the re-acquaintance of Nature with the anarchist spirit—its social spontaneity to release the potential of society and humanity, to give free and unfettered rein to the creativity of people—emancipating society from its domineering bonds and opening the way to a libertarian one. He underlined that "The explosive implications of an ecological approach arise not only from the fact that ecology is intrinsically a critical science—on a scale that the most radical systems of political economy failed to attain—but it is also an integrative and reconstructive science" (Bookchin 1964, p. 3).

Herbert Marcuse can also be considered a precursor of the emergent field of political ecology: his critical theory on technology and the workings of the capitalist mode of production gave significant ground for understanding the social conditions that catalyzed the destruction of Nature. Marcuse's reflections on Nature in his final writings align within the currents of political ecology. Thus, in *Counterrevolution and Revolt*, at the outburst of the environmental crisis and in a vein that echoes Bookchin, he asserted that, "What is happening is the discovery (or rather, rediscovery) of Nature as an ally in the struggle against the exploitative societies in which the violation of Nature aggravates the violation of man. The discovery of liberating forces of Nature and their vital role in the construction of a free society becomes a new force of social change" (Marcuse 1972, p. 59). Nature is then integrated into the emancipatory process of liberation. Marcuse privileges sensibility and the aesthetic quality of liberation over Bookchin's claim for ecological rationality and a dialectical naturalism to free society from its bonds. Through these critical views emerging from political ecology, the core of the ecological

[1] For a discussion of Bookchin's social ecology see Light 1998; for a critique of Bookchin's ontological monism and dialectical naturalism, see Leff 1998a and Clark 2008. Cf. Chap. 4.

question shifts the problem of abundance—of liberation from need and subjection of hierarchical and capitalistic domination—to the imperatives of survival and the responsibility of humanity for the destiny of life on planet Earth.

Political ecology emerged as a social response to the oblivion of Nature by political economy. In the transition from structuralism—focused on the determination of language, the unconscious, ideology, discourse, social and power structures, mode of production and economic rationality—to postmodern thinking, the discourse on liberation (of the subject, the individual, the nation) shifted to the *sustainability of life*. While inquiring into the root causes of environmental decay, political ecology is inscribed in the power relations that traverse the emancipatory process towards sustainability based on the potential of Nature. In this context, the political ecology debate gave way to the emergence of eco-socialism and eco-Marxism (Leff 1993, 1995; Benton 1996; O'Connor 1998; Foster 2000). By surfacing Marx's concept of Nature (Schmidt 1971) and analyzing the capitalistic causes of ecological decay, eco-Marxism uncovered a *second contradiction of capital*, the self-destruction of the ecological conditions of sustainable production (O'Connor 1998). Furthermore, a new paradigm of production was conceived, integrating the eco-cultural and technological conditions of production as an environmental potential for a sustainable world order with political power emerging from the environmental movements guided by environmental rationality (Leff 1995, 2004; see Chap. 7).

Political ecology emerged as a field of theoretical inquiry and political action in response to the environmental crisis: that is to the destruction of the conditions for sustaining human civilization caused by the reductive domination of the human logos over the emergency and *generative nature of physis*, and of the *logocentrism of science* over *subjected savoirs* of the people, and of privilege of techno-economic rationality over the conditions of life in the planet. Departing from a radical critique of the metaphysical foundations of modern epistemology, political ecology goes beyond the proposals for conservation of Nature—promoted by the International Union for the Conservation of Nature since its creation in 1948—and policies of environmental management—launched after the World Conference on Human Environment, Stockholm, 1972, *Our*

Common Future in 1987 and *Agenda 21* after the Rio Summit on Environment and Development in 1992—to inquire about the conditions for a sustainable life in the ecological stage of economic and technological hegemonic domination: in a world where—quoting Karl Marx and Marshal Berman—"all that is solid melts into air", generating global warming and the entropic death of the planet.

Political ecology is the study of power relations and political conflict over ecological distribution and the social struggles for the appropriation of Nature. It is a field of encounters, where the ontology of life meets the ontology of the techno-economic world order, where controversies on the ways of understanding the relations between humanity and Nature, and the history of exploitation of Nature grapple with the submission of cultures, their subsumption under capitalism and the rationality of the global world-system; where power strategies within the geopolitics of sustainable development engage in the construction of an environmental rationality. Conceived in this way, Michel Foucault (1980) appears as a fundamental precursor of political ecology by providing the insights that can untangle the power relations embedded in knowledge and in the institutional frameworks that have constrained, repressed and subjugated knowledge for alternative ways of conservation and construction of sustainable livelihoods. In Foucault's views, power is not only a relation of domination and an agency for repression. Eros, a desire for life, mobilizes emancipation processes from the will to power embedded in power strategies of knowledge to produce new forms of knowledge arising from the will to live. Political ecology is the field where power strategies are conceived and social struggles deployed to open new pathways for survival and for building a sustainable future. It involves the deconstruction of modern rationality and the construction of alternative environmental rationality.

The field of political ecology has emerged from cultural ecology, geographical studies, political economy, and critical rationalism, spreading out to neighboring disciplines: overlapping with environmental sociology and ecological economics; expanding from political economy of the environment to post-development and post-colonial studies; blending with eco-Marxism, social ecology, and eco-feminism; fusing with theories of complexity and with post-structural and post-constructivist approaches to Nature. Its scientific status and research approaches are still being debated and defined, as are its frontiers and alliances with other disciplines, and its theoretical genealogies, epistemological framings, and practical strategies.[2]

Establishing the field of political ecology in the geography of knowledge is a more complex endeavor than just delimiting paradigmatic boundaries between neighboring disciplines; merging academic traditions, forming clusters of research topics, drawing typologies of Nature ontologies, conceptualizing problematic areas of intervention and mapping environmental thinking. It implies deconstructing theoretical fields, resignifying concepts and mobilizing discursive strategies to forge the identity of this new epistemic territory in the construction of environmental rationality for a sustainable future.

Much of the political ecology elaborated in the North in the past two decades focuses in agrarian Third World environments, including peasant and indigenous peoples' traditional practices, resistance, and activism in the reconstruction of their life territories. Political ecology emerges in the South from a politics of difference rooted in the ecological and cultural conditions of its people; from their emancipation strategies for decolonization of knowledge, the reinvention of territories and reappropriation of Nature (Porto-Gonçalves and Leff 2015; see Chap. 12).

Precursors of Political Ecology in Latin America

Political ecology addresses the social struggles and power strategies for reappropriating Nature. Its social sources emerge from resistance to deterritorialization of habitats, to the pillage of the natural resources and to the subjugation of the original cultures by dominating colonial-modern powers. We can trace these processes 500 years ago, from the conquest and colonization of the Third World regions to the present strategies of the global economy and the geopolitics of sustainable development.

[2] For an account of the Anglo-Saxon political ecology literature, see Peet and Watts 1996; Biersack and Greenberg 2006; Escobar 2010; Peet et al. 2010; Robbins 2012; for an overview of French contributions to political ecology, see Debeir et al. 1986; Ferry 1995; Latour 2004; Lipietz 1999; Whiteside 2002.

Political ecology is inscribed in the history of submission and emancipation of original peoples from the global economic system: from the disruption of the livelihoods and the ecological catastrophe produced by conquest, colonization and imperial domination (Cosby 1986), to the present struggles to reterritorialize cultural beings and open the paths to sustainability.

In the face of the colonizing and exploitative character of the new geopolitics of globalization and sustainable development in the conflictive field of political ecology, different critical and creative responses emerged from the South after the 1970s, and throughout a period in which the environmental crisis shook the modern-colonial world-system, in particular from Latin American peoples. From the first ecological debates, the critique to the society of waste and pollution, of consumerism and productivism, has been considered from the a perspective of concern for peoples living under poverty and hunger, unable to consume the minimum necessary to sustain their existence. The debate over Nature was mobilized by authors like Josue de Castro with his seminal lecture on "Underdevelopment: the primal cause of pollution", presented at the Conference on Human Environment held in Stockholm in 1972, underlining the social causes of environmental unsustainability (de Castro 2003); or by Celso Furtado (1974), that questioned the idea of development after having been one of its principal defendants.

Even though orthodox currents of Marxism were at first critical of the new ecologism, soon after, some of its political thinkers and different social movements in Latin America and the Third World began to assimilate the environmental question into their inquiries and political agendas. New theoretical–political currents were developed in the nascent field of political ecology to render an account of an emergent "popular ecology," "eco-pedagogy of liberation" and "eco-socialism." The contributions to political ecology in the last 30–40 years are as important today to understand the complex socio-environmental processes under way, as they were at the time of José Carlos Mariategui's thesis (1971) on the decolonization of indigenous peoples and that of Aimé Césaire (1955), who together with Franz Fanon (2004) founded the *negritude* movement in the South. In the 1960s–70s, contributors were the *Theory of Dependence and Internal Colonialism*, with authors like Ruy Mauro Marini, Theotonio dos Santos, André Gunder-Frank and Pablo González Casanova (among others), and the *Theology of Liberation* and the *Pedagogy of the Oppressed* of Paulo Freire (Marini and dos Santos 1999).

John Murra (1956) developed a rich analysis of the organization of the geographical space of the original peoples of *Tawantinsuyu* (Quechua and Aymara, among others) where he studied Andean ecological floors as interconnected systems from the West Pacific coast to the Chaco-Pantanal region, and interlinked to the Central Brazilian Plateau to the East. Different from the territorial division of labor and space imposed by capitalist agriculture, the principles of complementariness and reciprocity commanded the organization of geographical space in their production practices. These conceptions of the cultural occupation of space are being re-evaluated by new theoretical-political approaches of original peoples' movements to reappropriate their ancestral territories (Tapia 2009).

The cultural territories of Latin America are a patrimony derived from the legacy of their rich and diverse cultures, of their original and traditional knowledge that goes back to the ancestral forms of occupation of the continent and to the formation of its climatic and botanical domains housed in the natural heritage of tropical forests, savannahs, steppes, *punas*, moorlands, mangroves, and wetlands; that is, of the wealth of biological diversity of the continent (Ab'Saber 1970). The original populations that inhabited these areas co-evolved with the ecosystem dynamics of their territories developing a rich collection of knowledge that, together with their biological diversity, represents a biocultural patrimony constructed mostly under the conditions of Nature. This diversified patrimony of knowledge of indigenous peoples, peasants, and maroons, subjugated by colonial and capitalist domination, is forced today into an encounter with scientific knowledge that supports the techno-economic appropriation of Nature.

Traditional knowledge is often referred to as "local knowledge", "popular wisdom" or "folk science"; as *indigenous science* (De Gortari 1963), *macro-systems* (Lopez-Austin and Lopez-Lujan 2001), *native sciences* (Cardona 1986), *popular knowledge* or *people's science* (Fals Borda 1981, 1987). In English literature, they are named *traditional, non- Western*, or *traditional ecological knowledge*. In general terms, these sets of practical, experimental, experiential and reflexive knowledge represent a cultural patrimony transmitted from generation to generation. These *systems of indigenous knowledge* (Argueta et al. 1994), are embodied in practices for the sustainability of life, such as food production and health care; they are embedded in their territories conceived as spaces where identities are forged and renewed. These cultural identities include language and communication systems; history and collective memory; norms for conviviality among parents and neighbors; relations with other peoples and societies that are expressed in common customs and law, myths and rituals, religion and festivities, where the transcendental lives of the peoples are expressed. This patrimony of critical thinking in Latin America is the fertile soil where the seed of environmental thinking is emerging and flourishing.[3]

In this perspective, political thinkers and activists such as José Martí (1963), Jose Carlos Mariategui (1971), Franz Fanon (2004) and Aime Césaire (1955) are precursors of Latin American political ecology. In Marti's affirmations, "There is no battle between civilization and barbarism but between false erudition and nature," or "The trenches of ideas are more fruitful than those of stone" (Martí 1963) we find a critical response to European epistemological–political colonization. From Mariategui's Latin American Marxism—intended to root socialism in the traditions of indigenous peoples, in the restoration of their community life and their productive organization—(Mariategui 1971), to the *liberation pedagogy* of Paulo Freire and the *eco-pedagogy* of Leonardo Boff, we can trace a lineage of critical thinkers that have forged Latin American political ecology.

The writer Eduardo Galeano (1971) updated this history of exploitative colonialism in his book *The Open Veins of Latin America.* Galeano brought to light the production of poverty generated through the exploitation of the Earth's wealth, with the fever of gold and silver that seemed to have exhausted the abundance of metals on the Earth'crust of Latin

[3] For a compendium of Latin American critical political thinking, see Marini, R.M. and dos Santos, T. (Coords.) (1999), *El pensamiento social latinoamericano en el siglo XX,* Caracas: UNESCO, 2 Vols.; Leff, E. (2012), Latin American environmental thinking: a heritage of knowledge for sustainability, *Environmental ethics,* Volume 34:4, Winter, pp. 431–450; Svampa, M. (2016), *Debates latinoamericanos. Indianismo, desarrollo, dependencia y populismo,* Buenos Aires y Bolivia: Edhasa/CEDIB.

America's territories, only to see a resurgence of this exploitative colonialism in the recent years. Likewise, the poverty that was produced in the old agricultural latifundia—like that of sugar cane in Cuba, rubber in Brazil, banana in Ecuador and Colombia—reappears today with monocultures of foodstuff, forest plantations, and transgenic crops.

Political ecology in Latin America was nourished by a rich tradition of anthropological and ethnoecological research, such as the studies on the Incas' *ecological floors* (Murra 1956), the cultural and ecological potentialities of Mesoamerica (Wolf and Palerm 1972), or the roots of *Profound Mexico* (Bonfil Batalla 1987). The *Geography of Hunger* (de Castro 1946) was a precursor of a legion of political ecologists that addressed the critical problems of hunger and poverty of Latin American populations generated by ecological degradation of their territories. New approaches in cultural anthropology and environmental geography are emerging together with the forging of a politics of territoriality and difference that is developing from socio-environmental movements guided by principles of political autonomy and cultural identity for the reappropriation of their bio-cultural patrimony of natural resources. The field of political ecology is establishing itself by welding critical theoretical thinking with case studies and political action. This dialogue of theory and practice is exemplified by the defense of the subsistence ecology of the Miskito Indians in Nicaragua (Nietschmann 1973), the *extractive reserves* of the *Seringueiros* (rubber tappers) in Brazil (Porto Gonçalves 2001a) and the *Process of Black Communities* in Colombia for the appropriation of their territories of biodiversity (Escobar 2008). A working group in political ecology was established in 2000 within the Latin American Council of Social Sciences (CLACSO) to develop this field of critical inquiry (Alimonda 2002, 2006).

One crucial question for political ecology in Latin America is the clash of strategies between the techno-capitalistic exploitation of Nature and the cultural reappropriation of the ecological patrimony and ethnic territories of the peoples. Today, this confrontation is exemplified by the invasion of transgenic crops through ethno-bio-prospecting and intellectual property rights of transnational enterprises that transgress the collective property rights and the natural resources of nations and peoples in the South. In the view of indigenous peoples, biodiversity represents their patrimony of natural and cultural resources, with which they have co-evolved throughout history, the habitat where their cultural practices are forged and embedded. Their ecological potentials and cultural meanings are incommensurable with economic values. These criteria differentiate what is negotiable and interchangeable in a debt for Nature equation, and the ethical–political principle that puts into question how to settle conflicts of ecological damage and distribution through economic compensations, establishing the threshold that separates ecological economics from political ecology.

Deconstruction/ Decolonization of Knowledge, Recognition of Nature and Reinvention of Life Territories

Political ecology is the field of the encounter of power strategies for the distribution of ecological costs and environmental potentials in the construction of sustainability. In the crossroads towards a sustainable future, the crucial point is the clash of views and ontological regimes to attain its objectives, traversed by economic, political, and ethical interests. Sustainability entails the deconstruction of unsustainable rationalities—of the theories that support them, the discourses that intend to legitimize them and the institutions that establish their function in the social order—as well as the construction of alternative rationalities and strategies to open paths towards sustainability. One primary objective of sustainable societies is to breach inequalities in economic and ecological distribution, which is the outcome of a history of metaphysical violence, of cultural colonization, territorial domination, and displacement under unequal power relations. Political ecology traces the construction and institutionalization of hierarchical social structures and domineering powers that are rooted in thinking and production modes that have deterritorialized primitive cultures.

Modern rationality built an unsustainable world whose signs are visible in the planet's environmental crisis. The ecological destruction generated by the exploitative appropriation of Nature during the colonial regime and the present world economic order was accompanied by the exclusion and oblivion of traditional practices and the imposition of Western knowledge for domination of territories in the conquest of the Third World. Indigenous peoples claim that their struggles for emancipation are political and epistemological: decolonization of knowledge becomes a condition for their cultural–political emancipation and for building their sustainable life-territories.

The claim for decolonization of knowledge has deep historical roots in critical thinking in Latin America. It follows the theories on unequal exchange, underdevelopment and dependency of the Third World from the global economy as organizing center of the world-system (Gunder-Frank 1966; Amin 1976; Dos Santos 1978; Cardoso and Faletto 1979; Wallerstein 1974, 1980, 1989, 2011). These theories set up the background for present political ecology theory insofar as they conceived dependency and underdevelopment as a structural state of world affairs where poor nations provide the natural resources and cheap labor in an unequal interchange for capital and technology from "developed" nations; that is, the hegemonic world order where the unequal "ecological distribution" within the geopolitics of "sustainable development" is inscribed. This structural domination was further developed by "internal colonialism," where hierarchies and inequalities were internalized and constructed within the class structure of poor dependent countries (González Casanova 1965; Stavenhagen 1965).

A critical inquiry has emerged in recent times with the works *Decolonial Thinking* and *Coloniality of Knowledge* (Lander 2000; Mignolo 2000, 2011; Mignolo and Escobar 2009; Quijano 2008) and the acknowledgement of the *Epistemologies of the South* (Sousa Santos 2008). The work of these scholars and of other intellectuals such as Catherine Walsh, Ramon Grosfogel and the Aymara-Bolivian sociologist Silvia Rivera Cusicanqui, seek to understand the historic-socio-political condition of the Latin American people. Decolonization of knowledge leads to questioning how Eurocentric ideas—from Greek philosophy to modern science and technology– were introduced to traditional societies and cultures through conquest, colonization, and globalization, invading indigenous ways of thinking and being, their traditional practices and cultural life-worlds, generating as reaction political resistance and purposive actions for the decolonization of knowledge and *rexistence* as a condition for the reappropriation of their natural and cultural patrimony.[4]

The emancipation purpose in political ecology implies deconstructing metaphysical thinking and *logocentric* science instituted in the hegemonic power relations by modern economic, scientific and technologic rationality. Beyond the need to understand the epistemological foundations, the ontological regimes and the power–knowledge strategies that dominated peoples and despoiled their territories, the construction of sustainable societies rooted in the ecological potentialities and cultural identities of the Third World peoples requires a strategy for decolonization of knowledge to liberate from exploitation, inequality, and subjugation.

Beyond an hermeneutic deconstruction of domineering knowledge, decolonization of knowledge implies the recognition and re-evaluation of traditional and "other" knowledge—"local knowledge", "popular wisdom" or "folk science"– unknown and negated by domineering paradigms. These non-Western "traditional knowledge from the South" are "other" ways of understanding the world, sets of imaginary–reflexive and practical–experimental knowledge that represents a cultural patrimony transmitted from generation to generation. This *savoir* and wisdom are embodied in practices for the sustainability of life, such as food production and health care; it is embedded in their territories conceived as spaces where cultural identities are forged and renewed, including signification codes and communication systems; collective history and memory; myths and rituals, religion and festivities. They are social norms for conviviality among families and neighbors; relations with other peoples and societies that are expressed in common customs (Thompson 1991). The biocultural patrimony of Latin American Peoples is the fertile soil where the seed of environmental thinking is emerging and flourishing (Boege 2008, 2018). This environmental knowledge/savoir/wisdom is fundamental to construct alternative rationality capable of uncoupling from the globalized world-system and building other possible life-worlds. The construction of a global world order founded in differences and specificities of various territories is supported by peoples' knowledge embedded in their

[4] For a compendium of Latin American critical social thinking see Marini and dos Santos 1999 and Svampa 2016.

ecological conditions and embodied in their cultural beings. *This traditional ecological knowledge* and *cultural imaginaries of sustainability* (Leff 2010, 2014a) are the roots and sources from where Latin American thinking offers new perspectives for sustainability.

Colonization of knowledge has been a fundamental instrument for cultural submission and appropriation of Nature, from the conquest of original peoples and their territories to the present strategies within the geopolitics of sustainable development. Third World territories are being revalued as areas for unrestrained exploitation of non-renewable resources (oil, coal, minerals), for biodiversity conservation to absorb greenhouse gases and for biotechnological prospection, or as natural resources—cellulose, transgenic crops, foodstuff– to be exploited and exchanged to fuel the continuing growth of developed and emerging economies. Resisting this exploitative conquer of Nature, Third World and Latin American peoples are claiming their rights to decolonize knowledge and emancipate from the global economic order.

Decolonization of knowledge implies the deconstruction of theories embedded in the ontology of the modern world order—the *logocentrism of science* and techno-economic rationality– and embodied in the life-worlds of the people, to disarm the institutionalized structures that constrain the world to unsustainable rationality. Deconstructionism unveils the ways that knowledge was constructed, inscribed and instituted as the hegemonic world order. Deconstructionist political ecology inquires the point in which ontological difference turned into social inequality by twisting the diverse modes of being-in-the-world into an objectified one world order, ruled by the abstraction of Nature and human labor into generalized monetary value. Therefore, decolonizing knowledge is an epistemological condition for deconstructing the exploitative trends of the global economy and recognizing the ecological and cultural potentials of the people to give life to alternative modes of production, of thinking and being.

Decolonization of knowledge as a condition and process towards the reappropriation of Nature and the reinvention of sustainable territories becomes a complex and challenging political task. Beyond the study of the colonization process, the environmental history of cultural subjugation and exploitation of Nature, the emancipation from subjection to central and external powers and the imposition of modern thinking over traditional worldviews and practices demands new ways of thinking that arise from these subjected places. In a globalized world, the social reappropriation of Nature is rooted in the reinvention of cultural identities. Rescuing and reconstructing traditional knowledge occurs upon encountering opposing and conflictive rationalities, intercultural hybridization and dialogue among forms of knowledge; it's the result of the clash between thoughts and actions, of reidentifications and negotiations, in the social construction of sustainability.

Alternative *environmental rationality* for sustainability (Leff 2004) is configured in the field of political ecology by rooting deconstructive thinking in ecological and cultural territories. This is not merely the application of deconstructive theories, complex sciences and sustainability blueprints to the design of new cultural territories; it goes beyond the purpose of adapting technologies to the ecological and social conditions of the South, building a new dominion of knowledge subject to the comparative advantages of the ecological conditions and endowments of Third World countries in the globalized world. Deconstructing theory and decolonizing knowledge in the perspective of political ecology implies politicizing the ontological concepts of diversity, difference, and otherness to construct sustainability rooted in specific cultural territories. This requires establishing and enforcing rights for cultural diversity to build *territories of difference* (Escobar 2008), and the deployment of political ethics of otherness. This process opens new perspectives in the deconstruction of the unitary hegemonic global world, to build a new world founded on different ecological potentials and the creativity of diverse cultural beings. Beyond the tolerance of cultural diversity and adaptation to different ecological contexts of a unitary world order, it reorients the destiny of humanity guided by the creative evolution of life, from the heterogenesis of natural and cultural diversity arising from eco-cultural co-evolution in opening ways to a future global world integrated by different cultural projects of sustainability, by different ways of being-in-the-world.

Geography has provided insightful spatial metaphors for the analysis of power strategies in knowledge. As Foucault had expressed, "once knowledge can be analyzed in terms of region, domain, implantation, displacement, transposition, one can capture the process by which knowledge functions as a form of power and disseminates the effects of power" (Foucault 1980, p. 69). Territorializing knowledge goes beyond the epistemic-psycho-ecological question of new cartography of knowledge to that of the embodiment and embedding of knowledge (Deleuze and Guattari 1987). The power strategies to reappropriate Nature and reterritorialize knowledge imply the restoration of subjugated and subaltern knowledge to generate alternative environmental rationality construed by the encounter of different cultural meanings: a critical resignification and reappropriation of "universal knowledge" from local cultural identities and the production of knowledge by different cultural beings.

The political philosophy that guides cultural emancipation and social actions towards sustainability emerge from the radical epistemological concept of *environment*, conceived as the limit of modern hegemonic rationality that has led to the environmental crisis of civilization and the source of an alternative sustainable world. *Environmental knowledge* (*saber ambiental*) emerges in the margins of *logocentric* science, from the periphery of central powers, in the externalities of hegemonic scientific and economic rationality. It is knowledge forged and rooted in the ecological potentials and cultural creativity of the peoples that inhabit the territories in the South (Leff 1998a).

Counter-hegemonic globalization—deconstruction of the one-dimensional oppressive force over diversity, difference, and otherness globalized under the dominance of techno-economic rationality– demands an epistemological decentering from modern rationality. The concept of environment is the point of anchorage outside the global economic order that deconstructs unsustainable knowledge. However, environmental rationality is not founded in a virgin territory untouched by the institutionalized global rationality that has negated other possible worlds. Environmental rationality is forged in the crossroads of the deconstruction of metaphysical and scientific thinking and the territorialization of the ontology of diversity, difference, and otherness. This critical concept of environment is the cornerstone of Latin American Environmental Thinking (Leff 2012).

Decolonizing from domineering knowledge involves the responsibility for the future of humanity and of the planet. Beyond prospective sciences that pretended to foresee the future in order to reorient present tendencies, political ecology constructs the future from the understanding of present unsustainable processes and the projection of social actions based on environmental rationality to harness unsustainable tendencies and trigger newly sustainable potentials.

Political ecology challenges the metaphysical duality of immutable space and transcendental time. In this perspective, historical time is not a uniform flow of events. Sustainability will not be attained through the optimization of present means oriented towards a prefixed end, but the outcome of diverse processes under their conditions and timings; with their uncertainties, encounters, convergences, and alliances. The paths opened by this purpose are defined by unexpected events that might trigger, accelerate, or hinder and block the paths towards sustainability. The construction of a negentropic sustainable future runs against historical entropic trends. Sustainability is built in the encountering of these conflictive processes; in the confluence of synergies of differentiated natural and social forces; in the negotiation of diverse interests and meanings that trigger counteractive economic, ecological and technological processes that in the coalescence of their different ontological regimes, will determine the future of life in the Planet.

Decolonizing the rich diversity of peoples/cultures discovered a new theoretical perspective of historical time and space as the expression of the complex integration of the social processes of appropriation of Nature and the way they have enacted the metabolism of their ecosystems, as the imprint of human culture in the sustainability of different territorialities. Thus, Milton Santos (1996) argued that different temporalities cohabited in geographical space challenging the coloniality of knowledge imposed by a modern culture that overvalues time in the detriment of space. The Eurocentric vision of cultural evolution was imposed on the world as the only possible universality. Thus, traditional peoples became backward societies, as if they were only a stage in the way of human development and economic growth. Thus traditional cultures were quieted and remained invisible. The simultaneity of different temporalities that forge cultural territories was occluded by the hegemonic temporality that orders the world, secluding other cultures.

Thus, both the Kantian conception of universal aprioristic categories of reason, of historical idealistic transcendence and geographic determinism have been contested, abandoning the Eurocentric linear and progressive conception of time as civilization stages, incorporating the incommensurability of time of different processes involved in building cultural territories. This conception of historical time and space has important political implications for social movements, such as the actuality of ancestry invoked by indigenous peoples, the reversal of internal colonialism through the political construction of pluri-nationality, the co-evolution of peoples/cultures and Nature/territories, and the social imaginaries of sustainability (Leff 2010, 2014a). Political ecology disentangles the historical and geographical inscription of conflicts of territoriality (Maier 2006; Haesbaert 2004).

Coloniality of knowledge has also been contested from the standpoint of ecofeminism, claiming that knowledge has been coded and molded as a masculine inscription in Western culture by hierarchical dualisms—particularly Cartesian dualism– (Merchant 1992), by "transcendent objectivity of male-dominated science" (Haraway 1991) and "monocultures of the mind" (Shiva 1993), in their intent to control Nature and dominate women. Decolonization from the South—emancipation of subjugated knowledge embodied in cultural beings and embedded in their life territories– demands the deconstruction of knowledge established from the North, to unearth alternative—diverse, different, other–ontological grounds and open other epistemological perspectives to guide the construction of sustainable societies.

Ecological Episteme and Political Ecology

The environmental crisis is the manifestation of a crisis of knowledge. Environmental degradation is the result of the forms of knowing the world that grew in the oblivion of being and Nature, away from the conditions of life and human existence. It is a crisis of civilization that results from the *ignorance of knowledge* (Leff 2000). From this perspective, political ecology explores the power strategies of the knowledge that traverse individual interests, social imaginaries, and collective rights, which are woven into the life-worlds of people in the globalized world. At the same time, it envisions new power strategies capable of deconstructing the unsustainable modern rationality and mobilizing social action for the construction of a sustainable future.

Political ecology constructs its territory of knowledge in the encounter of different systems of thought, ethics, practices and social action. It debates with *ecosophies* that responded to the first signs of ecological breakdown offering an ecological understanding of the world—the *Ecology of the mind* (Bateson 1972, 1979), *Gaia Theory* (Lovelock 1979), *Deep Ecology* (Naess 1989), the *Web of Life* (Capra 1996) and *Complex Thinking* (Morin 1990)—with their explicit and unintended political consequences. Political ecology responds to different ecological problems: population growth, human health, resource scarcity, deforestation, loss of biodiversity, pollution, climate change, peoples' livelihoods. It argues with different theories, discourses and policies, and socio-environmental conflicts: ecological distribution, dematerialization of production, geo-politics of sustainability, autonomy and self-reliance of the people. It is a field of confrontation of different approaches to sustainability: economic rationality–environmental rationality; ecologism–environmentalism; strong–weak sustainability; de-growth–sustainable development. Political ecology dialogues with other emergent ecological disciplines: cultural ecology, ecological economics, and bioethics; environmental anthropology, sociology, geography, history, law. Distinctive approaches within radical ecology—deep ecology, social ecology, eco-feminism, eco-Marxism, eco-socialism (with their internal polemical controversies)—converge and collide in the field of political ecology.

Notwithstanding its alliances and resonances with other eco-disciplines, political ecology is not an interdisciplinary paradigm that embraces them all. What is common to these new inquiries is the fact that they all are "post-normal" disciplines that do not have an established place within traditional and mainstream of science. Its post-normal character does not derive only from being applied domains of an ecological paradigm or approach based on the interrelatedness, feedbacks, and complexity of processes. Post-normal sciences contest the principles of epistemological representation—the identity of theory and reality—to incorporate "quality of knowledge" from "emergent complex systems" (Funtowicz and

Ravetz 1993, 1994). However, the specific trait of political ecology is the power relations that tense and cut across bio-cultural, socio-environmental and techno-economic processes, where it defines its proper identity, by borrowing conceptual metaphors from other disciplines to describe the socio-environmental conflicts derived from the unequal ecological distribution and the appropriation strategies of ecological resources, natural goods, and environmental services.

Political ecology, as well as other ecological disciplines, is forged within the emergent *ecological episteme* diffused to the social sciences in the transition from structuralism to post-structuralism. Although some authors assign an intrinsic political character to an ecological inquiry—i.e., Bookchin's ecological dialectics of Nature—power relations are not immanent to an ecological approach to reality. Political ecology is not a "normal" emergency within the realm of science resulting from the transition from the structural episteme—prevalent through the 1970s and 1980s– to a post-structural approach to the "politics of ecology" (Walker 2005, pp. 74–75). Political ecology informs environmental policies but focuses on social conflict regarding the distribution of environmental potentials and ecological costs, rather than in policy-making for environmental planning. The politicization of ecology is the expression of power struggles and strategies for the reappropriation of Nature.

Political ecology is not the amalgam or synthesis of differentiated positions and social responses to environmental problems. Conversely, it is the field of dispute of different visions and understandings of the environmental crises: pollution, resource scarcity, limits to growth, territoriality conflicts. In the inception of these discussions, the primal causes of ecological breakdown were debated between population growth (Ehrlichc 1968) and industrial development in capitalism (Commoner 1971, 1976) as the main causal triggers. Multivariable modelling projected current trends in population growth, economic development, technology, and pollution forecasting an ecological collapse. For the first time in modern history, the ideology of progress was contested, stating the limits to economic growth (Meadows et al. 1972). This scenario was reinforced by theoretical inquiries on the relations between the entropy law and the economic process (Georgescu-Roegen 1971) and research on dissipative thermodynamic processes (Prigogine 1961; Prigogine and Nicolis 1977). It surfaced then that economic growth had become the primary cause of ecological decay and environmental pollution leading to the "entropic death of the planet."

From the first moment that the environmental crisis gained worldwide awareness in the 70, a critical movement in Latin America got involved in these debates. Regarding the disputes on the "population bomb" and the "limits to growth," a seminal study conducted by Amilcar Herrera et al. (1976) asked: *Catastrophe or New Society?* In a similar vein as the critical economic and sociological thinking in Latin America—theories of economic dependency, underdevelopment and internal colonialism– they stated that environmental degradation was not fundamentally determined by population growth, nor directly by economic growth; instead, ecological decay was associated with poverty and unequal wealth distribution resulting from the imposed and adopted development model. From this conception, eco-development strategies (Sachs 1980) found fertile soil in Latin America.

The environmental crisis was associated with the fragmentation of knowledge in modern science that impeded the analysis of complex socio-environmental processes. Thus, a problem-solving approach to applied sciences emerged, positing interdisciplinary methods and complex thinking as essential tools for environmental management. From being the object of scientific research and economic domination, Nature became an object of theoretical inquiry, a political dispute, and social appropriation. Outside the field of science, various interpretative currents developed where Nature was no longer an object to be dominated and fragmented, but rather an entity to be redefined, re-embodied and re-embedded. This gave birth to a myriad of ecosophies—from deep ecology to eco-socialism; from eco-feminism to eco-anarchism– that nurtured the cradle of political ecology. Ecology became an encompassing paradigm based on a holistic vision of reality as systems of interrelations that orients thinking and action in a reconstructive path. Thus a *method* based on *generalized ecology* (Morin 1980) was promoted, where systems theories and interdisciplinary methods, complex thinking and the new sciences of complexity converged for the reordering and reintegration of knowledge.

This paradigm shift was epistemological and social, moving from mechanistic to a more organic and complex understanding of processes, that confronted the fragmentation of reality and knowledge in the classical science with a holistic view of the World. This view is understood as an interrelated and interdependent system that evolves through feedback loops as a cybernetic system, enabling novelty in knowledge, as well as chaos, indeterminacy and uncertainty, in order attain consciousness and creativity. Notwithstanding these paradigmatic changes in the understanding of things, the *ecological episteme* did not renounce to its quest for objectivity and its drive for totality. With ecology, a new theoretical centralism emerged: ecological thinking confronted the fragmentation of knowledge and the autonomy of self-centered paradigms, but it did not challenge the *logocentrism* of sciences or the totalitarian purpose to reintegrate knowledge in an all-encompassing paradigm. The ecological episteme did not dissolve the power structures of one-dimensional thinking installed in the generalizing-unitary *logos* and the globalizing law of the market.

Notwithstanding the usefulness of systems theories and the need for integrated approaches, *environmental epistemology* emerged as a critical understanding of the epistemological obstacles for building new interdisciplinary environmental paradigms (Leff 1986). Environmental epistemology revealed that what is at stake in knowledge for sustainability is not a neutral articulation of sciences but a reconstruction of knowledge from the critical exteriority of the environment—the otherness to the *logocentrism* of science of the concept of the environment– that challenges normal sciences and its ecological approaches. Sustainability is being built in the encounters and interplay of diverse and often incommensurable, and non-integrable paradigms; of conflicting modes of thinking and systems of knowledge. Moreover, environmental knowledge mobilizes social actors from other stands of reason and feeling for the social construction of their sustainable life-worlds. Political ecology is the field of an environmental, political epistemology, of power–knowledge strategies that open alternative paths towards sustainability (Leff 2001). Thus, environmentalism comes to challenge ecologism in the foundation of political ecology as a critical politics of difference.

The struggles for sustainability are epistemological and political. Ecology is politicized by opening the systemic vision of reality, and the symbolic and cultural ordering of Nature, from the ontology of life towards the domains of human ethics and social justice. What is at stake in the field of political ecology is not so much ecologizing the social order as the encounter of alternative and conflicting cultural and economic rationalities over the appropriation of Nature. The identity of political ecology in Latin America arises from the political–epistemological definition of the environment, differentiating ecology in affluent societies from *the environmentalism of the poor* (Guha and Martínez-Alier 1997). One fundamental trait of this epistemological difference is the conception of the environment as a potential for alternative sustainable developments. Thus a paradigm of eco-technological-cultural productivity can be constructed. The concept of *environmental complexity*—beyond complex thinking, the sciences of complexity, systems theory and interdisciplinary methods– and the category of *environmental rationality* emerge from a radical epistemological perspective (Leff 1995, 2001, 2004, 2006a).

Political Ecology/ Environmental Epistemology

Political ecology is the politics of the social reappropriation of Nature and the conflicts of human intervention in the courses of life on the planet. As in all politics, its practice is not just mediated by discursive strategies but is a struggle for the production and appropriation of concepts that orient social actions. This holds, not only because critical environmentalism confronts the ideologies that support unsustainable modernity (Leis 2001), but because the efficacy of any strategy for social reconstruction leading towards a sustainable future implies the deconstruction of theories and ideologies that have institutionalized the social rationality that generated the present environmental crisis.

The strategies for the construction of sustainable societies are configured by theoretical struggles and the politicization of concepts. Concepts such as Nature, biodiversity, territory, autonomy, identity, self-management, development, and sustainability, of being and life, are redefining their meaning in the conflictive field of political ecology, where different strategies for the appropriation of Nature are confronted. Thus, the concept of territory in the field of political ecology differentiates and goes beyond anthropological concepts related to the cultural definition of space or the State delimitation of Nation. Territoriality is the social construction of space arising from the encounter of conflictive rationalities and power strategies; likewise, the discourse and the geopolitics of sustainable development are confronted by the concept of sustainability configured under the broader scheme of environmental rationality (Leff 2004).

Beyond these theoretical debates, ecological emancipation in the globalized world is mobilized by concepts that gain significance, legitimacy, and power within peoples' radical imaginaries. Thus, the quest for sustainability is fused with cultural rights and civil society demands for decolonization, autonomy, and dignity as human beings: for their rights to reinvent their identities, to define their modes of being-in-the-world and to construct their life-territories. Politics of difference opens to the proliferation of existential meanings and civilization paths that are nurtured by the encounter of diverse modes of signification of reality, which establishes the territory of a *political epistemology*. Beyond the epistemology of normal science, *environmental epistemology* transcends complex thinking, system theories, and interdisciplinary methodology in the will to reintegrate, complement, reunify and totalize knowledge (Leff 2001). The construction of sustainability is traversed by *power strategies in knowledge* (Foucault 1980), redirecting environmental conflict and the fragmentation of knowledge to new political ethics: the *dialogue of knowledge and wisdoms* (*savoirs*). This implies the deconstruction of the epistemology of representation—the identity between the real and the concept, and of objective truth—in order to rethink the relation among the Real, the Symbolic and the Imaginary of different cultural beings.

Deconstruction of modern rationality goes beyond a paradigmatic shift from mechanistic and structural science to a new episteme of generalized ecology and complex thinking. The epistemology of "normal science" is decentered by *environmental rationality*. The environment is not the milieu that surrounds material and symbolic processes centered on their internal organizing principles: it is not only an "externality" of the economic system and of *logocentric* sciences that can be internalized by a holistic view, a systemic approach or an interdisciplinary method (Canguilhem 1971, 1977; Leff 2001). The environment as an epistemological category emerges as the exteriority of scientific and economic rationalities, as the "other" of totalitarian knowledge; it calls to rethink the relations between the Real and the Symbolic in order to create power-strategies to weave sustainable futures through the encounter and dialogue of knowledge/wisdom/*savoirs*. Environmental epistemology goes beyond a hermeneutics of Nature in order to deconstruct and resignify Nature through language games, symbolic codes, and power strategies, involving visions, feelings, reasons, and interests that are debated in the political arena. Thus, environmental epistemology provides an understanding of the conditions of life guiding socio-environmental movements for the social reappropriation of Nature and the construction of a sustainable future.

The epistemology of political ecology is sustained in the deconstruction of the ideological–scientific–discursive notion of Nature in order to rearticulate the Real in the biophysical order of life with the symbolic order that signifies Nature, where cultural worldviews and social imaginaries are embodied in the practices of their life-worlds and livelihoods. Environmental epistemology renews the debates over monism/dualism that confront radical ecologism—deep ecology, social ecology, and eco-feminism– in the perspectives of existential ontology, environmental rationality and the ethics of otherness; in the reconstruction / reintegration of the natural and the social, of ecology and culture, of the material and the symbolic. This is the core of algid disputes in environmental thinking and its political strategies, the point of confrontation of the theoretical dichotomy between the naturalism of physical–biological–mathematical sciences, and the anthropomorphism of cultural–social–human knowledge: the first attracted by positivistic logic and empiricism; the other by relativism, constructivism, and hermeneutics.

In the wreck of thought and the crisis of reason of the present *society of knowledge*, many scientists have jumped onto the life-raft offered by ecology as the "science par excellence" for the study of complex thinking and the interrelations of living beings and its environments, leading to generalized ecological thinking that maintains the will to embrace the wholeness of knowledge and reality in a method of complex thinking (Morin 1990). This holistic view intends to reunite all entities divided by dual metaphysical thinking—object–subject; body–mind; Nature–culture; reason–feeling– not as dialectical synthesis but by evolutionist creationism: by the emergence of an ecological consciousness that would reconcile and solve the debts of an anti-ecological rationality in the history of metaphysics following the idea of the *generativity of physis* to the *emergence of the noosphere* (de Chardin 1961). To dissolve the Cartesian dualism that lies in the roots of scientific and modern rationality, a philosophy of social ecology, based on principles of ontological monism and ecological dialectics proposes the reunification of Nature and culture (Bookchin 1971).[5] This philosophy does not offer a solid epistemological basis for an ontology of life rooted in a *politics of diversity and difference*—that recognizes the difference between the Real and the Symbolic– in the social construction of sustainability (Leff 2000, 2001, 2004, 2018).

Efforts to reunify Nature and culture arise as well from recent phenomenological perspectives in anthropology that claim that worldviews of traditional societies do not recognize a distinction between the human, the natural and the supernatural. These "matrixes of rationality"—to be understood in a metaphorical sense as the maternal womb where new rationalities and forms of being are conceived and fertilized from new ways of thinking– are not commensurable with, and translatable to, the epistemology of modernity. Politics of difference within environmental epistemology brings into new light the controversies of radical ecologism with dualist thinking as the source of hierarchical, domineering, exploitative and unsustainable societies. The principles of a *reflexive modernization* (Giddens et al. 1994) cannot dissolve at will the foundations of the dualism of modern rationality. If dualist thinking is responsible for the destruction of Nature, the solution does not arise from the epistemological reform of modern rationality but in opening scientific rationality to a dialogue of knowledge/*savoirs* with other cultural rationalities and traditional wisdom, under a politics of difference. The epistemology that sustains the geopolitics of economic–ecological globalization must not only coexist with other knowledge systems but must be deconstructed from its foundations to build sustainability on environmental rationality, where diverse cultural beings and different life-territories might coexist in a globalized world (Leff 2014a).

[5] For a critique of Bookchin's ecological dialectics, see Chap. 4.

Postmodern philosophy has come to question universalism and essentialism in theory as well as autonomous "regional" ontological regimes and discrete epistemological orders. Knowledge does no longer have the sole function of knowing the real. There is no longer an ontological principle of the Real that governs reality: knowledge denaturalizes Nature to generate *hyperreality* (Baudrillard 1983). Knowledge has produced a trans-ontological order where new hybrid entities emerge—*cyborgs*–made of organisms, symbols, and technology (Haraway 1991), in the encounter and blending of the Real and the Symbolic, of the traditional and the modern. Yet, it is necessary to differentiate this "hybridizing" of Nature, culture and technology that emerges as *environmental complexity* (Leff 2000) from the intervention of knowledge in the Real and of technology in Life, from the life-worlds of traditional peoples living "within Nature", where the separation between soul and body, life and death, Nature and culture, is absent from their imaginaries. The continuity and blending of the material and the symbolic in traditional people worldviews, cognition, and practices belong to a different ontological register from that of the relation between the Real, the Symbolic and the Imaginary in the techno-economic regime in modernity.

Political ecology faces the essentialism of Western ontology and the principle of universality of modern science, that through metaphysical thinking generated the *a priori* judgments of pure reason as well as a generic concept of man and the individual that constructed humanism and gave ideological support for the cultural domination of the *other* (Heidegger 1946; Levinas 1977/1997). Thus, universal human rights unify the rights of individuals while segregating, ignoring and discriminating the collective rights of other different cultures (Leff 2014b). Political ecology deconstructs the universal concepts of being, man, Nature, culture, identity, individual and subject—of power and knowledge—not with the purpose to pluralize them as "beings", "men", "natures" and "cultures" with differentiated "ontologies" and "epistemologies", but in order to construct the concepts of their differences.

Environmental epistemology thus transcends the interrelations and interdependencies of complex thinking and generalized ecology (Morin 1980) going beyond dialectical naturalism (Bookchin 1971). It emerges from the symbolic order and the production of meaning inaugurated by language; it is rooted in cultural significations, imaginaries, practices, and *habitus*, and is expressed in the confrontation of power strategies in knowledge for the reappropriation of Nature. In this perspective, political ecology is not inscribed in an ecological ordering of the world that would bring about a new consciousness-truth capable of overcoming anti-ecological interests; it is instead a new political space where the destiny of Nature and humanity are forged by the creation of new meanings and the construction of "truths" through power strategies in the interrelation culture–nature: in the interplay of a dialogue of knowledge/ *savoirs*.

Political ecology becomes a field where the Real, the Symbolic and the Imaginary converge and hybridize in environmental complexity. Entropy as the limit-law of Nature encounters the theories that support scientific–technological–economic rationality and the imaginaries of traditional cultures expressed in the controversial discursive field of sustainability. This epistemological question is not settled by scientific knowledge but is debated in the arena of *political ontology*, where other symbols and other imaginaries assign different meanings to the Real. Nature is thus "reconstructed" and "reconducted" from the powerful drive of symbolic and discursive strategies that are confronted in the geopolitics of sustainability.

Embodied/ Embedded Knowledge/ *Savoirs*

The epistemological project of modernity stands on the imaginary of representation, on a dualist separation of object and subject, body and mind, Nature and culture, reason and feeling, logos and writing. Knowledge is a relation with the Real that remains outside the knowing subject; it is knowledge "extracted" from Nature that does not belong to Nature. After four centuries of modern philosophy and science founded in this dualist principle—from Descartes, Bacon, Locke and Spinoza, Kant, Hegel and Marx—the environmental crisis have questioned the ontological and epistemological basis of a *res cogitans* outside space, and *res extensa* existing outside thinking. Phenomenology, hermeneutics, and constructivism problematize the existence of an intrinsic Truth of the Real. Psychoanalysis has uncovered the effects of the unconscious in the somatization of desire and showed that mental processes are symbolic and not mere organic expressions.

In reality, there is no pure thinking floating above the ordering of Nature, the bodies of individuals and the organization of society: philosophies, ideologies, and theories are embodied in beliefs and imaginaries, in worldviews and existential meanings that motivate gestures, postures, practices, and actions. Holistic views of ecology and phenomenological approaches in anthropology have stressed the close relations of culture and Nature and underlined the positioning of "living within Nature" in the life-worlds of the people. However, what brings political ecology to question the epistemology of modernity is not only the disembodiment of knowledge as a means to appropriate Nature but the fact that knowledge has penetrated life: the genetic structure of organisms and the biosphere's and ecosystems' organization, triggering the entropic decay of the planet.

The monist–dualist ontological–epistemological debate is transposed to the relations between life and knowledge in terms of the embodiment and embedding of knowledge. From Wittgenstein to Foucault and Derrida, research has shown how the structure and forms of language, speech, and discursive formations mold thinking and thus open different meanings that condense in social organization, are rooted in territories and orient political actions that mobilize the metabolism of the biosphere. For Castoriadis (1998), social imaginaries are embodied significations that have the potency to institute and alter; as *habitus* (Bourdieu), they are not always expressed as explicit representations that assign meaning to phenomena *a posteriori*, but constitute implicitly "sense in act." Knowledge is expressed through the body. Levinas pointed out that,

> Merleau-Ponty [...] showed that disembodied thinking that thinks the word before speaking thinking that forms the world of words and then adhere it to the world—previously made of significances, in a transcendental operation—was a myth. Thinking consists of elaborating the system of signs, in the language of a people or a civilization, to receive the signification from this same operation. The thinking goes to the adventure, in the sense that it does not start from a previous representation, neither from those significations nor from phrases to articulate. Thinking almost

> operates in the "I can" of the body. It operates in it before representing or forming this body. Signification surprises thinking [...] It is not the mediation of the sign that makes signification, but signification (whose original event is the face-to-face) which makes the function of the sign possible [...This] "something" that is called signification emerges in being with language because the essence of language is the relation with the Other. (Levinas 1977/1997, pp. 218–220)

Today, theory and knowledge have intervened Nature and are constructing new beings, entities, bodies, and organisms. Science is "embodied" in technology, and through technology in living beings. Science does not only "know" reality; it penetrates the real denaturalizing Nature, de-essentializing ontological orders, technologizing life, destroying the complex web of life of the biosphere. The identity between the concept and the real in the dualist relation of knowledge—as the correspondence between signifier and reality—turns into an instrument of knowledge that dissects, clones, and bursts the essence of being, from sameness to difference. Horkheimer and Adorno had rightly pointed out the paradox that,

> There is no being in the World that can avoid being penetrated by science, but that which is penetrated with science is not being [...] with this operation the step from mimetic reflection to controlled reflection is accomplished. In place of the physical adequacy to nature stands the 'recognition through the concept,' the assumption of the divers under the identical. [...] In the impartiality of scientific language, impotence has lost its expression force entirely, and only the existent finds there its neutral sign. This neutrality is more metaphysical than metaphysics. Ultimately, the Enlightenment has devoured not only all symbols, but also [...] the universal concepts, and from metaphysics, it has left nothing but the fear to the collective of which it was born. (Horkheimer and Adorno 1944/1969, pp. 41, 214, 37–38)

The epistemological inquiry on knowledge about the conditions of truth shifts to the problem of the effects of knowledge in the construction of reality; from the theoretical relation between knowledge and the real, the relation between being and knowledge is disclosed as the effects of alternative wisdom and truths in the social construction of life-worlds and the reappropriation of Nature: of *truth as cause* (Lacan), in the sense of the *production of existence* that generates the human rights to differentiated ways of being-in-the-world. In this new context emerges the question of the embodiment and embedding of knowledge in the biosphere, in new life-territories, in human bodies and social imaginaries. Thus, political ecology addresses the "mechanisms of power which have invested human bodies, acts and forms of behaviour [...] as a productive network which runs through the whole social body, much more than as a negative instance whose function is repression" (Foucault 1980, p. 61, 119).

Knowledge and wisdom are rooted in the living organism of the biosphere and the vital soil of human existence. Instrumental and technological knowledge—nuclear, medical and agricultural technologies, agrochemicals and toxic wastes– pollute the earth, air, and water, as well as the bodies of living beings through transgenic products and greenhouse effect gases; they invade human existence, rationalize thinking, reshape bodies and configure institutions; they codify the *self* through ideologies that mold feelings, orient behaviors and drive motivations through a process of rationalization that yields socio-environmental degradation and fuels the entropic death of the planet. Countering these tendencies environmental knowledge (*savoirs*) is embodied in new ethics and embedded in ecosystems through new social and productive practices oriented by environmental rationality; new identities are being reconfigured and embodied in cultural beings, unfolding in ecological practices embedded in new life-territories.

Political ecology embraces the purpose of reconstructing the world "from the perspective of multiple cultural, ecological and social practices embodied in local models and places" (Escobar 1999). This objective poses a radical question: Can the theory and practice of political ecology deconstruct the unsustainable world order and mobilize thinking and social action towards building a new global cosmopolitanism, conducting the destiny of humankind (and of the planet Earth) on the basis of a politics of difference and a strategy for the coexistence of diverse local environmental rationalities, where ecological potentials and cultural diversity become the basis of a new sustainable economy? Political ecology opens a new theoretical–practical field to embody environmental wisdom in cultural beings fertilizing new life territories.

Social imaginaries register the encounter of the Real and the Symbolic recorded in human existence throughout history. They are footprints of the conditions of life embodied in social beings in a lived world. Imaginaries of sustainability confront the rationalization of the world, especially the practices induced by the theoretical and instrumental rationality of modernity. Thus, radical social imaginaries become power strategies for emancipation (Leff 2010, 2014a). They are not only trenches of resistance to the rationalization of life, but potentials of creativity for the *rexistence* of the people and rebuilding their sustainable life-worlds.

Ecological Economics/ Political Ecology

Political ecology opens new horizons of social action and historical construction that go beyond the intention of ecological economics to internalize environmental externalities, to constrain economic performance or adapt economic mechanisms to ecological conditions of sustainability. Political ecology establishes its territory in the environmental hinterland, beyond the enclosure of economic rationality, of that which can be recoded and internalized in the realm of economics to value natural resources and environmental services. Political ecology is rooted in a space where the social conflicts for the appropriation of Nature and culture manifest their power strategies, where Nature and culture resist the homologation of different ontological orders and the reduction of symbolic, ecological, epistemological, and political processes to market values. This is the *polis* where cultural diversity acquires "citizenship rights" within a *politics of difference*: a radical difference, as what is at stake is beyond the equitable distribution of costs and benefits derived from the economic value of Nature.

The questioning of the *limits to growth* triggered a fierce debate worldwide that lead to a confrontation of diagnosis and perspectives, and yielded into a politics of discursive strategies to respond to the environmental crisis. Political ecology emerged in the margins of ecological economics to analyze the non-chrematistic value, the cultural meanings and the power struggles in the social appropriation of Nature that cannot be understood, nor solved, through the economic value of Nature nor by ecological norms imposed on the economy. These socio-environmental conflicts are expressed as controversies derived from diverse—and often antagonistic– meanings of Nature, where ethical, political and cultural values overflow the field of political economics, including the political economy of natural resources and environmental services. Thus, political ecology emerges in the exteriority of ecological economics.

In the interplay of concepts that define the difference of these neighboring fields of inquiry, the notion of "ecological distribution" has gained significance. Ecological distribution refers to the extra-economic—ecological, cultural and political– processes that link ecological economics with political ecology, in analogy with the concept of economic distribution that turns economics into political economics. Ecological distribution is expressed as power conflicts involved in the social strategies for survival and sustainable production alternatives in the political economy of the environment, as well as to struggles for the social appropriation of Nature and the distribution of the costs and damages from different forms of ecological destruction and environmental pollution. Ecological distribution embraces criteria and values that overflow economic rationality and contest the intention of reducing such values to chrematistic costs and market prices, mobilizing social actors for material and symbolic interests—identity, autonomy, territory, quality of life, survival– that are beyond strict economic demands for land property, the means of production, employment, income distribution and development.

Ecological distribution refers to the unequal repartition of the environmental costs and potentials, of those "economic externalities" incommensurable with market values, but that appear as new costs to be internalized through economic instruments and ecological norms, if not by the effect of social movements that emerge and multiply in response to ecological damage and the struggle for the social appropriation of Nature.

In this context, the notion of *ecological debt* has permeated the political discourse, as a strategic concept that mobilizes resistance against globalization of the market and its coercive financial instruments, questioning the legitimacy of the economic debt of the poor countries, as well as the capitalistic appropriation of their natural resources and the historical dispossession of their patrimony of natural resources. The ecological debt surfaces the largest—and until now submerged– part of the "iceberg" of the unequal exchange between rich and poor countries, that is to say, the appropriation and destruction of the natural resource base of the "underdeveloped" countries in the construction of a global capitalist system. The state of poverty of their peoples does not derive from their cultural condition or natural limitations—from geographical determinations, ecological conditions, or worse, cultural "backwardness"– but rather from its dominated insertion in a global economic rationality and internal colonization processes that has overexploited their natural resources, displaced their cultural territories and degraded their environments.

Notwithstanding the environmental injustice behind these social inequalities, this historic ecological debt is incommensurable and unquantifiable in economic terms, as there are no standards to measure it, nor discount rates to update the historical processes of exploitation of Nature and cultural colonization and submission. The ecological debt uncovers the history of dispossession, the pillage of Nature and cultural subjugation that has been masked by the economic principles of the endowment of natural resources, comparative advantages and efficient use of productive factors leading to—and pretending to justify– unequal exchange in the free market global economy.

Political ecology as a theoretical discipline and field of inquiry has the objective of analyzing the historical power struggles and appropriation strategies over Nature among nations and peoples, as well as present distributive conflicts of ecological resources. These inquiries are triggered by the pressing imperatives of the environmental crisis: scarcity of natural resources, climate change, environmental degradation, survival needs, emancipation desires, existential rights and the quest of a sustainable future. Political ecology becomes a field of political ethics, for the deployment of power strategies (in knowledge, economy, politics, social relations, common property, and cultural rights) that have denaturalized Nature and deterritorialized cultures, mobilizing social actions towards building a sustainable world.

Denaturalization and Reconstruction of Nature

In the course of history, Nature was "constructed" as an ontological order. Nature as *physis* embraced the emergence potency of the Real. Further on, the naturalness of reality became a fundamental argument to legitimate the "real existing order." "Natural" were the entities that had the "right of being to exist." This naturalness of the order of things—that of the ontology and the epistemology of Nature– was the metaphysical foundation of anti-nature rationality, based in the unassailable, ineluctable and immovable laws of Nature. In modernity, Nature was converted into the object of inquiry of science, the object of labor and the raw material for production; economic theory ignored the complex ordering of the living and the ecosystem organization of Nature. From classic economics on, capital and labor became the fundamental production factors; Nature was input for production but did not determine the value of commodities. Nature affected decreasing yields but was ignored as a condition and potential for sustainable production. Furthermore, Nature was externalized from the economic system. Nature was *denaturalized*; it became an object to be appropriated by knowledge and the economy; a "resource" that was consumed in the flow of value and economic productivity.

In the early sixties, Nature regained its status as an ontological order and political referent, a subject of philosophical and ethical inquiry, and soon after a standpoint for criticism of the established economic order. The first signs of concern for Nature appeared somehow before, leading to the establishment of the International Union for the Conservation of Nature in 1948. However, the seminal works of authors like Rachel Carlson (1962), Paul Ehrlichc (1968), Barry Commoner (1971) and Arne Naess (1989) raised the ecological alarm. The study of the Club of Rome, *The Limits of Growth* (Meadows et al. 1972) disseminated worldwide the questioning of the economic system and its catastrophic effects in the destruction of Nature and pollution of the environment. The environmental crisis gave way to the rise of consciousness of the destruction of the ecological bases and conditions for sustainability of the planet, leading governments to design policies for the conservation of Nature.

To be sure, the mainstream thinking that guides ecological actions—from critical ecosophies to complex thinking, as well as the domineering ecological schemes and economic instruments that guide the geopolitics of sustainable development– have complexified the social understanding and interventions on Nature. However, they have not yet deconstructed a naturalist view that from bio-sociology to system's approaches and generalized ecology, have been unable to dissolve the techno-economic siege of the world, where natural law becomes the support of power strategies that denaturalize Nature.

If Nature was denaturalized when metaphysical thinking separated Nature and culture, the reconstruction of Nature does not imply the restoration of the essentialist ontology of Nature. Political ecology is not only the hermeneutic and deconstructive inquiry on the history of denaturalization or a constructivist approach to resignify Nature but rather the politics of cultural appropriation and territorialization of Nature. The re-evaluation of Nature involves the reconstruction of the concept of Nature: from resignification of the *reél hors signifié* and *physis* as the emergency potency of everything existent to the "natural" conditions of existence and demystification of "natural" disasters, and to an ecological view of cultural, social and political relations. This deconstruction of Nature goes beyond the hermeneutics of Nature, environmental history, and postmodern constructivism. Against ontological realism, political ecology stresses the power relations that tense all social relations: relations of human beings with Nature; power relations in knowledge, in production and the appropriation of Nature; it is the field where discourses, behaviors, and actions attracted and embedded in the concept of Nature are contested.

Beyond the ecological approaches that dominate environmental thinking, new constructivist and phenomenological insights are contributing to deconstruct the concept of Nature (Rorty 1979), stressing the fact that Nature is not merely an objective entity in the realm of the Real, but is always *meaningful*: a signified, geo-graphed, territorialized, politicized entity. This paradigmatic change is supported by recent studies in environmental anthropology (Descola and Pálsson 1996) and environmental geography (Santos 1996; Porto Gonçalves 2001a). Its approaches and findings demonstrate that Nature is not the product of biological evolution, but instead of the co-evolution of "nature" as *Physis* guided by cultures that have signified and inhabited Nature. In the field of political ecology, "organic/cultural Natures" encounter "capitalized Natures," that is Nature intervened by the globalized techno-economy that impose its hegemonic and homogenizing dominance through technologic breeding and market mechanisms.

Nature is thus being reconstructed in the hybridization of different ontological and epistemological orders: physical, organic, symbolic, techno-economic; in the encounter and confrontation of heterogeneous rationalities that redesign Nature through social knowledge and practical appropriation strategies. Following a long historical process of resistance, which origins can be traced back in the colonial and imperialistic domination of the original Peoples of the Earth, their cultural identities are being reinvented and reaffirmed in their present struggles to defend, revalue, design collective rights and assign new cultural meanings to Nature: to design and legitimize new productive strategies for the conservative and sustainable use of their cultural patrimony of natural resources.

An emblematic example of these cultural innovations of Nature is the identity invention of the *Seringueiros* and the invention of their extractive reserves in the Brazilian Amazonia (Porto Gonçalves 2001a), as well as the more recent *Process of black communities* in the Colombian Pacific (Escobar 2008). Identities are being configured through struggles for the affirmation of cultural beings that confront the domination/appropriation strategies promoted and imposed by economic globalization. These political actions are more than processes of resistance: they are movements for *rexistence* of peoples and Nature (Porto-Gonçalves 2002).

Cultural Politics, Politics of Cultural Difference and Otherness

Politics of difference is founded in ontological and symbolic roots—the continuing and evolving differentiation of *Physis*; the infinite signification and complexifying of being– which destiny is to diversify, to ramify, to redefine (Derrida 1978, 1982; Deleuze and Guattari 1987): to manifest in *distinction* (Bourdieu 1984); to radicalize in *otherness* (Levinas 1969, 1991). Postmodern thinking on difference—*differance*– (Derrida 1978, 1982) is the project to deconstruct the unitary thinking of Logos in metaphysics and *logocentrism* of science, with their will to subsume diversity in universality, to subject heterogeneous being to the measure of a universal equivalent, to close the circle of science in a unifying system of knowledge; to reduce ontological diversity to the structural homologies of system theory, to pigeonhole ideas in one-dimensional thinking, or worse, to reduce all beings to economic values and market prices. Political ecology roots theoretical deconstruction in the political arena; beyond recognizing cultural diversity, traditional knowledge, and indigenous peoples' rights, environmentalism contests the overwhelming unification power of money in the exchange market as the fate of human history.

Political ecology contests the essentialist ontological conception of Nature while acknowledging that there is nothing intrinsically political in original Nature or ecological organization. The relations between living beings and its environing Nature and food chains—even depredation and domination among them and the territoriality struggle of species—are not political in any sense. Politics is drawn into Nature not only in response to the fact that economic rationality and the social sciences have negated the ecosystems organization of Nature. Nature becomes political by the fact that power relations are established in the symbolic order of human beings in their radical difference with all other living creatures. The political engage Nature in power relations through symbolic, cultural, economic and technological interventions of Nature.

From this perspective, Arturo Escobar refers to *ecologies of difference*, underlining the notion of *cultural distribution*, to address the conflicts that emerge from different cultural meanings assigned to Nature, as "power that inhabit meanings is a source of power" (Escobar 2006). As cultural meanings become means to legitimize human rights, they mobilize discursive strategies for the claim of cultural values; it is as such human rights that cultural values enter the power field of political ecology to confront "intellectual property rights" and the "rights of the market" in the social struggle for the appropriation of Nature.

However, the notion of "cultural distribution" can become as fallacious as that of ecological distribution if submitted to homologation and homogenization. Incommensurability does not only apply to the difference between economy, ecology, and culture, but within various cultural orders, where there are no equivalencies, no possible translation between different cultural codes and meanings. Distribution always appeals to a homogeneous object: income, wealth, employment, matter, energy, Nature and power. However, being, as the subject of rights, is inherently heterogeneous. Political ecology is forged in the realm of otherness. Cultural difference implies shifting from the generic and abstract concept of being conceived by essentialist and universal ontology to the politics of difference, as specific and localized *rights to the otherness of cultural beings*.

Political ecology in Latin America is operating a similar procedure as the one achieved by Marx with Hegelian idealism, turning the philosophy of post-modernity (Heidegger, Levinas, Derrida) on its own feet, territorializing thinking on being, difference and otherness in *environmental rationality*, rooted on the *politics of cultural diversity, territories of difference and ethics of otherness* (Leff 2004, 2014a, 2018). Cultural diversity and ontological difference nested in the symbolic order becomes the core of a *politics of difference*. Otherness becomes the original root of diversity and difference that dissolves the unitary and universal ontological/epistemological conception of being, reality, world, and knowledge. Political difference is the right to be different, the right to differ: to contest the already existent reality. Otherness radicalizes difference beyond dialectic contradiction—the *alter ego* that mirrors identity; the alternation of powers—as the expression of an *absolute Other*: the *Other* as something else than the new and unknown that emerges from the *generativity of physis* and transcendent dialectics. The *Other* is incommensurate and untranslatable; it does not assimilate to a consensus of conflictive differences or common knowledge through communicative action (Habermas 1984, 1987).

Beyond various and different paradigms of knowledge that can be integrated into a holistic view and an interdisciplinary paradigm, the political ethics of otherness opens different modes of cognition and knowledge, intuition and intelligibility, of imagination and creativity. The *dialogue of knowledge* is the encounter of different cultural beings in their non-synthetic, untranslatable ways of being (Leff 2004, 2020a). If the ethical politics of otherness searches the pacific coexistence of different ways of being-in-the-world, the varieties of ways in which human cultures construct Nature open political ecology to conflicts of *equality in difference* arising from different cultural visions and valuations of Nature, as well as the confrontation of cultural/economic rights to appropriate Nature and to territorialize cultural diversity.

Cultural ecology, ethnoecological studies and environmental anthropology converge in the field of political ecology to understand the different ways of constructing Nature, involving different ways of knowing, attracting the "rationality debates" in anthropology and philosophy, calling traditional ecological knowledge and ethnosciences (Fals Borda 1981, 1987; Lopez-Austin and Lopez-Lujan 2001) and inviting non-Western science (Needham 1954) to a dialogue of knowledge, the encounter—the alliances and conflicts– of the different forms of *being–savoir* as the creative source of a sustainable cosmopolitan world.

However, the difference in cultural values and visions does not become a political force only by their ontological and ethical principles. The legitimization of difference that codes new values and empowers cultural beings and their subjugated principles of life and existence—i.e., the *living well* of Andean indigenous peoples (Huanacuni 2010)—emerges from the saturation effects of the forced homogenization of life induced by metaphysical thinking and modern rationality. Politics of difference emerges as the resistance of cultural beings to the dominance of global hegemonic homogeneity, to the objectifying of beings and unequal equality. The strife for equality within the scope of human rights and its juridical procedures based on individual rights ignores the political principle of *equality in difference* that claims its rights in a culture of diversity and otherness. As stated by Escobar,

> It is no longer the case when one can contest dispossession and give arguments in favour of equality from the perspective of inclusion in the domineering culture and economy. The opposite is happening: the position of difference and autonomy is becoming so valid, or more, in this contestation. Appealing to the moral sensibility of the powerful is no longer effective [...] This is the moment to test [...] the power strategies of cultures connected by networks and *glocalities* in order to be able to negotiate

contrasting conceptions of the good, to value different forms of life and to reaffirm the pending predicament of difference-in-equality. (Escobar 2006)

The rights to difference are forged in the encounter with otherness, in the confrontation of the domineering rationality with everything external, which has been excluded, breaking the metaphysical identity of equality and the unity of the universal. In this tension, political ecology transgresses one-track thinking and one-dimensional reason, to open history to the difference of being immersed in a field of power relations and political forces. To be sure, "the struggles for cultural difference, ethnic identities and local autonomies over the territory and resources are contributing to define the agenda of environmental conflicts beyond the economic and ecological field", valuing and claiming the rights of "ethnic forms of otherness committed to social justice and equality in difference" (Escobar 2006). Peoples' cultural emancipation is not a claim for ethnical essence or universal rights of the individual, but the collective rights of cultural beings—including the intrinsic values of Nature as cultural rights—together with the rights to dissent from established meanings and present hegemonic power structures, and to weave different life-worlds. Thus indigenous peoples are offering alternative views to the environmental crisis, to solve climate change and to construct "other" possible worlds based on their worldviews and the ecological conditions of their life-territories.

Politics of difference goes beyond the recognition of different views, interests and political positions in a plural world. Difference is understood in the sense that Derrida (1978, 1982) assigns to his concept of *differance*, which not only establishes difference here and now, but opens being to time, to becoming, to the events and the advent of the unexpected, to the creative evolution of life, to the eventuality of the yet not thought and inexistent, of the yet to come into being: to a sustainable future. Facing the *end of history*—conceived as the siege and sealing of cultural evolution by the ineluctable domination of technology and the globalized market—politics of difference reopens history to utopia, to the construction of differentiated and diverse sustainable societies. The right to differ in time opens the meanings and the senses of being that weave live-times that which is possible from the *Real of Life*—the emergential potentials from Physis; the complexifying evolution of Life—the creativity of life arising from the Symbolic order, and the drive... is not (Levinas 1969; Leff 2020a).

Political ecology embraces the power struggles for the production and distribution of use values; but above all, of meanings–values assigned to needs, ideals, desires, and forms of existence that drive the transformation of culture and Nature. From the incommensurability of cultural rationalities, the politics of cultural difference stresses the rights of existence of different values and meanings assigned to Nature that configure diverse identities and life-worlds. Thus, politics of difference leads sociological imagination to design power strategies capable of building a cosmopolitan world based on cultural diversity and political plurality as the conviviality of different cultural rationalities. Cultural politics is the quest of "other possible worlds" claimed by the World Social Forum: a world that embraces many worlds (Subcomandante Marcos *dixit*); a New World constructed by the encounter of different rationalities through a dialogue of knowledge/*savoirs*.

Un-difference of Ecological Consciousness

Political ecology is not politics merely informed by ecology. Ecological awareness that emanates from the narratives of different ecosophies or the discourse of sustainable development is not a homogeneous understanding shared by different cultural worldviews, social imaginaries, and interest groups. Thus, ecological consciousness has not gained in clarity, consistency, legitimacy, and force to reorient criteria towards the construction of sustainability. Techno-economic rationality prevails over the ontology of life as the driving force of history and ecological downfall. Decision-making regarding the environment continues to be geared by economic interests rather than prioritizing ecological balance, human survival and weaving alternative sustainable worlds, to the point of negating scientific evidence on the risks of climate change. The principles of *sustainable development* (polluter pays, previous and informed consent, common but differentiated responsibilities) have become *slogans* with limited effect in decision-making criteria, in changing the trends of ecological degradation and in building a sustainable world order. The environmental movement is a disperse field where various social actors intervene, often confronted by their different views, interests, claims and political strategies, rather than a space for consensus and solidarity of common objectives.

The idea of an emergent "species consciousness" that would safeguard humanity from ecological catastrophe is a problematic illusion. The *Economics of Spaceship Earth* (Ward 1966; Boulding 1966) is an ideology that veils the social differences of the fellow passengers; just as that of *Our Common Future* (WCED 1987) that through the principle "thinking globally and acting locally" reinforces the trends and strategies established by the domineering global thinking—the views on "sustainable development" within the hegemonic economic order—blurring other approaches to construct a sustainable future. Environmental consciousness would seemingly emerge from the deep sources of being and in the realm of the noosphere to restore the conditions of life in our unsustainable world. However, for such generalized and unified consciousness to emerge as an existential condition it would be necessary for humanity as a whole to share the experience of a common threat or a shared destiny in equal terms; as when the generalization of plagues (sent by the gods) turned the symbolism of the Aristotelian syllogism on the mortality of all men into self-consciousness of humanity through a lived experience, transforming the axiom of logic into the production of meaning in the social imaginary. From the Aristotelian statement "all men are mortal" does not follow a generalized meaning that nested in consciousness. Only once the pest spread in Thebes and society as a whole felt concerned by the threat of real death, pure symbolic form turned into a social imaginary (Lacan 1974–1975). The same applies to a sufficient scale to the generalized experience that since the origins of humanity established the imaginary of the prohibition of incest. The symbolism of the Oedipus complex and the meaning of the Greek tragedy had been already internalized as a lived cultural law; it was not instituted by Sophocles nor by Freud, but by lived experience.

Environmental consciousness is not a unifying imaginary of different individuals and cultures that integrate humanity. The deconstruction of the modern subject, from Nietzsche and Freud to Heidegger, Levinas, and Derrida, has surfaced the fact that the subject fails to establish himself as the source and foundation of his thoughts and acts. The interiority of the subject is exposed to the infinity of otherness previous to any consciousness of his being. If otherness in the field of political ecology implies a radical difference in cultural beings, it follows that there are no foundations to postulate a unified trans-individual and trans-cultural ecological consciousness of the human species.

In the *risk society* that we presently live, the imaginary of insecurity and terror is drawn to the threats of war and generalized violence and insecurity rather than to the imminent dangers of climate change and ecologic collapse. Even traumatic human experiences like the Holocaust and genocides along history have been unable to give preeminence to an ethic of life over the will to empower the human death drive. It seems vain to posit a certain consciousness that could respond effectively to ecological risk and guide social actions towards sustainability when the environmental crisis that looms the World is still perceived as false consciousness, as a misguiding uncertain premonition by science and by the prevailing economic and political interests that dominate Nature and society. The threat that has penetrated the collective imaginary is that of *ontological insecurity*—the fear of war and terrorism; the collapse of basic social rules of human coexistence—rather than consciousness of the revenge of subdued and overexploited Nature and to orient actions towards an ecological reordering of the world founded in the conditions of life.

No doubt, today everybody has a certain awareness of the environmental problems that affect their quality of life; but this consciousness appears as fragmented and diverse perceptions depending on the specificity of diverse ecological, geographic, economic, social and cultural contexts and conditions that configure a variety of environmentalism (Guha and Martínez-Alier 1997). Not all forms of awareness and consciousness become "ecological cases" that generate social movements (Yearly 1991). The more widespread that these expressions are worldwide—like global warming—the less clear and more general that the perception of ecological risks becomes: not only because their occurrence varies in different latitudes, but because they are sensed through different visions and conceptions: from God's will and the fatality of natural phenomena to the expression of the law of entropy and the effects of the global economy. Environmentalism is thus a kaleidoscope of theories, ideologies, strategies, and actions that are not typified as class consciousness nor unified by a species consciousness, lest for the fact that ecological narratives have already penetrated all languages, discourses, theories, and imaginaries of our globalized world.

The entropy law—which gives scientific support to such previsions– and the evidence of "natural" disasters that have developed and proliferated in the last years, have not yet dissolved the certainties of the economy with the uncertainties and probabilities of climatic events. What prevails is a dispersion of visions and previsions on the conditions of human survival and existence and their relation to the environmental crisis, where class consciousness boundaries become diffuse but not erased, divided by differentiated values, understandings, and interests. At the same time, the political rights for cultural diversity are generating new ways of thinking and positioning of social groups that contest the possibility of a unitary vision to save the planet, biodiversity and the human species. These emergent cultural and environmental common rights confront the current juridical framework constructed around the principle of individuality, of positive law and private rights similarly as economic rationality is being questioned by environmental rationality.

Changing our minds about life, survival, and existence is not primarily a matter of consciousness, but instead of constructing alternative rationality and modes of understanding through a politics of knowledge. As viewed by Foucault, "the genealogy of knowledge needs to be analyzed, not in terms of types of consciousness, modes of perception and forms of ideology, but in terms of tactics and strategies of power [...] deployed through implantations, distributions, demarcations, control of territories and organization of domains which could well make a sort of geopolitics" (Foucault 1980, p. 77). The geopolitics of sustainability involves a "new politics of truth [...] the political, economic, institutional regime of the production of truth" (Ibid., p. 133).

If environmental consciousness arises from human awareness on the limits of existence that today face the entropic death of the planet, environmental rationality is built by the relation of being with life, of the Real with infinity and with its limits, in the encounter of the *possible* with the objectified world, in the interconnection of the Real, the Imaginary and the Symbolic that obliterates the subject in the *lack of being* of human existence.[6] The "subject" of political ecology is not the man of humanism constructed by metaphysics, phenomenology, and anthropology, nor the generic *Dasein* of existential ontology (Heidegger 1927/1962). Diverse human beings forged by their wisdom and practices weave their life-worlds as *production of existence* (Lacan 1974/1975). Mobilized by the desire for life, they build their future by forging the relation of being with *savoir* through history, with the present and with the becoming of other possible worlds: with a sustainable future beyond any transcendence prescribed by ecological evolution, historical dialectics, economic rationality or intentionality of an enlightened subject. Environmental rationality is configured in a politics of difference, in the construction of the rights of being and the reinvention of identities constituted through power relations.

Ecofeminism and Gender: Phallocracy, Difference, and Otherness

In recent years, the upsurge of gender issues and the legitimization of women's rights have converged with environmental concerns and struggles. From radical feminism to ecofeminism, the domination of women and the exploitation of Nature appear as the result of hierarchical social structures established since patriarchy and gerontocracy in traditional cultural formations, to class division and domination processes in modern societies. Ecofeminism has become a diverse and polemic field of inquiry and social action. The first expression arose from women's responses to the effects of environmental degradation on their labor place and their living conditions. Women appeared as one of the most vulnerable social groups as a result of the social functions inherited by patriarchy and the modern social/gender division of labor.

In a first approach, ecofeminism associated the life-giving, caring and nurturing sensibility of women with Nature conservation, linking

[6] The meaning of the syntagm *lack in being* is knitted into the whole of Lacanian psychoanalytic discourse. Regarding my own appropriation and use of this concept, see Leff, E. (2020a), *El conflicto de la vida*, Mexico City: Siglo XXI Editores.

feminist and environmental struggles. The Chipko movement became one of the most emblematic eco-feminist movements in the South (Anand 1983; Shiva 1988). Transcending a naturalist and essentialist vision, ecofeminism developed and contrasted its stands from deep ecology and social ecology within radical ecology (Zimmerman 1994). Following radical feminism, ecofeminism viewed in the patriarchal social hierarchy and ontological dualism the primary sources of ecological destruction and women's domination through male social formations that organize thinking, culture and gender relations.

Political ecology includes eco-feminist inquiries and struggles within its broad scope of politics of difference. Eco-feminism is not only a claim for distributing roles to women in environmental matters or granting new civil and gender rights opened by a democratic culture in the perspectives of sustainable development. It further implies the inquiry of the specific difference from which new perspectives can be opened for sustainability. Beyond emancipation from all masculine forms of domination, feminism faces the challenge of deciphering the enigma of the *difference opened by the division of sexes* within the various dualisms that cross and tense the ontology of difference. Feminism entails the inquiry of the socially constructed difference that has divided humanity between *mankind* and *woman being*; ecofeminism enlarges the political perspectives opened by a feminist and gendered vision of power, culture and social organization, to the relations to Nature and sustainability. This inquiry goes beyond establishing the place and roles of women in a social structure and their claims for equal rights under the privileged status of men that govern the established social order.

Within the complex scope of feminism, ecofeminism embraces the ideas, theories and practices that in a different perspective and from other strands of radical ecologism search to identify the specificity of sexual and gender relations in the genesis of the environmental crisis, as well as the status of sexual difference within power structures in the present social, economic and political order, that offsets environmental degradation (Mellor 1997). In this perspective, the eco-feminist movement inquiries from sexual division and gender difference, the specific standpoint from where women—from their being and condition– understand the environmental crisis and offer a specific feminine vision for the construction of sustainable societies.

Besides including gender differences and sexual rights in the progress of democratic societies new questions arise from ecofeminism: Is there a natural affinity of women to Nature that would legitimize their social claims and turn them into privileged spokespersons of the rights of Nature? How do cognition and sensibility vary with sexual difference and gender identities? How does this difference complexify the approaches to the deconstruction of the logic of domination? How do different gender visions open alternative cognitive, sensitive, epistemological and ethical perspectives on sustainability?

After Simone de Beauvoir (1968) stated that no revolution could dissolve social structure in the way that social revolution changes class differences, eco-feminism has opened a debate on the place of gender difference and social hierarchy in phallocentric societies, in the historical division of labor and its environmental effects. In the beginning, much of the debate turned around the biological and physiological condition of women in the sexual–social division of labor, within the relations of domination of patriarchal, hierarchical structures. However, a deeper quest lead to inquiry the *lack in being* set off by the *difference of sexes*: the primary difference produced by sexual otherness, not as biological and physiological difference, but as that established through symbolic structures and signification by language.

Eco-feminist thinking takes a similar stand as other radical ecologies in assigning ontological dualism one of the primal causes of Nature's objectification and women's domination that has to lead to environmental crisis, extending gender difference from its biological and symbolic origin, up to its socio-historical construction (Merchant 1992; Haraway 1991). The gender debate in ecofeminism goes beyond any natural causes derived from sexual difference, to explain the inequalities and domination of women. It opens the inquiry about the processes of signification in the symbolic order and its effects in the forms of identification of subjects, in social hierarchies and domination relations arising from gender difference as a social-symbolic construction. Beyond essentialist and naturalist approaches,

> The difference is always in the order of the signifier, in the symbolic order, from where it distributes gender emblems and attributes. These attributes will be re-signified as sexual difference in the way of identifications that will lead the subject to be a man or a woman, or any combination of both [...], because the content of what can be masculine or feminine has no natural essentiality; it acquires different modalities depending on a socially determined historicity [...] phallocracy emanates from a totally different order: it is the way in which difference is organized as the differentiated appropriation of privileges and powers. From this difference derives a hierarchical ordering of domination and submission. (Saal 1998, pp. 24, 33)

Thus, nor biology, nor the symbolic order—the oedipal structure and castration complex– can adequately determine the sexual difference and explain the places that men and women occupy in the social order. It is not a difference of constitutive essences that would determine man to be the congener of culture and woman of Nature: man's subjectivity deriving from its place in production and women in reproduction. Ecofeminism leads to inquiring about the role played by the interdiction of incest in a particular oedipal structure, in establishing certain relations of domination between men and women and how *phallocracy* organizes power relations. The fact that always and in every culture, there are laws that allow the access to certain women while prohibiting others, and that men have always occupied the higher ranks in social structure, would seem to confirm the universality of Oedipus. However, as Safouan (1981) has proposed, the Oedipus is not universal. If phallic domination is in no way natural, it is not determined either by a universal symbolic order. The social rules for the exchange of women have varied with the evolution of the economic process (Meillassoux 1977). As George Bataille explained,

> By being sexual, prohibition underlined the sexual value of its object [...] Erotic life could only be *regulated* for a particular time. In the end, these rules expelled eroticism outside the rules. Once eroticism was dissociated from marriage, it acquired a more material meaning [...]: rules pointing to the distribution of women-object of greed were those that secured the distribution of women-labour force. (Bataille 1957/1997, pp. 218–219)

From the *lack in being* (Lacan) that results from being inscribed in the symbolic order, and in its search for completeness, human desire opens its way to *will to power* (Nietzsche). Thus, the man takes resources from his physical strength to gain supremacy in the social order, developing power strategies—physical–gestural, economic–political, juridical–institutional– as instruments of domination. From a position of power in his relation to women, man has constructed discursive strategies that operate as power devices. However, nothing legitimizes such claims of superiority. Feminist politics emerges from those pre-established places set in the symbolic and economic structures that find their origins in the gift-exchange of women: in their functions of production and reproduction.

For Moscovici (1972), domination of men underpinned in his use of the law of prohibition of incest, clinging to it as a trans-historic symbolic law established for any social order. From a Freudian-Marxist feminist vision, women find their way to emancipation by moving away from their reproduction function and the places assigned to them by the economic division of labor. Furthermore, women have to deconstruct the imageries built by psychoanalytic theory—the Oedipus complex and the law of prohibition of incest—to delink from economic rationality and from *rationalizations of the unconscious* (Deleuze and Guattari 1983).

Together with deep and social ecology, ecofeminism agrees that cosmogonies and use-practices of Nature in traditional cultures are more "ecological" than in modern societies. However, women have not been less submitted by gerontocracy and patriarchy in traditional societies. Feminist claims are induced to traditional cultures from modern democratic culture. Gender identities and emancipation arise in the encounter of cultural differences. Gender politics poses the question of a radical but non-essential sexual difference where the symbolic order configures the identities of human beings (men, women or any gender construction) and assigns their places in social structures, attributing forms of thinking, feeling and being-in-the-world. From sexual division, cultural gender differences are constructed: the domineering reasoning and objectifying will of men; the caring sensibility of women in Modern Western culture; their contrast with more spiritual, holistic, ecological and non-possessive oriental and traditional cultures. Ultimately, culture distributes social roles and configures different forms of gender-beings in their relations to

Nature. Gender/culture identity in the order of being and meaning denaturalizes the sexual question to view the conflicting interests that arise from the disjunction of sexual difference in the symbolic order, within power relations and social hierarchies.

Politics of difference inquires gender identity and sexual division in their relations with thinking and the construction of reality; it searches to understand the relation of sexual difference with the ontological disjunction between being and entities (Heidegger 1957/1969; Derrida 1983/2008) that developed in the history of metaphysics in the Cartesian dualisms of object and subject, mind and body, Nature and culture, man and woman, that lead to the objectification of the World, to the construction of hierarchies and institutionalization of relations of domination of women and Nature in modern societies. Ecofeminism complexifies power relations in the field of political ecology by inquiring the links between Nature, language, the unconscious, sexual difference and social structure as conjugated agencies in the construction of Nature–culture–gender relations.

In this perspective, what distinguishes women from men is not their affinity with Nature or the organic functions of women (pregnancy, progeny, maternity, care), buy their resistance to submit their being in a totalitarian rational order. Gender equity demands human rights beyond claims for a better distribution of functions, privileges, and rights established by modern society. By forging new meanings, ecofeminism claims gender rights as *rights to otherness*. Gender difference emerges from the sources of desire that disjoints the metaphysics of the One into the ontology of difference and ethics of otherness, where masculine/feminine positions collide. Within a politics of difference, eco-feminist and gender claims overflow the scheme of economic or ecologic distribution as a way of reassigning property and appropriation rights to women in their socio-ecological roles, functions, and relations with Nature; ecofeminism opens new ways to dissolve hierarchy, oppression, and domination arising from power relations originated by the division of sexes and constructed by masculine power strategies.

If ecofeminism is called to think the deconstruction of the theoretical and social structures in which men forged domination powers, it must arm itself with strategies that, without being exclusive of women, are more "feminine" in the face of *macho* forms of domination. The power of seduction is wiser than the imposition of power through knowledge (Baudrillard 1990). Seduction reorients the power of desire—the Nietzschean will to power– to the *will of power to desire life* (Leff 2006b) opening history into forging new rationality through relations of otherness in an emancipatory process where men and women will reconstruct their rights of being.

However, political ecology inquires: is there a specific speech of women; do different ways of reasoning and feeling concerning Nature arise from gender variances that far from justifying any dualism founded in sexual or gender difference could open new ways of building a sustainable world? Can ecofeminism offer to political ecology new thinking, new grammars for culture–Nature relations: a strategy of seduction, solidarity, reciprocity, and emancipation of being as an alternative to strategies of domination?

These questions lead to a more fundamental inquiry on the difference of sexes. Beyond biological and symbolic (*phallic*) determinations, an inquiry arises on the difference in gender positions in the face of different *modes of jouissance* (Lacan 1998). This implies thinking the relation *being/savoir* within the structure of *jouissance*, searching the possibility of being in "other knowledge," or in a *savoir Other*, knowing that it is impossible to *know the other*. In the incompleteness of being, in the unknowing of the other, in the void that organizes the modes of *jouissance*, different positions and perspectives of *being/savoir* can come to existence. Here a womanly mode of *jouissance* is speculated beyond the frontiers of language, symbolic law, and phallic legislation. At stake are the different modalities of the relation of gender identities with *jouissance*. In waiting for these varieties of relations of *jouissance* and knowledge to be disclosed and to emerge to existence, what is speculated is a manly way of knowing, in close relation to positive knowledge, to the present, to reality, to the truth as the identity of thought and reality.

Conversely, a womanly *savoir*, in her relation to *jouissance*, convenes "*Other knowledge*", a no-knowledge, in her "letting be" into the realm of the unknown, in the horizon of what is not, in the obscurity of nothingness. The woman would be molded by a *jouissance* "Other" beyond knowledge organized by signifiers—by the Phallus as signifier—beyond consciousness and will:

> Woman inaugurates a new time by presenting her *jouissance* in the field of knowledge, not a knowledge that does not know itself, but a no-knowledge, knowledge that obliterates the Other. It is not an unknown knowledge that refers to the place of the Other, but the new face that woman presents of this Other as no-knowledge […] *a metamorphosis unknown for normal pathways of understanding*. (Morales 2011, pp. 210, 50)

In this perspective, sexual difference opens a new inquiry in the ways of knowing, very much in the vein of Emmanuel Levinas who stated: "The caress does not know what it seeks" (Levinas 1993, p. 133). In their relation to *jouissance* and knowledge, *the man comes, while the woman goes*! Women would be prone to being more "cosmic" and "oceanic" in character, more disposed to letting themselves be within the unknown, to restrain from totality, to float over the uncertainties of life and to fly towards the future; while men would be more predisposed to objectify being in present entities, to be driven by the ambition of totality and will to power to grasp reality and control the world.

The above speculation opens an ontological-anthropological inquiry into the relation of Being with *Geschlecht*. If there is an original split in the sexual condition of human beings—an otherness more original than the difference between Being and entities derived by Ancient Greek thinking (Heidegger 1957/1969; Derrida 1983/2008) it opens the question about the male character of metaphysical thinking that derived in modern societies governed by men. However, things are even more complex: if the Oedipus is not universal and traditional cultures are not cut by homogeneous patriarchal social structures, anthropological studies should provide evidence of different ways of understanding the world and organizing the world-lives of traditional cultures governed by different patriarchal/matriarchal social relations emerging from different "modes of *jouissance*", by different cultural/Oedipal forms/ways of being-in-the-world. Womanly and gendered *savoir* arise from their secluded unrealized potentialities and encounter/blend with other constellations of "*savoirs* without knowledge" that call the yet unknown sustainable future into being.

Women—and men– will not regain their rights to being from an equalization of power in the order of rationality that has dominated and subjected them. To emancipate from that oppressive order, men and women are forging new gender identities, restoring their being through Other power–knowledge strategies, merging the realm of desire for a life with new forms of cognition and thinking, meaning and feeling; reweaving and fertilizing the social fabric with new forms of being-in-the-world. Thus, ecofeminism claims its transcendent otherness to emancipate from established power relations.

In this sense, ecofeminism is not only a standpoint to criticize the places assigned to women in the economy, in politics and the family. Its substantive difference is not only established by the different and subjugated roles determined by a hierarchical, patriarchal and phallocentric culture but in stating sexual and gender difference in new languages, concepts and sensibilities, other to male construed rationality. In this perspective, political ecology opens an inquiry on the ways gender difference generates other forms of identification, distinct forms of knowing and feeling in which life is embodied and being comes to life in the midst of savoir emerging from nothingness.

Ethics, Emancipation, Sustainability: Towards a Dialogue of Knowledge/ *Savoirs*

Political ecology constructs its theoretical and political identity in a world in mutation, driven by an environmental crisis: a crisis of *being-in-the-living-world*. The concepts and understandings that had guided our intelligibility of the world until then, the meaning of our life-worlds and the intentions of our practical actions, seem to vanish from our everyday language. Yet, the established world order holds onto a dictionary of signifiers and discursive practices that have lost their capacity to sustain life: dialectic logic, universal principles, unity of sciences, essence of things, eternal truths, transcendence of thought, and intentionality of actions or deeds, resonate and echo the nostalgic reminder of a world forever gone. *Something* new is emerging in this world of uncertainty, chaos, and unsustainability. Through the interstices opened by the cracking of monolithic rationality and totalitarian thinking, environmental complexity sheds new lights on the future to come. This "something" is expressed as a *need for emancipation* and a *will to live*.

While language games keep proliferating and revolving around this false and unsustainable world, they also serve to envision possible alternative futures, to design utopias that redirect the course of life. If this process is not to succumb to the "fatal strategies of *hyperreality*" (Baudrillard 1983) generated by the "simulacra and simulation" of the discourse of "sustainable development", and guided by the power strategies of an unsustainable rationality that drifts the world into the entropic death of the planet, one basic principle must continue giving support in reason to human existence: the coherence of thinking with life conditions on Earth, knowing that the world will never be totally known nor controlled by thought.

Environmental crisis expresses the limits of growth, the unsustainability of economic rationality and technological reason. These are the effects of the history of metaphysics and Western knowledge: of *logocentrism* of theory, universality of science and one-dimensional thinking; of instrumental rationality between means and ends; of the law of economic value as universal equivalent to measure all things, that under the sign of money and the laws of the market have recoded all things and ontological orders in terms of exchangeable and tradable market values. Human emancipation arises from the deconstruction of knowledge and declamping from the iron cage of modern rationality. It implies giving new meanings to the emancipatory concepts of modernity—liberty, equality, and fraternity– as principles of a political ethics that ended up being co-opted and corrupted by economic and juridical liberalism—by the privatization of individual rights and the coercion of economic interests over other human values—in order to legitimize the values of a politics of difference and an ethics of otherness: of conviviality in diversity and solidarity among human beings with different cultures and collective rights.

Political ecology is a politics for cultural diversification. Cultural diversity is the standpoint to deconstruct the unitary logic and universal equivalence of the market and to reorient being through the diversification of ethnic–eco-cultural paths for building sustainable societies. Political ecology roots the deconstructionist spirit of postmodern thinking in a politics of difference activating an abolitionist agenda for direct democracy and sustainability:

> The abolitionist agenda proposes self-managing communities established according to the ideal of a spontaneous *organization*: personal links, creative work relations, affinity groups; community and neighborhood councils based in respect and sovereignty of human persons, environmental responsibility and the exercise of direct democracy "face to face" for decision making in matters of collective interest. *This agenda intended to change our course towards a civilization of diversity, ethics of frugality and a culture of low entropy, reinventing values, untying the knots of the mind, avoiding cultural homogeneity with the force of a planet of diverse peoples, villages and cities.* (Borrero 2002, p. 136)

Political ecology is a conceptual texture that weaves material Nature, symbolic meaning and social action with emancipatory thinking and political ethics to renew the sources and potentials for the sustainability of life (Leff 2002b; PNUMA 2002). Environmental rationality constitutes its theoretical core and its strategic actions. It entails the deconstruction of totalizing knowledge—of established paradigms and instituted rationalities– to open up new paths for sustainability built on the potentials of Nature, cultural creativity and the actualization of identities that open being to becoming of that which still-is-not. From a drive for life, from the intimacy of existence that was reduced by totalitarian theories, emerges the emancipatory power for the sustainability of life:

> A certain fragility has been discovered in the very bedrock of existence—even, and perhaps above all, in those aspects of it that are more familiar, more solid and more intimately related to our bodies and our everyday behaviour. However, together with this sense of instability and this amazing efficacy of discontinuous, particular and local criticism, one also discovers […] something one might describe as precisely the inhibiting effect of global, *totalitarian theories*. (Foucault 1980, p. 80)

In deconstructing totalitarian theories Foucault foresaw a *return of knowledge* where "it is not theory but life that matters"; the *genealogies* and *insurrection of subjugated knowledge*; the re-emergence of disqualified knowledge in the struggle for truth and legitimacy of "particular, local, regional knowledge, of differential knowledge incapable of unanimity and which owes its force only to the harshness with which it is opposed by everything surrounding it [...] by the effects of the centralizing powers which are linked to the institution and functioning of an organized scientific discourse within a society such as ours" (Ibid.: 81, 85, 82, 84).

The insurrection of subjugated knowledge drives emancipation from the dominant regime of modern rationality that has marginalized and exterminated other cultures; that has occluded other knowledge and impeded other possible worlds to come into being. Beyond the deconstructive intentionality of postmodern thinking that has mobilized epistemological debates over scientific knowledge, decolonizing knowledge encompasses a broader historical struggle for legitimizing other knowledge/savoir/wisdom, alternative ways of understanding reality, Nature, human life and social relations; different ways of constructing human life in the planet.

What is at stake in the emancipatory ethics of environmentalism is the legitimization of the different popular and traditional knowledge in their encounter with erudite and formal knowledge. Political ecology encompasses such historical struggles and their present power strategies; it embraces the genealogy of environmental knowledge and extends it to consider not only present clashes of knowledge involved in the geopolitics of sustainable development, but also in the power strategies involved in the present processes of hybridization of scientific knowledge and renewed traditional practices; in the construction of new cultural identities through the embodiment of knowledge and its embedding in new life territories and territorialities, in present struggles for the appropriation of Nature.

Environmental ethics in the perspective of the social construction of sustainability projects genealogy of knowledge to a prospective horizon. The *ethics of otherness* (Levinas) is rooted in the field of political ecology as a *dialogue of knowledge/savoirs*. Sustainability is envisioned as the historical outcome of the emancipation of subjugated knowledge, of new understandings of life in the planet and human life, for the construction of *negentropic societies* that internalize the entropic conditions of living. This entails the invention of different economic rationalities: other modes of sustainable production and consumption. Political ecology addresses the power relations involved in the paradigm shifts and social changes in the construction of environmental rationality to build a sustainable world.

Political ecology renews the reflection on ethics for emancipation. Emancipatory needs are not limited to "reducing alienated labor," generating "autonomous free time," "ending role-playing" and promoting receptivity, tranquility and abounding joy instead of the "noise of production" (Marcuse 1992, p. 35). Emancipation from our convulsed globalized world and risk society goes beyond the search for the "ontological security" of the ego. Emancipation of life implies the affirmation of new identities, the rights of cultural beings and new forms of knowledge/*savoir* to uncouple from constrictive hegemonic rationality. Political ecology opens new pathways to sustainability through a dialogue of knowledge, to build a global world where diverse forms of being and living can coexist supported by a politics of difference and ethics of otherness.

This emancipation process from the subjection of being by the hegemonic rationality imposed on the world cannot be the agency of the individual, a rational choice among the alternatives set up by the rationalized world. Emancipation from the present unsustainable world demands the deconstruction of modern techno-economic rationality. It implies rethinking, reknowing and reapprehending the conditions of living, the ecological organization of life on the planet and the conditions of human existence. This civilization transition is a task that cannot be achieved by individual subjects in the process of *reflexive modernization* (Giddens et al. 1994). The construction of a sustainable world demands the social control of environmental degradation: slowing down the trends towards the entropic death of the planet, enhancing the principles and securing the conditions of life. It implies the reinvention of common identities, collective forms of being and cultural world-lives to empower the negentropic processes that sustain life in the planet.

Sustainability is the horizon of such purposive living, an objective not attainable by the restoration of the unsustainable hegemonic rationality, the enlightenment of reason and scientific truth. Travelling towards the horizon of sustainable life guided by environmental rationality, opens the world to the reconstruction of different cultural beings, of beings reconstituted by *other* forms of knowledge, and by their *environmental savoirs* and *social imaginaries of sustainability* (Leff 2010). Sustainability will be the outcome of a dialogue among forms of knowledge: of the encounters of cultural beings instituted by their *savoirs*, with techno-scientific-economic powers and their strategies for the capitalistic appropriation of the planet; of the alliances among *beings/savoirs*, with their differences and their unknowns. Political ecology is the field for the deployment of this odyssey towards a sustainable future, crossed by power strategies for survival and the sustainability of life; for the reinvention of human modes to inhabit our living planet in compliance with the conditions of life on Earth.

The perspectives of political ecology are not only opened to understanding the ontological and political nature of socio-environmental conflicts and the power strategies involved in social struggles over ecological distribution, but also to envision new potentials that emerge from *other* forms of knowledge—from social imaginaries, the reinvention of identities and renewal of traditional productive practices– through the rights of existence of diverse cultural beings, a politics of difference and a dialogue of knowledge, to open new paths towards sustainability; to liaise with the organization of emergent social movements for the reappropriation of Nature and to design and institute political ethics and juridical procedures for the pacific solution of such conflicts and for the conviviality of different modes of being in a sustainable and pacific world.

Bibliography

Ab'Saber, A.N. (1970), "Províncias geológicas e domínios morfoclimáticos no Brasil", *Geomorfologia* 20:1–25, São Paulo.

Alimonda, H. (Ed.) (2002), *Ecología política. Naturaleza, sociedad y utopía*, Buenos Aires: CLACSO.

———. (Ed.) (2006), *Los tormentos de la materia: aportes para una ecología política latino-americana*, Buenos Aires: CLACSO.

Amin, S. (1976), *Unequal development: an essay on the social formations of peripheral capitalism*, New York: Monthly Review Press.

Anand, A. (1983), “Saving trees, saving lives: Third world women and the issue of survival”, in L. Caldecott & S. Leland (Eds.), *Reclaim the Earth*, London: The Women’s Press.

Argueta, A. et al. (1994), *Atlas de las plantas de la medicina tradicional mexicana*, 3 volumes, México: Instituto Nacional Indigenista.

Bataille, G. (1957/1997), *El erotismo*, México: Tusquets. [1962. *Death and sensuality; a study of eroticism and the taboo*, New York: Walker and Company].

Bateson, G. (1972), *Steps to an ecology of mind: collected essays in anthropology, psychiatry, evolution, and epistemology*, Chicago: University Of Chicago Press.

———. (1979), *Mind and nature: a necessary unity (advances in systems theory, complexity, and the human sciences)*, Hampton Press.

Baudrillard, J. (1983), *Les stratégies fatales*, Paris: Éditions Grasset & Frasquelle.

———. (1990), *Seduction*, New York: St. Martin’s Press.

de Beauvoir, S. (1968), *The second sex*, London: Jonathan Cape.

Benjamin, W. (1936), “The work of art in the age of mechanical reproduction”, and (1940), “Theses on the philosophy of history”, in Arendt, H. (ed.) (1968), *Illuminations*, London: Fontana.

Benton, T. (Ed.) (1996), *The greening of Marxism*, New York: Guilford Press.

Biersack, A. & Greenberg, J.B. (Eds.) (2006), *Reimagining political ecology*, Durham/London, Duke University Press.

Boege, E. (2008), *El patrimonio biocultural de los pueblos indígenas de México. Hacia la conservación in situ de la biodiversidad y agrobiodiversidad en los territorios indígenas*, México: INAH/Comisión Nacional para el Desarrollo de los Pueblos Indígenas.

———. (2018), “Hacia una antropología ambiental para la apropiación social del patrimonio biocultural de los pueblos indígenas”, in Toledo, V.M. & Alarcón-Cháires, P. (Eds.), *Tópicos bioculturales: reflexiones sobre el concepto de bioculturalidad y la defensa del patrimonio biocultural de México*, México: UNAM.

Bonfil Batalla, G. (1987), *El México profundo, una civilización negada*, México: Grijalbo.

Bookchin, M. (1964), “Ecology and revolutionary thought”, *Comment* (republished in the British monthly *Anarchy* in 1965).

———. (1971/1990), *Post-scarcity anarchism*, Montreal/New York, Black Rose Books, Second edition, 1990.

———. (1982/1991), *The ecology of freedom. The emergence and dissolution of hierarchy*, Montreal/New York: Black Rose Books, Second edition, 1991.

———. (1990), *The philosophy of social ecology. Essays on dialectical naturalism*, Montreal/New York: Black Rose Books.

Borrero, J.M. (2002), *La imaginación abolicionista. Ensayos de ecología política*, Cali: PNUMA/CELA/Hivos.

Boulding, K.E. (1966), "The economics of the coming spaceship earth", in Jarrett, H. (ed.), *Environmental quality in a growing economy*, Baltimore: Johns Hopkins University Press, 1966: 3–14.

Bourdieu, P. (1984), *Distinction: social critique of the judgment of taste*, Boston: Harvard University Press.

Canguilhem, G. (1971), *La connaissance de la vie*, Paris: J. Vrin.

———. (1977), *Idéologie et rationalité dans l'histoire des sciences de la vie*. Paris: J. Vrin.

Capra, F. (1996), *The web of life*, New York: Anchor Books.

Cardona, G.R. (1986), *La floresta di piume. Manuale di etnoscienzia*, Roma: Editore Laterza.

Cardoso, F.H. & E. Faletto (1979), *Dependency and development in Latin America*, University of California Press.

Carlson, R. (1962), *The silent Spring*, Boston: Houghton Mifflin.

Castoriadis, C. (1998), *The imaginary institution of society*, Cambridge: MIT Press.

de Castro, J. (1946/1975), *The geopolitics of hunger*, New York: Monthly Review Press.

———. (2003), "Subdesenvolvimento: causa primeira da poluição", in: *Fome: um tema proibido—últimos escritos de Josué de Castro*, Rio de Janeiro: Civilização Brasileira, pp. 135–137.

Césaire, A. (1955), *Discourse on colonialism*, New York: Monthly Review.

Chardin, P. Teilhard de, (1961), *The phenomenon of man*, San Francisco: Harper & Row Publishers.

Clark, J. (2008), "Domesticating the dialectic: a critique of Bookchin's neo-Aristotelian metaphysics", *CNS*, Vol. 19, No. 1, March, p. 95.

Commoner, B. (1971), *The closing circle: nature, man, and technology*. New York: Knopf.

———. (1976), *The poverty of power: energy and the economic crisis*. New York: Random House.

Cosby, A.W. (1986), *Ecological imperialism. The biological expansion of Europe, 900–1900*, New York: Cambridge University Press.

Debeir, J.C., Deléage, J.P. & D. Hémery (1986), *Les servitudes de la puissance. Une histoire de l'énergie*, Paris: Flammarion.

De Gortari, E. (1963), *La ciencia en la historia de México*, México: Fondo de Cultura Económica.

Deleuze, G. & Guattari, F. (1983), *Anti-Oedipus*, Minneapolis: University of Minnesota Press.

———. (1987), *A thousand plateaus. Capitalism & schizophrenia*, Minneapolis/London: University of Minnesota Press.

Derrida, J. (1978), *Writing and difference*, Chicago: The University of Chicago Press.

———. (1982), *Margins of philosophy*, Chicago: Chicago University Press.

———. (1983/2008), *Geschlecht: différence sexuelle, différence ontologique*, *Cahier de l'Herne* consacré à Heidegger, dirigé par Michel Haar, Paris, in *Psyche. Inventions of the Other*, Vol. II, Edited by Peggy Kamuf and Elizabeth Rothenberg, Stanford: Stanford University Press, pp. 7–26.

Descola, Ph. & G. Pálsson (Eds.) (1996), *Nature and society: anthropological perspectives*, London: Routledge.

Dos Santos, Th. (1978), *Imperialismo y dependencia*, México: Ed. Era.

Enzensberger, H.M. (1974), "A critique of political ecology", *New left review*, 84 (3–31).

Ehrlichc, P. (1968), *The population bomb*, New York: Ballantine Books.

Escobar, A. (1999), "Gender, place and networks. A political ecology of cyberculture", in Harcourt, W. (ed.), *Women@Internet. Creating new cultures in cyberspace*, London: Zed Books, pp. 31–54.

———. (2006), "An ecology of difference: equality and conflict in a glocalized world", in *Focaal, European Journal of Anthropology* 47: 120–140.

———. (2008), *Territories of difference. Place, movements, life, redes*, Durham/London: Duke University Press.

———. (2010), "Post-constructivist political ecologies", in Redclift, M. & Woodgate G. (Eds.), *International handbook of environmental sociology*, 2nd edition, Cheltenham/Northampton: Edward Elgar Publishing Limited, pp. 91–105.

Fals Borda, O. (1981), "La ciencia del pueblo", in *Investigación participativa y praxis rural. Nuevos conceptos en educación y desarrollo comunal*, Lima, Perú Editorial Mosca Azul.

———. (1987), *Ciencia propia y colonialismo intelectual. Los nuevos rumbos*. 3a. edición, Bogotá: Carlos Valencia Editores.

Fanon, F. (2004), *The wretched of the earth*, New York: Grove.

Ferry, L. (1995), *The new ecological order*, Chicago: The University of Chicago Press.

Foster, J.B. (2000), *Marx's ecology: materialism and nature*, New York: Monthly Review Press.

Foucault, M. (1980), *Power/knowledge*, New York: Pantheon.

Funtowicz, S. & J. Ravetz (1993), *Epistemologia política. Ciencia con la gente*, Buenos Aires, Centro Editor de América Latina.

———. (1994), "Emergent complex systems", *Futures*, 26 (6): 568–582.

Furtado, C. (1974), *O mito do desenvolvimento econômico*, São Paulo: Ed. Paz e Terra.

Galeano, E. (1971), *Open veins of Latin America: five centuries of the pillage of a continent*, New York: Monthly Review Press.

Georgescu-Roegen, N. (1971), *The entropy law and the economic process*, Cambridge, Mass.: Harvard University Press.

Giddens, A., Beck, U. & S. Lash (1994), *Reflexive modernization. Politics, tradition and aesthetics in the modern social order*, Stanford: Stanford University Press.

González Casanova, P. (1965), "Internal colonialism and national development", in *Studies in comparative international development*, Vol. 1, Num. 4:27–37.

Gorz, A. (1989), *Critique of economic reason*, London/New York: Verso.

———. (2006), "Gorz. Ecologie une éthique de la libération", interview in *EcoRev*, January.

———. (2008), *Ecologica*, Paris: Galilée.

Guha, R. & Martínez-Alier, J. (1997), *Varieties of environmentalism. Essays North and South*, London: Earthscan.

Gunder-Frank, A. (1966), *The development of underdevelopment*, New York: Monthly Review Press.

Habermas, J. (1984), *Theory of communicative action, Vol. I: Reason and the rationalization of society*, Boston, Mass.: Beacon Press.

———. (1987), *Theory of communicative action, Vol. II: Live-world and system: a critique of functionalist reason*, Boston, Mass.: Beacon Press.

Haesbaert, R. (2004), O mito da desterritorialização: do "fim dos territórios" à multiterritorialidade, Rio de Janeiro: Bertrand Brasil.

Haraway, D. (1991), *Simians, cyborgs and women. The reinvention of nature*, New York, Routledge.

Heidegger, M. (1927/1962), *Being and time*, trans. John Macquarrie & Edward Robinson, San Francisco: HarperCollins Publishers.

———. (1935/1977), "The question concerning technology", in *Heidegger. Basic writings*, trans. D. Farrell Krell, New York: HarperCollins, pp. 283–317.

———. (1946/1977), "Letter on humanism", in *Heidegger. Basic writings*, trans. D. Farrell Krell, New York: HarperCollins, pp. 189–242.

———. (1957/1969), *Identity and difference*, Chicago: Chicago University Press.

Herrera, A.O. et al. (1976), *Catastrophe or new society: a Latin American model*, Ottawa: IDRC.

Hetman, F. (1973), *Society and the assessment of technology*, Paris: OCDE.
Horkheimer, M. & Th. Adorno (1944/2002), *Dialectics of Enlightenment*, Stanford: Stanford University Press.
Huanacuni, F. (2010), *Vivir bien/Buen vivir: Filosofía, políticas, estrategias y experiencias regionales*, La Paz: Convenio Andrés Bello/Instituto Internacional de Integración.
Illich, I. (1973), *Tools for conviviality*, New York: Harper & Row.
Kropotkin, P. (1902/2005), *Mutual aid: a factor of evolution*, Boston: Extending Horizons Books.
Lacan, J. (1974–75), *Séminaire XXII RSI (Réel, Symbolique, Imaginaire)*, mimeo.
———. (1998), *On feminine sexuality the limits of love and knowledge*: The Seminar of Jacques Lacan, Book XX *Encore* (edited by Jacques-Alain Miller), New York: Norton.
Lander, E. Ed. (2000), *La colonialidad del saber: eurocentrismo y ciencias sociales. Perspectivas latinoamericanas*, Buenos Aires: Clacso/Unesco.
Latour, B. (2004), *Politics of nature. How to bring the sciences into democracy*, Cambridge, Mass.: Harvard University Press.
Leff, E. (Ed.) (1986), *Los problemas del conocimiento y la perspectiva ambiental del desarrollo*, México: Siglo XXI Editores.
———. (1993), "Marxism and the environmental question: from critical theory of production to an environmental rationality for sustainable development", *Capitalism, nature, socialism*, Vol. 4 (1): 44–66, Santa Cruz, California.
———. (1995), *Green production: towards an environmental rationality*, New York: Guilford.
———. (1998a), "Murray Bookchin and the end of dialectical naturalism", *Capitalism, nature, socialism*, Vol. 9 (4):67–93.
———. (2000), "Pensar la complejidad ambiental", en Leff, E. (Coord.), *La complejidad ambiental*, Colección "Aprender a aprender", México: Siglo XXI Editores (2nd revised edition, 2003).
———. (2001), *Epistemología ambiental*, Sao Paulo: Cortez Editora.
———. (Ed.) (2002b), *Ética, vida, sustentabilidad*, México: PNUMA, Serie Pensamiento Ambiental Latinoamericano No. 5.
———. (2004), *Racionalidad ambiental: la apropiación social de la naturaleza*. México: Siglo XXI Editores.
———. (2006a), *Aventuras de la epistemología ambiental: de la articulación de las ciencias al diálogo de saberes*, México: Siglo XXI Editores.
———. (2006b), "Ética por la Vida. Elogio de la Voluntad de Poder" en *Polis*, Revista de la Universidad Bolivariana, vol. 5, núm. 13, Universidad de Los Lagos Santiago, Chile.

———. (2010), "Imaginarios sociales y sustentabilidad", in *Cultura y representaciones sociales*. Num. 9, México, pp. 42–121.

———. (2012), "Latin American environmental thinking: a heritage of knowledge for sustainability", *Environmental ethics*, Volume 34:4, Winter, 2012, pp. 431–450.

———. (2014a), *La apuesta por la vida: imaginarios sociales e imaginación sociológica en los territorios del Sur*, México: Siglo XXI Editores.

———. (2014b), "Los Derechos del Ser Colectivo y la Reapropiación Social de la Naturaleza", en Pereira de Cunha, Belinda; Agustín, Sergio; Araújo da Costa, Nálbia Roberta (Org.), *Saberes Ambientais, Sustentabilidade e Olhar Jurídico: visitando a obra de Enrique Leff*, Joao Pessoa e Caxias do Sul, Brasil.

———. (2015), "Encountering political ecology: epistemology and emancipation", in Bryant, Raymond & Soyeun Kim (Eds.), *The International Handbook of Political Ecology*, Cheltenham, UK: Edward Elgar, pp. 44–56.

———. (2018), *El fuego de la vida: Heidegger ante la cuestión ambiental*, Mexico: Siglo XXI Editores.

———. (2020a), *El conflicto de la vida*, Mexico: Siglo XXI Editores.

Leis, H. (2001), *La modernidad insustentable: las críticas del ambientalismo a la sociedad contemporánea*, Montevideo PNUMA/Nordan.

Levinas, E. (1977/1997), *Totalidad e infinito: ensayo sobre la exterioridad*, Salamanca: Sígueme, 4th edition [*Totality and infinity: an essay on exteriority*, Pittsburgh: Duquesne University Press, 1969].

———. (1993), *El tiempo y el otro*, Barcelona: Paidós [*Time and the Other*, Pittsburgh: Duquesne University Press, 1987].

Light, A. (Ed.) (1998), *Social ecology after Bookchin*, New York/London: The Guilford Press.

Lipietz, A. (1999), ¿Qu'est-ce que l'écologie politique? La grande transformation du XX Siècle, Paris: La Découverte.

Lopez-Austin, A. & Lopez-Lujan, L. (2001), *Mexico's indigenous past*, Norman: University of Oklahoma Press.

Lovelock, J. (1979), *Gaia. A new look at life on Earth*, Oxford: Oxford University Press.

Maier, C. (2006), "Transformations of territoriality 1600–2000", in G. Budde, S. Conrad & O. Janz, Eds., *Transnationale Geschichte: Themen, Tendenzen und Theorien*, Göttingen: Vandenhoeck & Ruprecht.

Marcuse, H. (1968), *L'homme unidimensionnel*, Paris: Ed. Minuit [*One-dimensional man*, Boston: Beacon Press, 1964].

———. (1972), *Counterrevolution and revolt*, Boston: Bacon Press.

———. (1992), "Ecology and the critique of modern society", *Capitalism, nature, socialism*, Vol. 3 (3):29–38.
Mariategui, J.C. (1971), *Seven interpretive essays on Peruvian reality*. Texas Pan American Series, Austin: University of Texas Press.
Marini, R.M. & dos Santos, T. (Coords.) (1999), *El pensamiento social latino-americano en el siglo XX*, Caracas: UNESCO, 2 Vols.
Martí, J. (1963), *Obras completas*, La Habana: Editorial Nacional de Cuba.
Meadows, D.H., D.L. Meadows, J. Randers & W.W. Behrens III (1972), *The limits to growth*, New York: Universe Books.
Meillassoux, C. (1977), *Mujeres, graneros y capitales*, México: Siglo XXI Editores.
Mellor, M. (1997), *Feminism & ecology*, Cambridge: Polito.
Merchant, C. (1992), *Radical ecology*, London: Roultledge.
Mignolo, W. (2000), *Local histories/global designs: coloniality, subaltern knowledges, and border thinking*, Princeton: Princeton University Press.
———. (2011), *Modernity and decoloniality*, Oxford bibliographies.
Mignolo, W. & A. Escobar, eds. (2009), *Globalization and the decolonial option*, London: Routledge.
Morales, H. (2011), *Otra historia de la sexualidad*, México: Palabra en Vuelo.
Morin, E. (1980), *La méthode. La vie de la vie*, Paris, Editions du Seuil.
———. (1990), *Introduction à la pensée complexe*, Paris: Éditions du Seuil.
Moscovici, S. (1972), *La société contre nature*, Paris: Union générale d'édition Collection 10/18.
Mumford, L. (1970), *Pentagon of power: The myth of the machine*, Vol. II, New York: Harcourt.
Murra, J. (1956), *The economic organization of the Inca state*, Chicago: University of Chicago.
Naess, A. (1989), *Ecology, community and lifestyle*, trans. D. Rothenberg, Cambridge: Cambridge University Press.
Needham, J. (1954), *Science and civilization in China*, Cambridge, UK: Cambridge University Press.
Nietschmann, B. (1973), *Between land and water: the subsistence ecology of the Miskito Indians, Eastern Nicaragua*, New York: Seminar Press.
O'Connor, J. (1998), *Natural causes: essays in ecological Marxism*, New York: Guilford.
Peet, R. & Watts, M. Eds. (1996/2004), *Liberation ecologies: environment, development, social movements*, London: Routledge, 2nd edition.
Peet, R., Robbins, P. & Watts, M., Eds. (2010), *Global political ecology*, London: Routledge.

PNUMA (2002), *Manifesto for life. For an ethic for sustainability*, www.rolac.unep.mx.

Polanyi, K. (1944) *The great transformation: the political and economic origins of our time*, Boston: Beacon Press.

Porto-Gonçalves, C.W. (2001a), *Geo-grafías. movimientos sociales, nuevas territorialidades y sustentabilidad*, México: Siglo XXI Editores.

———. (2002), "O latifúndio genético e a r-existência indígeno-camponesa", *Geographia*, Ano 4, No. 8, Niterói: Universidade Federal Fluminense, pp. 7–30.

Prigogine, I. (1961), *Introduction to thermodynamics of irreversible processes* (Second ed.). New York: Interscience.

Prigogine, I. & Nicolis, G. (1977), *Self-organization in non-equilibrium systems*, New York: John Wiley & Sons.

Quijano, A. (2008), "Coloniality of power, eurocentrism, and social classification", in Moraña, M., Dussel, E. and Jáuregui, C.A. (Eds.), *Coloniality at large. Latin America and the postcolonial debate*, Durham, NC: Duke University Press, pp. 181–224.

Robbins, P. (2012), *Political ecology: a critical introduction*, Chichester, UK: John Wiley & Sons, 2nd edition.

Rorty, R. (1979), *Philosophy and the mirror of nature*, Princeton: Princeton University Press.

Saal, F. (1998), *Palabra de analista*, Mexico: Siglo XXI Editores.

Sachs, I. (1980), *Ecodesarrollo: desarrollo sin destrucción*, México: El Colegio de México.

Safouan, M. (1981), "Is the Oedipus complex universal?", in *M/F*, No. 5/6, pp. 83–90.

Santos, M. (1996), *A natureza do espaço: técnica e tempo; razão e emoção*, São Paulo: Hucitec.

Schmidt, A. (1971), *The concept of nature in Marx*, London: New Left Books.

Schumacher, E.F. (1973), *Small is beautiful. Economics as if people mattered*, London: Blond & Briggs.

Shiva, V. (1988), *Staying alive: women, ecology and development*, London: Zed.

———. (1993), *Monocultures of the mind*, London: Zed Press.

Sousa Santos, B. (2008), *Conocer desde el Sur. Para una cultura política emancipatoria*, Buenos Aires: CLACSO/CIDES-UMSA/Plural Editores.

Stavenhagen, R. (1965), "Classes, colonialism, and acculturation. Essay on the system of inter-ethnic relations in Mesoamerica", *Studies in Comparative International Development*, 1 (6): 53–77.

Svampa, M. (2016), *Debates latinoamericanos. Indianismo, desarrollo, dependencia y populismo*, Buenos Aires y Bolivia: Edhasa/CEDIB.
Tapia, L. (2009), *La invención del núcleo comum—ciudadanía y gobierno multisocietal*, La Paz, Bolívia: Ed. Muela del Diablo.
Thompson, E.P. (1991), *Customs in common: studies in traditional popular culture*, London: Merlin Press.
Throne, F. (1935), "Nature rambling: We fight for grass", *The science newsletter* 27, 717, Jan. 5: 14.
Walker, P.A. (2005), "Political ecology: where is the ecology?", *Progress in human geography* 29 (1):73–82.
Wallerstein, I. (1974), *The modern world-system, vol. I: capitalist agriculture and the origins of the European world-economy in the sixteenth century*, New York/London: Academic Press.
———. (1980), *The modern world-system, vol. II: mercantilism and the consolidation of the European world-economy, 1600–1750*, New York: Academic Press.
———. (1989), *The modern world-system, vol. III: the second great expansion of the capitalist world-economy, 1730–1840*, San Diego: Academic Press.
———. (2011), *The modern world-system, vol. IV: centrist liberalism triumphant, 1789–1914*, Berkeley: University of California press.
Ward, B. (1966), *Spaceship earth*, New York: Columbia University Press.
WCED (1987), *Our common future*, Oxford: Oxford University Press.
Whiteside, K (2002), *Divided natures*, Cambridge: Massachusetts Institute of Technology.
Wolf, E. (1972), "Ownership and political ecology", *Anthropological quarterly* 45 (3): 201–205.
Wolf, E. & A. Palerm (1972), "Potencial ecológico y desarrollo cultural en Mesoamérica", in *Agricultura y civilización en Mesoamérica*, No. 32, México: SepSetentas.
Yearly, S. (1991), *The green case. A sociology of environmental issues, arguments and politics*, London: Harper Collins Academic.
Zimmerman, M.E. (1994), *Contesting earth's future. Radical ecology and postmodernity*, Berkeley: University of California Press.

11

Power–Knowledge Relations in the Field of Political Ecology

1. Critical Epistemology of Political Ecology: Power in Knowledge

The environmental crisis irrupted in the 1960s and 1970s as a manifestation of a crisis of civilization: a crisis of the hegemonic ways of understanding and constructing the world in which we live. The environmental question interrogates scientific knowledge and the techno-economic reason that have been institutionalized in the globalized world, which interfere with the natural courses of life and destroy the essential bases for the sustainability of life. For political ecology, the construction of a sustainable world raises the challenge of questioning ontological regimes, paradigms of science and the power strategies at work in the knowledge that underlies the techno-economic rationality that degrades the sustainable conditions for life by appropriating and transforming Nature. So

Published in *Ambiente & Sociedade*, 20(3), November 2017, pp. 225–254. This text is a revised, extended version of my more synthetic original article published as Leff, E. (2015), The power-full distribution of knowledge in political ecology: a view from the South, in Perreault, T., Bridge, G., and J. McCarthy (eds), *Routledge Handbook of Political Ecology*, London & New York: Routledge, pp. 64–75. Permission for publication granted by *Ambiente & Sociedade*.

E. Leff, *Political Ecology*, https://doi.org/10.1007/978-3-030-63325-7_11

building sustainable future calls for the *deconstruction*[1] of economic rationality and market logic (Marx 1965), of the *logocentrism of science* (Derrida 1976, 1978, 1982), and of the juridical principles and procedures (Foucault 1998)—the power devices in knowledge that give support to the hegemonic rationality of modernity—which have unraveled the unsustainable world that we inhabit. Moreover, political ecology moves from this critique of the metaphysical causes of unsustainability to derive theoretical–political strategies capable of bringing about a process of social reorganization in line with the conditions of life on the planet.

The environmental crisis calls for thinking the state of the world: the thermodynamic-ecological and the symbolic-cultural conditions of organic and human life in our living planet. It requires understanding both the trends imposed by techno-economic rationality on the entropic destination of life and the conditions imposed by the ontology of life on the *sustainability of life on Earth*. Thus, the environmental crisis prompts us to think about something that has so far been unthought: the disregard for the unsustainability of life inadvertently produced by humanity and the conditions of life on the planet.

Political ecology has taken up this call and intends to respond to the challenge of this civilization crisis. Anglophone political ecology has opened up a critical space in the American and Anglo-Saxon academy to attempt a *deconstruction* of those theories that seek to understand the relations between culture and Nature—geographical and anthropological

[1] *Deconstruction* designates the critical purpose of postmodern philosophy to discover and unravel the conceptual framework and the social effects of metaphysical thought (Derrida 1976, 1978, 1982). One of these cores consists of the whole set of binary oppositions (object–subject; reason–intuition; Nature–culture) established by the Cartesian division between *res cogitans* and *res extensa* that affect the way we talk and write and understand sexual, racial and social differences. This controversy is not solved by ontological monism pretending to dissolve the radical difference between the Real and the Symbolic, from where the human condition emerges in the order of Life. Deconstruction is the task of questioning metaphysics. The deconstructionist perspective in political ecology allows us to analyze the pathways which, through the history of metaphysical thinking, have led to the formation of frames of thought and theoretical paradigms, legitimized by the dominant rules of science, establishing hegemonic ways of understanding the world and determining the approaches for intervening in Nature. In this context, theoretical deconstruction in political ecology is called upon to unravel and reveal the strategies of power in knowledge (Foucault 1980) that are intertwined with the logic and rhetoric of the discursive development of the scientific disciplines from which political ecology distinguishes itself to establish its own disciplinary identity and political objectives with a study program on the power relations that permeate the socio-environmental field.

traditions, political economy, and agrarian and peasant studies—and those that have ignored the epistemological causes of the environmental crisis: the economic theories that drive economic decisions and the theory of evolution that naturalizes "adaptive" human behavior. Thus, it seeks to understand the social processes that affect, condition and determine the unsustainability of life that are inducing the destruction of ecosystems as life-support systems, triggering global change and the entropic degradation of the planet driven by techno-economic rationality. Thus, political ecology unravels the power relations—and the power in knowledge—that underlie the socio-environmental crisis that currently afflicts and challenges humanity and determine the ways of intervening in, appropriating and degrading Nature.

Today, Anglophone political ecology is undergoing a process of self-criticism, reflecting on its *positioning* in the face of global challenges, both in science and politics. The construction of political ecology as a disciplinary field[2] of power relations in knowledge (Foucault 1980) requires thinking critically about the supremacy of the Anglophone school—and more generally about the perspectives from the North—in the field of political ecology: its hegemonic control over knowledge and the way it has shaped and weaved its knowledge into the structure of this emerging social science—namely, its frames of intelligibility, research agenda, strategies for social activism, and impact on public policies—over and above the interpretations that emerge in other geographical latitudes and other economic, social, and cultural contexts.

There are currently different "regional political ecologies" across the planet. They erupt from the depths of the environmental crisis and are rooted in ontologies of diversity and difference, counteracting attempts to confine them in a unifying paradigm or seal them in the hegemonic process of globalization. The call to think about the geography of power in the field of political ecology is an invitation to engage in a conversation that instigates the main forces of discursive strategies to deconstruct the logic that lead the world toward socio-environmental degradation,

[2] I adopt the concept of *field*, both in the epistemic field of political ecology and in the more general field of socio-environmental struggles and conflicts, as well as in relation to the behaviors and practices of social actors as expressed by Pierre Bourdieu (Bourdieu and Wacquant 1992).

fostering innovative theoretical capacities, stimulating sociological imagination and en-living social imaginaries to lead the way toward the construction of a sustainable world that embraces diverse, different, and "other" possible worlds (Leff 2014a).

Political ecology assigns itself the task of responding to the call of thinking and conceptualizing "the geographies of knowledge production in political ecology" by mapping the "geographical imaginary of political ecology." But more than being a call to rearrange the puzzle of environmental disciplines in the world of sciences, it summons sociological imagination to rethink the world from the perspective of the politics of knowledge in which political ecology is inscribed and, in this way, it can facilitate the dialogue between different modes of knowledge-*savoirs*, providing consistency to the thinking that can steer the construction of a sustainable world, based on its theoretical, geographical and cultural diversity.

Here, I will limit my argument to problematize and radicalize some of the concepts and primary lines of inquiry that collide in the field of political ecology. I intend to open a dialogue and to contextualize some of the fundamental principles and proposals of political ecology practiced in Latin America and contrast them with the contributions of Anglophone political ecology. In doing so, beyond exploring and displaying the political socio-geography of environmental conflicts, I intend to question the epistemic core of this discipline and stimulate a more cosmopolitan and critical reflection to strengthen its theoretical consistency and strategic effectiveness and, thus, enable this social science to confront the hegemonic powers that are leading the world toward socio-environmental degradation.

2. Origins and Foundations of Political Ecology

Political ecology emerged as a new discipline within the social sciences between the 1960s and 1970s, driven by the eruption of the environmental crisis. It launched an inquiry on the socio-environmental conflicts generated by the capitalist appropriation of Nature, strongly influenced by a Marxism that was very much in vogue during these times of theoretical and cultural revolutions, and by the work of pioneering authors such as Murray Bookchin (1962), Eric Wolf (1972a), Hans Magnus Enzensberger (1974), and André Gorz (1975).

The field of political ecology evolved from critiques of the paradigms in geography and anthropology and spread to new disciplines such as cultural ecology and ethno-ecological and ethnic-geographic studies, influencing political economy, and merging with post-development studies, eco-Marxism, social ecology, and eco-feminism. It engaged in dialogue with other theories of complexity that have given rise to new ontological and epistemological lines of inquiry on Nature. Political ecology is thus forged from the melting pot of post-structural, deconstructionist, post-constructivist, post-colonial and anti-essentialist studies (Escobar 2010). The intention here is not to synthesize nor produce a comprehensive map of the discipline's genealogy, sources, interdisciplinary odysseys, and mainstays of its relations with other disciplines, nor will this chapter address its application to different socio-environmental issues and case studies. The aim is merely to think about the boundaries and cross-relations of diverse regional political ecologies in different territories around the world.

In an analysis of the development of the Anglophone field of political ecology, Peet and Watts (1993) identified its origins in the politicization of the earth sciences and agricultural culture and practices. Political ecology was forged from an ecological critique of economic rationality (Gorz 1989) and the unraveling of the second contradiction of capital (O'Connor 1998), in the critical margins of ecological economics (Martínez-Alier 1995a). It came into being together with other strands of thought and socio-environmental activism such as social ecology, eco-feminism, and eco-Marxism. Another fertile field opened with the demarcation of areas of study more closely associated with disciplines whose objects are clearly defined in a direct relationship between social practices and Nature. Thus, it detaches itself from an ecologist-evolutionist-adaptive view of human geography, cultural ecology, ethnobiology, sociobiology, and human ecology, to consider the power relations that have resulted in theoretical, technical and discursive devices that steer and define the means of human intervention in Nature.

One of the founding principles of political ecology is the critique of the theoretical effect that led to the emergence of the *ecological episteme* and *post-structural thought* in the social sciences. This colonizing effect of the life sciences reifies social relations and is expressed both in functionalist sociology and in complex thought when cybernetic models and generalized ecological frameworks are adopted as methods for the interdisciplinary integration of the sciences (Morin 1980). Similarly, the sciences of complexity seek to operate as a trans-disciplinary framework, spreading into and over the social sciences (Prigogine 1955, 1981; Prigogine and Nicolis 1977; Prigogine and Stengers 1984; Urry 2003, 2005). By contrast, economics applies its own instrumental devices to merchandise Nature and regulate the environment, disregarding the ontology of life, Nature, and culture.

From its inception, the Anglophone school of political ecology emerged from a critique of adaptive theories that originated in cultural ecology, upheld by authors such as Julian Steward (1972), Roy Rappaport (1968, 1971) and Peter Vayda (1969, 1983), in conjunction with other ecological paradigms—i.e., Bateson's ecology of mind (1972), Holling's ecological systems (1973) and Wilson's bio-sociology (1975)—that intended to found a biological, ecological and organic framework for understanding society. New functionalist paradigms of sociology and anthropology were constructed in the scheme above, assigning an adaptive meaning to the social order, whose effects were expressed in the ecological ontology that underlie the understanding of environmental problems, that oriented the social responses to the environmental question and shaped the biological bias of the nascent discipline, veiling its political Nature.

However, the problems of colonization of knowledge are not confined to masking the symbolic organization of cultures or power relations, obscured by the naturalization effects of biological theories in ecological anthropology, social ecology, and cultural geography. They also permeate all interventions—from anthropology to the social sciences—in the life-worlds of traditional societies and in their attempts to understand their cultural organization through scientific concepts. This situation requires political ecology to adopt a strategy encompassing knowledge deconstruction, epistemological vigilance and the ethics of otherness to understand, to intervene and to open a dialogue with the different cultural contexts of socio-environmental processes in which it is involved for academic and political purposes.

3. The Regionalization of Knowledge in Political Ecology: *paterfamilias*

By applying the principles of critical deconstruction of scientific paradigms to itself, to reflect on the regionalization of political ecology's theories and practices, how are the theoretical difference, the demarcation of the discipline's boundaries and the central core of its discursive formation conceived? If one of political ecology's study objects is the different modes of deterritorialization produced by capital, then this must be founded on the political geography of knowledge, that is, on the reasons why ideas, concepts, and theories flow, acquire meaning, interconnect, and become territorialized as power strategies, devices, and practices. Political ecology is regionalized from different theoretical perspectives, social meanings, and ecological conditions: from the frameworks of the disciplines that converge in this field, to the diverse ways in which the environmental crisis and socio-environmental conflicts are expressed and the social responses in different regions of the planet. They diverge in the manner they are perceived by the different social imaginaries that motivate and steer the actions of social agents, be they local populations or academics. Sensibilities, reasons, and practices that guide the theoretical interests of the *Homo academicus* (Bourdieu 1984) lead to the formation of epistemic communities that promote the thinking–acting alliances made by the intellectual actors of political ecology.

This epistemological reflection on the field of political ecology opens up a new research program on the environmental sociology of knowledge (Leff 2014a). Political ecology emerged from the flow of reasons and motivations that influenced the founding authors of this field. It is the result of trajectories and theoretical interests associated with their academic development in disciplines such as anthropology, geography, Marxism, and political economy. The discipline's academic lines, affinities, and alliances formed the basis for constructing a theoretical framework on conceptual principles that allowed for the definition of a new disciplinary field. However, before attempting to assign the approaches inherent to political ecology to specific *epistemes*, paradigms or theoretical frameworks, it is important to note its emergence from the "disciplinary schools" and "academic niches" that developed research programs based on particular theoretical legacies, patriarchies and knowledge lineages.

Marxism and its different strands shaped the lines of argument adopted in eco-Marxism, eco-anarchism, social ecology and radical ecology that molded a political ecology in the margins of economics and political economy (Zimmerman 1994). Anglo-American authors acknowledge that the origins of political ecology lie in Carl Sauer's geography school and Julian Steward's cultural ecology school (Watts 2015).

Along with a constellation of factors derived from the institutional geography of knowledge, it is possible to identify certain subjective conditions that have influenced the development of political ecology and its inscription within the Anglophone academy. One question often evaded by the sociology of science, but of particular importance, relates to the personal and academic interests and subjective sensibilities that, while leading some authors to be echoed by others and become theoretical references, makes some academics ignore others despite their intellectual affinities and convergence of objectives. Indeed, at play here is the recognition or not of academic peers, the attractiveness of some theories and disciplinary engagements that lead researchers to explore and accept some sources of inspiration and reject and disregard others when establishing their academic identity.

This may explain the lack of dialogue between political ecology and neighboring fields that emerged at the same time, driven by similar interests, such as environmental sociology[3]. Also, its distant relations with eco-Marxism and ecological economics; as well as a lack of dialogue and a more consistent theoretical interaction with the sources of philosophi-

[3] The dialogue between political ecology and sociology of knowledge is essential for establishing differences, convergences, and affinities in the approach to the objects of study, between causal or comprehensive, realist and constructivist frames, within the different schemes of intelligibility of the social processes that drive the dynamics, transformations, and degradation of Nature. An example is the study of these processes as social rationalities, as in Max Weber's (1922) comprehensive sociology that "reveals the development of a given phenomenon based on a conflictive behavioral game between actors driven by different rationalities [...] from the global to the individual level and from causal laws to the logic of action" (Berthelot 1998, p. 33). The political ecology approach emerges precisely from this framework founded on environmental rationality (Leff 2004). The constructivist debate in the field of political ecology cannot be solved as an epistemological attempt to discover how far and by what means Nature is socially constructed. Constructivism becomes political by the fact that construction/ deconstruction of Nature is the result of the conflicts between social imaginaries, interests, beliefs, and practices invested in Nature, the environment and sustainability; by discursive strategies and the confrontation between capitalist and environmental rationalities in the appropriation of Nature (Leff 2014a).

cal inspiration, from which political ecology adopts its conceptual references. The distance of Anglophone political ecology from authors such as Murray Bookchin, Barry Commoner, and Michael Zimmerman is symptomatic, considering the close associations between this discipline and Bookchin's eco-anarchism, social ecology, and communitarianism (Bookchin 1971/1990, 1982/1991, 1990a) as well as with Commoner's critical analysis of "the poverty of power" (Commoner 1971, 1976), the essential conversation with radical ecology,[4] Heidegger's existential ontology (1927, 1957, 1971) and postmodern thinking (Zimmerman 1994). This lack of dialogue is indicative of the way theoretical preferences guide our research, define our views and leave their mark on the theoretical frameworks, conceptual approaches and lines of thought that inform the objects of study and design the discursive strategies of the emerging political ecologies.

Anglophone political ecology has developed inside an academic cloister, insufficiently open to intercultural dialogue between regions and different types of knowledge or to the protagonists of political ecology. Though one of the core themes of political ecology is the unequal access of different populations to resources, asymmetry in the access to the means of disseminating ideas is also evident in this academic field. The result is that the discipline is heavily weighted toward Anglo-Saxon authors who control and have privileged access to publications and the dissemination of theories at a global level. Undoubtedly, language

[4] The distinction between political ecology, deep ecology and radical ecology is not only rooted in their different epistemological perspectives of the analysis of social conflicts and power strategies that occur in the fields of ecological distribution, social inequality, and sustainability, but also in political ecology's ontological and ethical commitment to resolve these conflicts. While social ecology and eco-feminism seek emancipation by suppressing the dualities that generate and underpin oppression and by freeing up the potential subdued by the dualisms of modern patriarchal and social power structures, political ecology is deeply rooted in the pre-ontological difference between the Real and the Symbolic and in the sexual difference expressed in the hierarchical dualisms and socio-environmental conflicts. Political ecology sees emancipation not as the elimination of ontological differences, but as the construction of environmental rationality that embraces these differences. Emancipation is not transcendence via ontological dialectics, subjective intentionality or the reflexive restoration of modernity. It is the reidentification, the repositioning of being-in-the-world. The construction of a sustainable world does not transcend sexual division, nor does it dissolve ontological differences, that is, the difference between Being and beings, between the Real and the Symbolic. It requires a reidentification within the environmental complexity of the actual state of the world, deconstructing the dominant rationality and constructing new environmental rationality (Leff 2014a, 2018).

barriers play a role in constraining the flow of communication and a more fruitful exchange of ideas between the different regional cultures of political ecology. It may be the case that Latin American thought and academy have been more affected, especially Latin American political ecology, where authors are less inclined to publish in English than their Asian, African and Australasian counterparts who are more likely to be educated in English-speaking universities and be affiliated to that intellectual culture.

However, the resonance of ideas and the control of the means of dissemination are not only influenced by language barriers. Personal disposition, intellectual cultures, and theoretical affinities also determine empathy and alignment to different trends of thought which can either stimulate or hinder academic exchange and dialogue of knowledge that foster the recognition or rejection of different modes of understanding produced in the social environments where ideas are debated, and concepts are applied to the construction of alternative worlds. These circumstances can either lead to a paradigmatic closure or open the interdisciplinary and intercultural dialogue between diverse epistemic communities and social actors.

The Politicization of Ecology and the Epistemic Regionalism of Political Ecology

Before starting an inter-regional dialogue, it is crucial to question anew the principles that found political ecology: In which way is ecology political? (Walker 2005), Can political ecology claim any general principles to construct knowledge within its field? Or, What is the "regional" and "localized" character of political ecology? We must acknowledge that ecology is in no way political if understood as a web of relations between non-human populations and their environment, such as the complex flows in matter, energy, and information that occur in the metabolism and organization of the biosphere—in depredation relations, trophic chains, and the ecological dynamics—not induced by human action. It becomes political through the effects of human intervention on ecological transformations when they cease to be driven exclusively by natural laws. Of course, ecology, understood as the web of life transformed by human action, is a field of co-incidence of heterogeneous, different, and complex orders of Nature, ontological regimes, and cultural meanings. This is where the effects of the human intervention are manifested, based on different existential ontologies and social rationalities that mobilize the metabolism of the biosphere through ecological flows and thermodynamic pathways of matter and energy resulting in environmental changes.

Ecology becomes political as a result of the *will to power* to which Nature is subjected, that is, the processes of appropriation steered by different, and often conflicting, interests and values. Indeed, it becomes political due to the way that human intervention, inscribed in a range of meanings and rationalities, leaves an ecological imprint on the environmental conditions of society through the pressure it exerts on the transformation of Nature. Thus, strategies for appropriating Nature mobilized by different rationalities and imaginaries in the different ecological contexts, whether capitalist or traditional, create politicized ecological processes that are the result of power strategies. Politics is the means by which the ontology of the Real is realized as the living conditions of the people, and by which is it possible to move from a global world, governed by the unifying power of the market, to constructing diverse life-world—different modes of being-in-the-world and of inhabiting the planet—steered by an ontology of diversity, difference, and otherness. This shift in the trends of the global world towards the sustainability of life requires the deconstruction of the rationality based on the metaphysical principles of unity, universality, totality, and generality. This path toward politicization of life is rooted on an ontology of diversity and difference: it unleashes the Real potentialities, steered by different existential meanings and mobilized by the rights of cultural beings to construct their sustainable life-worlds.

Robbins (2012) sees Kropotkin and Marsh as precursors of political ecology, given their interest in the impacts of human activities on soil degradation.[5] However, ¿what is the source of this politicization? If it

[5] Some authors consider Kropotkin (1902/2005) as a precursor of political ecology because he argued, against social Darwinism, that rather than competence or skills, cooperation and mutual aid are essential for evolution and are the basis of survival.

were the mere result of a universal law of matter degradation and life, there would be nothing essentially political about it, except for the conflicting effects on humanity and reactions of different social groups of the accelerating trends of the destruction of the planet, the deterritorialization and subjugation of peoples' livelihoods, and the defense of their territories. Following this line of argument, Claude Lévi-Strauss believed that no cultural regime could resist the ultimate principle of the law of entropy and that anthropology would eventually become an *entropology* (Lévi-Strauss 1955). However, we must search the political roots of ecology in its origins outside Nature. Ecology becomes political due to the effects of the history of metaphysics, that is, the effects of the ontological obsession on Being on the oblivion of life—the original intuition of Heraclitus's *physis* as the emergent potential of all entities and of life itself—because of the abstract principles of unity and universality in philosophical and scientific thought that have led to the techno-economic unification of the globalized world (Leff 2018). It has become political because of the lack of knowledge about the ontology of life's diversity and difference, because of the will to power (Nietzsche 1968) that acts as a compulsion to dominate Nature, and violence forged in metaphysical thinking (Derrida 1978); because of the supremacy of identity as sameness of being and the disregard for otherness (Levinas 1969).

The political denaturalizes ecology, given that the will to power that drives the process of social appropriation of Nature was forged originally from outside Nature, by the human *logos* that launched metaphysical thinking and through *ratio* that erected the objective measure of all things, and by the representation of the objectified world by the Cartesian *cogito*, that enlightened the economic, technological, and juridical rationality of modernity and its power devices. In the deployment of metaphysical thinking, *physis*, as the emergent potency that orders matter and life, derives in an objectified and measurable Nature, which with modern rationality is converted into a "scientific object" and "natural resources," disposed to economic appropriation and planning. The unconscious "*lack in being*" (Lacan) mobilizes the will to power to dominate Nature; that engendered the *logocentrism of science* and the theoretical paradigms of political economy, as well as those of the life and human sciences that form the basis of the hegemonic interpretation and appropriation of the world.

Political ecology emerged by dissociating itself from two dominant theoretical paradigms of modernity: 1) economic theory as the means of appropriating and transforming Nature that triggers entropic (ecological and environmental) degradation; 2) biological theory—in particular evolutionary Darwinism—which when applied to the social fields—cultural ecology, sociobiology or structural-functionalism—becomes the naturalizing standard of socio-environmental dynamics. Political ecology displaces the dominant paradigm of political economy; it is situated on its margins and focuses on the ecological distribution conflicts that cannot be absorbed, adjudicated or resolved by ecological economics (Martínez-Alier 1995a). In contrast to evolutionist theories, political ecology emerges by deconstructing the naturalizing effects of social Darwinism and the ecological framing of the ethnological order and anthropological practices (Watts 2015).

Within these limits, political ecology has sought to characterize and establish its identity vis-à-vis other neighboring scientific disciplines. This endeavor transcends the definition of a new discipline whose aim is to unravel and circumscribe the political character of ecology and study the political processes that leave their marks on Nature. Political ecology places ecology within the perspective of a political epistemology by deconstructing the theories that have disregarded the environment, by unraveling the power–knowledge strategies behind the human actions that mobilize environmental changes that can neither be attributed to Nature nor inscribed within the "naturalness" of the actual social order.

To be sure, the impacts of rural modernization—deforestation and soil degradation, the exclusion, and poverty of peasant populations—have had a political effect on the resistance of indigenous and peasant peoples, in their struggles for land and the defense of their traditional practices and modes of being. However, the exacerbation of these processes, reflected in the destruction of Nature, has also driven the emergence of other political, epistemological, and ontological factors, denaturalizing the scientific paradigms which have concealed the social effects and power relations that determine and condition socio-environmental degradation. These factors have also promoted processes of epistemic, political, and environmental reconstruction that, through social resistance, are engendering new ways of conceiving the process of production and the habitability of the world more in line with conditions of life in the biosphere.

The key to deconstructing the dominant theories of the environmental field—and political ecology—is the interpretation of the concept of *environment*. While in economic and development theories the environment is conceived as an externality and a cost, or as that which surrounds an organism or a cultural organization in biological and ethnological theories, in our *environmental epistemology* the environment is conceived as exteriority to the normal paradigms of science. However, more than merely being a space where externalities are reflected, the environment is the "other" of the *logocentrism* of sciences: the unknown Real, the subjugated knowledge, and the "no knowledge," that accompanies socio-environmental reconstruction. The concept of *environment* appears as a *productive potential* that opens up possibilities for the construction of other possible sustainable worlds. Thus, it becomes possible to conceive *other sources and means of production* and *other life-worlds* based on the ecological productivity and cultural creativity of the peoples of the Earth (Leff 1986, 1995, 2004, 2018).

This conceptual demarcation of the *environment* allows us to understand the "regional" character of political ecology. Blaikie and Brookfield define their research perspective as a "regional political ecology" whose aim is to analyze "different geographical and hierarchical scales of socio-economic organization [...] environmental variability, spatial variations in resilience and sensitivity of the earth, where different demands for land take place at different times" (Blaikie and Brookfield 1987, p. 17). In this way, regional factors influence different geographical scales and ecosystems resilience that determine the levels of degradation according to "demands" for land use. Thus, political ecology emerges from the impacts of the processes of techno-economic appropriation of various regions and the resistance of social groups affected by them.

The type of political ecology that results from the critical concept of environment brings into play its exteriority and potential; namely, the knowledge of the peoples and the ecological potential of their lands for the reconstruction of their life territories. Geography and anthropology, from which Anglophone political ecology is derived, acquire an ontological purpose and become politically more "active" as they move from social resistance to the processes of degradation, toward the reconstruction of sustainable eco-cultural territories. In the environmental rationality conception, political ecology does not only relate to phenomena such as the economic asymmetries of the globalized world and the unequal distribution of economic benefits and environmental costs. Ecological distribution plays a decisive role in the regions of the South when the *negentropic productive potential* (Leff 2019) of tropical ecosystems and the cultural creativity of their people enact the productive forces of their life-territories. Thus, political ecology can dramatically alter the reactive nature of the bio-power strategies inscribed in economic, ecological, and anthropological theories[6] moreover, unleash new potentialities of interpreting, appropriating, and constructing sustainable life-worlds.

The potential for thinking and putting into practice this productive ecological rationality is far more significant in the tropical ecosystems and ethnic territories of the South which have the most considerable ecological productivity on Earth and where creative cultural diversity is kept alive through the worldviews and traditional practices embedded in the ethno-ecological co-evolution of the bio-cultural heritage of its peoples (Boege 2008, 2018). Deconstructing the capitalist rationality based on the ontological (geographical, ecological and cultural) foundations of the environmental rationality, and the epistemological perspectives that consequently emerge are one of the main criteria for distinguishing Northern from Southern political ecology from a Latin American environmental perspective (Leff 2012).

In Latin American thought, political ecology is not only a field of research and social practices associated with socio-environmental conflicts and the unequal distribution of costs and benefits within the context of global change. Socio-environmental struggles lead to new ways of constructing diverse culturally and ecologically sustainable societies. The deconstruction of the ecological principles that have colonized the social

[6] Michael Watts argues that Foucault's theory of bio-power questions hegemonic concepts such as adaptation, security, risk management and resilience that organize life in accordance with the dominant neoliberal governance regime (Watts 2015).

sciences as a new *post-structuralist episteme*—the critique of the ecological paradigms of the cultural order and agro-productive practices, based on the principle of biological adaptionism (Watts 2008)—is one of the main lines that distinguish the development of Anglophone political ecology. By contrast, environmental rationality goes further and introduces new theoretical and social action strategies associated to the culture–Nature relationship based on the ontology of life and the existential ontologies of the people (Leff 2003, 2014a, 2018).

The environmental crisis has led us to question the "exceptionalism" of the social sciences (Dunlap and Catton 1979, 1994) in isolation from Nature's determinations and conditions on social actions. Thus, new environmental sociology is born, inscribed within the emergent paradigms of the complex sciences that have driven the anti-essentialist and post-structural movement, hermeneutics and the deconstruction of the epistemological crisis of representation, opening the way to constructivism and postmodern thinking (Leff 2011).

Today, the emancipation of the original and indigenous peoples, peasants and Afro-descents has resulted in the demand for rights to their ancestral territories and biocultural heritage. In face of the rational choice principles whose consequences are the so-called "tragedy of the commons" (Hardin 1968), the privatization of territories rich in biodiversity and the commodification of Nature (including environmental goods and services), the recognition of "customs in common" of "moral economies" (Thompson 1971, 1991), "common property rights" (Orstrom 1990) and "collective rights to the common goods of humanity" (Leff 2014b) are mobilizing the peoples of the Earth for the reconstitution of their life-territories. The environmental rationality questions the legitimacy of the paradigms that underpin dominant economic and juridical rationalities. It recognizes "other" knowledge and wisdom of various forms, as well as the effectiveness and significance of the traditional practices of peoples, the right to reinvent their cultural identities, build new production practices and renew their life-worlds in confrontation, resistance, and *rexistence* with/ against modernity.

5. The Epistemological Emancipation and Rights of Existence of the Peoples of the Earth

Indigenous peoples are claiming that their struggles are "political" and "epistemic." They argue that environmental problems were epistemologically caused by the hegemonic ways of understanding the world and, above all, the imposition of the scientific and economic rationality that objectified the world and colonized, marginalized, and exterminated the knowledge and practices of other peoples. The result is the suppression of alternative ways of conceiving life and the closure of pathways for constructing other forms of inhabiting the world. They also argue that emancipation is not possible without the deconstruction of the dominant forms of knowledge and the reconstruction, legitimation, and establishment of other ways of understanding life and alternative modes of being-in-the-world.

Ecology becomes political through the power relations that alter the ecological dynamics derived from the emergency and *generativity of physis*, driving it towards the techno-economic transformation of Nature. The political root of this historical process is both ontological and epistemological. It is ontological because it arises from the disjunction between the Real and the Symbolic, of the difference between Being and entities, that geared by metaphysical thinking, resulted in the separation between Nature and Culture in modern rationality, building up a techno-economic regime that exacerbates the entropic degradation of Nature. It is epistemological because power is also inherent in the forms of dominant global knowledge based on the *logocentrism* and the "normal" paradigms of science. It is forged in the Cartesian *cogito* and the *a priori* of reason that steers the production of knowledge that intervenes life whirls the world and degrades Nature. The potentialities of the Real are appropriated by technologies and rationalities that encounter the immanence of life, undermining the ecological potential and colonizing cultural organizations. Thus, modern epistemology frames the paradigms of knowledge that configure the ontological regimes that infringe people's life-worlds and their existential ontologies.

The order of rationality developed in the course of the history of metaphysics conflicts with the original ontology of life: the ontology of diversity, difference, and complexity, is denied by the rationality of the one, of unity, of identity and metaphysical universality. It is expressed in the oppression, subjugation, discrimination, marginalization, and exclusion of other modes of being-in-the-world. Thus, processes for resisting the invasion of modernity in traditional life-worlds are transformed into *rex-istence* movements. They are the expression of other existential ontologies manifested in the political arena. The restoration of the ontology of life demands reflexive thinking. This is not merely a theoretical shift. It implies the cultural *en-living and reappropriation of Nature* that occurs in the strategic field of political ecology. Socio-environmental struggles are the expression of the *will to the power of living* that, based on social imaginaries of sustainability, claim the rights of cultural beings to their autonomic modes of existence and their repositioning in the world in accordance to the principles and conditions of life (Leff 2010, 2014a, 2018).

Environmental justice is expressed in peoples' rights of being, to reconstruct their life-worlds and to plan their possible futures, over and above the internalization of ecological costs and the distribution of Nature's benefits and potentials. In this perspective, an essential ontological and anthropological question re-emerges: how to interpret the real conditions for human life in the living planet we inhabit? Radical socio-environmental movements resist the economic-ecological colonization of modernity. They are not willing to merely accept better ecological distribution, greater access to resources and a fairer distribution of the benefits of the capitalization of Nature. In line with their existential ontologies, they demand not only the right to survive but to "live well"—as expressed in the *Sumak Kawsai* and *Suma Qamaña* of the Quechua and Aymara Andean peoples—within the "meaning" of their cosmogonies, identities, and cultural imaginaries (Huanacuni 2010). These emancipatory processes question the domination of capital over their life-worlds and the capacity of the globalized economy to manage the sustainability of the planet. Here a fundamental question emerges: how to conceive the reconstruction of the world based on the cultural imaginaries of sustainability of the peoples of the Earth (Leff 2010).

Strands and Slopes in the Regional Geography of Political Ecology

The political ecology in Latin America is situated within this perspective. To be sure, both the Anglophone and the Latin American conception depart from the same understanding of the roots of political ecology, namely, the power relations that structure this field. They also drink from the same theoretical traditions, from Marx and Kropotkin to Thompson, Harvey, Deleuze, and Foucault. They coincide on the challenges in deconstructing the power devices and networks to which the world is subjected that degrade the biosphere, contaminate the environment, endanger ecological sustainability and hinder cultural diversity. These phenomena lead to deforestation, soil erosion, and the destruction of biodiversity in the planet, an increase in dispossession, deterritorialization of Third World peoples, as well as the exacerbation of poverty and social inequality worldwide. In general, political ecology research and case studies take place in the poor and developing countries in the South (Watts 1983/2008; Blaikie and Brookfield 1987; Robbins 2012). Studies by Raymond Bryant (1992), among other authors, have focused on the Third World, while political ecology has sought to distinguish between environmental movements from the North and the South (Redclift 1987; Guha and Martínez-Alier 1997). In contrast to the motivations of the actors from the North, Southern political ecology academics focus on the processes that affect socio-environmental conditions and the movements that resist, defend and reconstruct their livelihoods and life-worlds.

However, beyond political ecology's interest in the asymmetries between North and South, the socio-environmental impacts caused by the hegemonic power of globalization on the territories of the South, and despite the diversity of environmentalism, theoretical sources, disciplinary origins, and schools of thought that feed and form political ecology's diverse frameworks and programs, a fundamental question remains: Is there a clear regional division in the theoretical field of political ecology? Is it possible to classify theories, concepts, and methods to establish regional typologies of environmental thought and delimit their theoretical frameworks and practical strategies? Since its inception, Anglophone political ecology has been concerned with the effects of environmental changes on political relations.[7] However, it is just as important—or even more—to analyze the effects of power in knowledge on environmental change.

In each region discursive strategies are formed according to theoretical choices and the problematic that guide the different schools of thought and frameworks for analysis that orient social action. The Anglophone slope of political ecology has formed by deconstructing the biological, ecological, and "adaptation approaches" to the socio-environmental dynamics and their sources in ecological anthropology, cultural geography, and agrarian studies. By contrast, in Latin American political ecology, a particular conception derives from the emancipation of traditional *savoirs* from the coloniality of knowledge and from the more direct connection in which the peoples subject to the effects of power interpret and resist domineering knowledge. It emerges from its proximity to the processes of ecological degradation and environmental conflicts, from belonging to this same history, culture, and landscape; it springs from a disposition to open up a dialogue of knowledge with indigenous *savoir* and wisdom, to listen to the voices of the peoples that express the "cry of the earth" and the "voice of Pachamama," in order to weave the discursive web of political ecology (Boff 1996b; Rivera Cusicanqui et al. 2016).

While Anglophone political ecology considers the socio-environmental transformations of the Third World as critical topics and privileged objects of study, the political ecology of the South is embedded within the historical and existential roots of the disruption of the life-territories of traditional peoples' and in their emancipation processes. If the former manifests a discursive purpose of deconstructing the dominant rationality and conducting case studies on the expressions and manifestations of power in the socio-environmental processes of the South, Third World political ecology is a discursive amalgam between academic and political

[7] "By encompassing both the historical and the contemporary dynamics of conflict, this area of investigation illustrates how those 'with no' power fight to protect the environmental foundations of their ways of life. The latter part of this investigation considers the political ramifications of environmental change. Addressing questions related to the impact of socioeconomic and political processes, this element of the reference framework focuses on an area of human interaction with the environment that is often forgotten: the significant effects of environmental change in socioeconomic and political relations" (Bryant 1992, p. 14).

actors, between theoretical thought, participatory research, and the radical imaginaries of the peoples in alliance with resistance movements and their strategies for *rexistence* (Guha 1989; Guha and Gagdil 1992; Arnold and Guha 1995; Shiva 1988; Escobar et al. 1998; Esteva and Prakash 1998; Acosta 2010; Quintero 2014). Third World political ecology goes beyond the study and analysis of socio-environmental processes and conflicts or the sociology of resistance movements. It is historically, theoretically, and politically committed to the construction of a sustainable future and of other possible worlds.

Thus, political ecology acquires a *strategic and prospective purpose.* It ceases to be just a new discipline or a new epistemic field. The transition towards sustainability comprises a process that deconstructs the rationality of modernity and constructs environmental rationality that integrates cultural diversity with the evolving diversity of life and the environmental complexity of the world. Political ecology fosters the dialogue between different types of knowledge-*savoirs* (understood as the encountering of different cultural beings formed by their knowledge-*savoirs*), where different ways of being-in-the-world confront each other or make alliances in the reintegration of Nature and culture in the immanence of life. The political ecology of the South is built within a discursive web that facilitates communication between different modes of being-in-the-world—between theoretical frames of understanding, social imaginaries, and ways of living—of social actors. It is the confluence of ontologies and rationalities, of diverse ways of reconstructing worlds and appropriating Nature. It is the reinvention of identities, and the different modes of people's *rexistence* hand-in-hand with nature's *reliving.* In this context, political ecology moves from the conflicts over the social appropriation of Nature to the existential questions of the rights of existence—of diverse modes of being-in-the-world—in the more radical field of *political ontology* (Blaser 2009; Escobar 2013; Leff 2018).

Epistemic communities have developed their identities from different sources, from diverse strands and slopes of thinking that converge in the field of political ecology. Beyond the intention of establishing a universal paradigm and a new branch of science, political ecology is a way of understanding the socio-environmental movement, marked by the diversity of geographical and cultural contexts within which they are manifested.

This does not mean that differences in intelligibility and approach to socio-environmental processes have developed only as a reflection of geographical and cultural contexts. Diverse intellectual and academic environments, in different historical, political and cultural contexts, have played a significant role in the formation of perspectives of analysis, theoretical approaches and political strategies in Latin America.

Eric Wolf and Angel Palerm's research on the ecological potential and agricultural systems of the Mesoamerican peoples (Wolf and Palerm 1972), John Murra's analysis of the organization of geographical space and the ecological floors of the original peoples of Tawantinsuyu (Murra 1956), the *Geography of Hunger* by Josue de Castro (1975) and the *indigenato* studies by Darcy Ribeiro (1973) opened new study areas for anthropological, ethnological, geographical, and agricultural analyses associated to the knowledge and practices of traditional cultures in the territories of the South. Dependency and internal colonialism theories (González Casanova 1965; Stavenhagen 1965), the eco-theology of liberation (Boff 1996a), decolonization and the ethics of liberation (Dussel 1998; Quijano 2008; Lander 2000; Mignolo 2000, 2011; Mignolo and Escobar 2009), as well as other theories and agro-ecological practices are all the result of a demand for *knowledge from the South* (Sousa Santos 2008) and represent the emergence of *Latin American environmental thought* (Leff 2012). From the perspective of the ecology of difference and territoriality conflicts, they print their mark on the political ecology of Latin America (Leff 2003, 2014a).

Thus, the construction of the field of political ecology is open to diverse approaches and experiences of socio-environmental conflicts. It is open to "Southern environmentalism" and to "other knowledge." More than just the construction of an interdisciplinary paradigm, an academic exchange between North and South and an intercultural dialogue of knowledge/*savoirs*, political ecology weaves its comprehension frameworks and research practices in a *dialogue between different types and codes of the radical imaginaries and social struggles* peasant, and Afro-descendant populations, as well as the discursive practices and the political strategies of the protagonists and actors of political ecology in their struggles for the reappropriation of their bio-cultural heritage and the construction of new life-territories.

Regional political ecologies are more than just a mosaic or a map of different power relations occurring in different socio-ecological contexts. They are more than forms of analysis, negotiation tactics, and strategies for resolving environmental conflicts. Political ecology is where rationalities, logics of meaning, and political practices meet, where the will to power surfaces, expressing different, and often opposing, meanings. It is a space for struggling against and resisting the territorialization of the hegemonic global order and the geopolitics of "sustainable development" that invades and blocks alternative strategies for building sustainable worlds. This hegemonic sustainable development produces poverty, breaking the fragile equilibrium and the resilience of ecosystems as they seek to exterminate the traditional practices for living in harmony with Nature, invented for centuries and millennia by the peoples of the earth. The accumulation and expansion of capital lead to processes of deterritorialization, imposing a "green economy" rationale in the form of the production of agro-fuels, forest plantations, genetically modified crops, mineral extraction, shale gas and hydrocarbons through hydraulic "fracking" technology. It is within the field of political ecology that resistance movements to these processes and movements for the emancipation of the Peoples of the Earth unfold, reinventing collective identities and creating alternative strategies for *enowning* their bio-cultural heritage and constructing a sustainable future.

7. The Political Ecology of the North and the Winds from the South

Blaikie and Brookfield (1987) highlighted the concept of soil degradation as a fundamental topic for political ecology, following Marx's original ecological concern raised by Justus von Liebig and a generalized sign of the metabolic rift between capital and Nature.[8] However, assessing environmental degradation has become not only relative, depending on different theoretical perspectives and interests, with the ever more complex processes and variables involved in the planet's ecological degradation. The criteria to balance ecological costs and economic benefits have been superseded by the intensification of the technological intervention

[8] See Chap. 7.

on Nature. Beyond the discussion as to whether degradation is a question of subjective values or the effect of social processes, the discussion between realist or constructivist intelligibility schemes of the ecological processes and socio-environmental conflicts has shifted to the debate on the ontological condition of Nature so as to understand whether entropy is being driven by the cognitive interest of modern rationality and how economic processes and social practices enact the second law of thermodynamics, mobilizing the metabolism of the biosphere and triggering the entropic degradation of the planet. Although environmental degradation is empirically observable (in pollution of air, water, and soil; in deforestation and climate change), understanding these processes requires conceptualizing the way in which the economic process and social agency, inscribed in power relations, do not only mobilize the metabolism of the biosphere but trigger processes of environmental degradation; and how the effects of techno-economic rationality can be reconciled with the laws governing Nature.

Nicholas Georgescu-Roegen (1971) is here an essential reference for political ecology. He demonstrated how economic process triggers entropic degradation of the planet as the economic appropriation of Nature enacts the second law of thermodynamics. The consequences for political ecology, political epistemology, and environmental sociology are complex, as the relationship between economic processes as a social agency and the law of entropy is not simple, nor transparent or direct. Understanding how these entropic processes operate in the metabolism of the biosphere is fundamental to environmentalism's critique of the destructive effects of the dominant economic rationality and to open the way to constructing negentropic pathways to sustainability. Entropy used to be conceived as the limiting law of Nature. Today, a new theoretical venue for the construction of sustainability emerges through the intervention of different rationalities that enact the dialectics between the entropic and negentropic processes in the metabolism of the biosphere. This construction is both epistemological and political: it requires imagining other ways of understanding life, establishing sustainable means of production, instituting low-entropy practices and building a negentropic social order based on the thermodynamic and ecological conditions of life on the planet (Schrödinger 1944; Vernadsky 1998; Prigogine 1955,

1981; Prigogine and Nicolis 1977; Prigogine and Stengers 1984; Schneider and Sagan 2005; Kauffman 2003; Leff 2018).

Political ecology emerges closely associated with the condition of peasant societies, the agrarian question, and rural studies. It is in the rural space that environmental issues become radicalized as territorial conflicts, where the dispute for the ecological means of production and the rights of appropriating Nature takes place. Though currently, these processes also encompass the oceans, the atmosphere, and cities, conflicts are more acute in rural areas, where there is a potential for constructing new negentropic means of production and modes of living and inhabiting the world, based on ecological productivity and cultural creativity. If rural areas were the stage for the agrarian revolutions of the 20th-century struggles for land, today, they are the arena for processes for reappropriating and reconstructing life-territories. In addition to the struggles for land, the management of agrarian systems, land-distribution policies for developing agrarian subsistence economies and traditional ways of life, it is in the realm of political ecology that struggles take place between capitalist displacement and deterritorialization of traditional peoples and cultural reterritorialization and reappropriation of Nature that could result in the construction of sustainable alternatives and negentropic life-territories.

Political ecology thus, revolutionizes the traditional approaches to the agrarian question (Chayanov, Kautsky), reconceiving the reconstruction of rural areas on new ontological bases. The agro-ecological struggles for cultural autonomy and territorial rights are high on the Latin American political ecology agenda. The implementation of these practices could lead to new possibilities for constructing local sustainable economies, based on ecological productivity and the cultural creativity of the peoples of the Earth articulating in a global sustainable economy founded on the negentropic potentials and the ecological organization of the biosphere (Leff 1995, 2018).

Robbins (2012) argues that *ecological distribution* is an essential concept for contesting the geographical naturalness of socio-environmental processes and incorporating ecological factors into the economic rules of wealth distribution. Political ecology reveals the most critical points that escape the multi-criteria evaluations of ecological economics. Ecological distribution addresses issues related to the unequal sharing of the environmental costs and ecological benefits—the so-called "economic externalities"—that cannot be measured in market values and appear as new entities to be internalized by economic instruments and ecological norms: that is, the valuation of environmental goods and services. Ecological distribution also leads to the emergence of social movements against ecological damage and struggles for the social appropriation of Nature. Thus, it refers to the conflicts over social strategies for survival, to sustainable production alternatives and disputes over territoriality. It addresses the risks and damages of different forms of ecological destruction and environmental contamination. Ecological distribution encompasses criteria and values that are beyond economic rationality, question the aim of reducing values to financial costs and market prices, mobilizing social actors to defend their own material and symbolic interests—identity, autonomy, territory, livelihoods, and quality of life—that transcend the economic demands for land, means of production, employment, income distribution, and development.

Ecological distribution leads to consider how colonial conquest, capital domination, and economic rationality have deterritorialized cultures, altering the ecological distribution of the planet. These are the outcomes of the capitalization, commodification, and merchandizing of Nature, serving the growing demands of capital and the geopolitics of sustainable development: altering the climate, causing deforestation of the biosphere, eroding territories of biodiversity, overexploiting water sources, and aquifers and exhausting the resources of the subsoil. Moreover, unequal exchange promotes the transfer of resources from the poor (though rich in their ecological endowment) countries of the South to the countries of the North, altering the climate and metabolism of its ecosystems and reducing the resource base and the ecological potential and of Southern territories on an unprecedented scale.[9]

[9] As an example, large extensions of biologically diverse tropical forest is being cleared for monospecific forest planting as eucalyptus or transgenic crops to supply natural commodities to the Chinese market, transferring water and cellulose to more affluent countries, draining the soils, eroding ecosystems and altering the ecosystem's metabolism, and generating climate change through meteorological dynamics that affect larger areas that extend beyond the zones directly affected and throughout the whole South American region.

Emblematic examples of the barbarism of capital in the era of "ecological" destruction are health risks, biodiversity erosion, and socio-environmental impacts caused by phenomena such as extractive mining, genetically modified crops, forest plantations, and agro-fuels (Houtart 2010). The response from the South to the ecological destruction induced by the geopolitics of sustainable development is a politics of difference, the construction of means of production and life-worlds based on the conditions of life: on ecological potentials and cultural creativity.

8. Political Ontology: The Rights of Existence of Nature and the Peoples of the Earth

In this way, political ecology confronts alternative strategies for constructing sustainability. Within the framework of post-structural theory, Peet and Watts (1993) analyzed the makeup of the field of political ecology in terms of a "dialectical collusion" and the subjugation of otherness in universal and totalitarian Eurocentric discourses, opening a discussion on the discursive power relations between hegemonic and dominated regions. This criticism not only leads to the regionalization of political ecology, but to the radicalization of a politics of difference and otherness, where different rationalities converge, resulting in a critique of the hegemonic power of discourses in the field of political ecology itself, highlighting that:

> The connections between rationality, truth, discourse and the global system of power relations led post-structural discourse theory in interesting directions […] One, is the idea that regional discursive traditions are capable of capturing even oppositional modes of thought, so that the dialectic, perhaps the main logic of critical thinking, may be exposed as Eurocentric […] the argument is that Hegelian–Marxist dialectics […] expresses a self-searching for power over that which is the "other" […] Marxism's universalizing narrative of the unfolding of a rational system of world history is seen as a negative form of the history of European imperialism and hence a conceptual system that remains collusively Eurocentric. Thus, poststructuralist–postmodern thinkers distrust "totalizing" systems of knowledge. They emphasize the singular and the contingent and seek a knowledge that

> respects the other without absorbing it [...] The second theme of particular geographic interest involves the expansion of the social production of regional discourse, through reflection on the other to a critique of discursive relations between hegemonic and dominated regions. (Peet and Watts 1993, p. 229)

Thus, Peet and Watts present a crucial problem for political ecology: the deconstruction of the universal theoretical discourses from the perspective of an irreducible otherness and the task of putting into practice a politics of difference by which conflicts are not resolved within the framework of the prevailing hegemonic rationality. This perspective introduces the possibility of understanding the root of the problems and conflicts that emerge due to a lack of understanding of the principle of otherness in the dispute between theoretical-axiological meanings in the construction of sustainability, such as the imposition of economic reason (market logic and economic mechanisms for environmental management) in environmental and sustainable development policies, and economic compensation for ecological damages. At the same time, within this point of view, political ecology is associated with the construction of a sustainable world based on the political ethics of otherness.

In this way, political ecology becomes the arena where the world's social construction occurs through the encounter of different modes of understanding the world that confront each other through discursive strategies and intertwine in a dialogue of knowledge. In these processes, the subaltern populations of the Third World develop "semiotic resistance" strategies to maintain their "control over the meaning of their lives." The symbolic resistance and strategies of the people are leading to a new turn in post-structural political ecology relating "structural theories of global transformation to the 'subjective mapping of experience'" (Escobar 1992). Thus, political ecology, "Instead of talking on behalf of subaltern populations, help them to unravel resistance discourses [...] and allow discourses to speak for themselves" (Peet and Watts 1993, p. 247).

The democratization of the environmental question marked a shift in the political ecology discourse toward the theory and praxis of *political ontology*. This new scheme radicalizes political ecology, which ceases to be a discourse *about* the other, to become a dialogue between different beings and their constitutional knowledge, between techno-economic and environmental rationality, between postmodern theory and the emancipatory struggles of the peoples of the Earth through the reappropriation and legitimation of their own *savoirs* as constitutive of their rights to existence and modes of being-in-the-world. The environmental question ceases to be a critique by political ecology of the adaptation theories in geography and cultural ecology and environmental distribution conflicts and the unequal access to resources. Political ecology is radicalized: beyond the analysis of discursive strategies of resistance and conflicts under the rules of communicative rationality, economic instruments, and legal procedures, it positions itself as the field for emancipation: for the *rexistence* of subaltern populations through strategies for re-identification and reappropriation of Nature. It does not only open a broader range of "development styles," adapting and adjusting traditional practices to the framework of an ecological economy and green economy strategies. Political ecology goes beyond the resistance strategies employed by the original peoples in the last 500 years to foil attempts to take over their territories, preserving and reinventing their identities, modes of being and traditional practices (i.e. from their *milpas* and subsistence family orchards in face of the appropriation by large colonial estates and capitalist commercial crops and genetically modified plantations, to the invention of extractive reserves, as those of the *Seringueiros* in the Brazilian Amazon region).

The shift of *political ecology* toward *political ontology* is a quantum jump in a field of forces crossed by high voltage lines tensed by different and opposing ways of constructing the world. It is mobilized by distinct processes of territorialization, of alternative ways of producing and existing within Nature, in the face of the geographical, thermodynamic, and ecological potentialities and limits of the planet, resisted and built upon human, environmental, and cultural rights of different modes of being-in-the-world. Political ontology cannot be reduced to a politics of cultural difference. It introduces the existential ontologies of the peoples intertwined in the environmental conditions of their territories. That is the reappropriation of their biocultural patrimony of cultural meanings associated with the ecological potential and geographical conditions of their territories to reconstruct their sustainable ways of living.

The right to appropriate and transform territories bring into play power strategies embedded in the development and environmental justice discourses that do not dissolve under any knowledge, methods, logic, or ethics in common for resolving human, cultural and environmental conflicts. Political ontology introduces an issue that cannot be resolved within the framework of representative democracy, nor even deliberative democracy. It is not only a question of allowing all voices to be heard. It is the practice of environmental democracy, the right to inhabit the world through different cultural rationalities and territorial conditions. It summons the construction of a world where many other worlds can fit in, in an aphorism of the Zapatista movement. It is a world-to-be, constructed in the peaceful interaction and cohabitation between different life-worlds, in a planet subjected to natural laws that establish limits but opened to the negentropic potential of the biosphere and the creativity of cultural diversity.

In this way, political ecology not only questions the economic and ecological paradigms that monopolize the understanding of socio-environmental processes and conflicts; it exerts an epistemological vigilance and imposes the ethical principle of cohabitation in otherness (Levinas 1977/1997, 1999) on power relations of knowledge (Foucault 1980) across the socio-environmental field. It not only summons us to overcome the Hegelian–Marxist dialectic framework and the idea of developing a new (ecological) dialectics of Nature (Bookchin 1971/1990, 1982/1991, 1990a) as the basis for the practice and actions of political ecology (Leff 1998a). Political ontology challenges the construction of a juridical and legal rationality capable of resolving conflicts between different "logics of meaning" (Deleuze 1969), superseding the logic of communicative action (Habermas 1984, 1987) that aims to reach consensus through rational argument, recognizing the legitimacy of distinct social imaginaries, values, and modes of understanding, but seeking to settle differences within a "basic framework" through the ability to contest conflictive interests and different existential ontologies through rational argumentation.

A few examples will reveal the challenges faced by a politics of difference when dealing with the territorial conflicts and cultural rights put forward by the environmental question, and the incapacity of communicative rationality to settle these conflicts: the dialog between the Zapatista

National Liberation Army and the Mexican government were not resolved through legal procedures imposed by the State. The conflict between the rights of ownership and reconstruction of indigenous territories and the reappropriation of their bio-cultural heritage vis-à-vis the rights of companies and the state cannot be resolved by any means of economic distribution. Finally, ecological damage cannot be solved through financial compensations. The principle of environmental justice is not only challenged by the impossibility of equating the desire for ownership of different social groups under a common logic, but also by the forced cohabitation, within a single planet, of different logics of meaning and sustainability strategies.

Political ecology is the arena for the meeting of distinct rationalities and the dialog between different knowledge-*savoirs*, of ways of comprehension of the world and being-in-the-world under the conditions of life. It should be understood as the exercise of living in an ontology of diversity, a politics of difference and ethics of otherness that transcends a dialectics of opposites and calls for the deconstruction and the reconstruction of the economic and juridical rationality order of modernity (Leff 2004).

Raymond Bryant (1992) has emphasized the focus of political ecology in the conflict for access to resources. However, territorial conflicts are not just conflicts for the rights and conditions of access to resources. They also involve struggles associated with the impacts caused by the imposition of ecologically inappropriate and culturally alien models to local peoples. They are conflicts between the different modes of appropriating and transforming Nature. Conflicts of access become conflicts of the modes of territorialization and construction of sustainability, involving accumulation by dispossession, capitalization of Nature, impacts on ecosystems, displacement and marginalization of populations, environmental costs, damages, and risks. Thus, the field of political ecology is defined in terms of power struggles, that occur not only in present socio-political conflicts—including movements of resistance and negotiation—but of power strategies that will gear a transition toward a sustainable future, bringing together different visions–interests–rationalities.

Capitalist accumulation has spread across the planet, conquering and colonizing territories, marginalizing and excluding communities, deterritorializing people from ecosystems and detraditionalizing their cultures.

It has also generated processes of resistance and survival in which political ecology merges into environmental history. Today, accumulation by dispossession (Harvey 2003) is devastating the planet through the geopolitics of "sustainable development" (Leff 2002a), transforming and degrading their life-territories for the expansion of mining, oil extraction through hydraulic *fracking* of the geosphere, the implantation of genetically modified crops, and the production of agro-fuels. Mega infrastructure projects implemented to allow the free flow of commodities have also caused ecological and cultural impacts, fostering many resistance movements in defense of ancestral and biodiverse territories.[10] These phenomena are not only instigated by corporate, national, and transnational capital with the agreement of liberal or "progressive" governments but directly by the policies of the neoliberal state.[11] These movements are no longer simply resistance, but *rexistence* movements (Porto-Gonçalves 2002). They are processes in which the peoples of the Earth reinvent their identities and their conservationist modes of production in line with the sustainable productivity of ecosystems, their cultural values, their existential meanings and purpose for life. The rubber tapper movement in the Brazilian Amazon is emblematic in terms of the invention of their extractive reserves (Porto-Gonçalves 2001a). Another example is the Black Community Process in the Colombian Pacific (Escobar 2008).

Emancipation processes are thus resisting the process of capitalist dispossession and subjugation of traditional peoples: by the reterritorialization, reidentification, and reappropriation of their bio-cultural heritage. If in the first stage of accumulation capital expanded by colonizing the minds of the original peoples through the actions of missionaries who subjected these peoples' consciences to the design of their God; if they

[10] The recent TIPNIS (Indigenous Territory of the Isiboro Securé National Park) case, in Bolivia, is an emblematic case. The conflict was caused by the State's interest in building a road across the indigenous territory to promote the economic development of the region against the right of the indigenous people to their territory of biodiversity. This is an example of the confrontation between the development drive of the so-called "progressive" governments and the emergence of new regimes of ecologically sustainable production and cultural reappropriation of Nature (Porto-Gonçalves and Betancourt 2013).

[11] Secondary legislation established by the recent energy reform in Mexico adopted, in Article 91 of the Oil Law, the legal status of "expropriation by the Mexican State for social interest and public order purposes," in which oil exploitation and exploration is a priority.

were dehumanized, deprived of their knowledge to be converted into labor force to produce economic value, in an era that lasted to post-industrial times, today these peoples and their lands are deterritorialized so they can operate within the logic and interest of the geopolitics of "sustainable development." In response, these populations exercise their rights of being, repositioning themselves in the world in the face of the global changes engendered by the market logic and the rationality of modernity. They do so through "other rationalities," based on their cultural identity, in the conflicting and solidary reinvention of their imaginaries, knowledge, and practices, reopening the fate of humanity to a sustainable future, leading to the political repositioning of the peoples of the Earth in the reconstruction of their life-territories.

From this perspective, Latin American political ecology emerges as distinct from European and Anglo-American sociology. Latin American environmental thought feeds on selected sources of sociological and philosophical theories to bring critical thinking to their livelihoods that re-emerge from their diverse cultural roots to hybridize "universal" thought with the imaginaries that sprout from other geographies—from the ecological and cultural conditions of Latin America—with its political theories and its traditional knowledge. Through this process of deconstruction, hybridization, and reterritorialization of knowledge, theories, and practices, political ecology opens new perspectives for the reconstruction of a sustainable world.

By recognizing in the original *physis* the sources and conditions of life, an ontology of diversity opens the way to the politics of difference and the ethics of otherness that sustain environmental rationality. In a world order where the hegemonic power of the rationality of modernity dominates and exterminates difference, political ontology territorializes the principles of postmodern thinking dialoguing with the imaginaries of traditional cultures; socio-environmental movements erupt from the fire of life of Heraclitus and the bosom of Mother Earth, opening new pathways to building a sustainable world. Political ecology transcends theoretical regionalisms not only to become hybridized with other disciplines and generate a holistic vision of the complexity of the globalized and ecological social world. It aims to redefine and reterritorialize the knowledge that derives from other sources of thinking, social conditions and environmental contexts, as well as other political interests to weave the epistemic web of political ecology in its different ecological–cultural contexts.

Thus, the indigenous struggles for identity and land not only demand the right to existence of different cultures. The claim for autonomy and the demand for territories to maintain alive the existence of cultural beings transcends the proposals for a cosmopolitanism that tolerates racial and cultural differences while imposing its hegemonic power in the construction of a world driven by the unity of the market and the supremacy of techno-economic rationality. In the field of political ecology, different rationalities confront each other. The rationality that drives the modernization of the world as a process involving detraditionalizing of peoples and progress toward the entropic degradation of the planet is faced with the emancipation of life: of the construction of a negentropic world order founded on cultural diversity in the encounter of tradition and modernity.

Political ecology is the field where the socio-environmental conflicts generated by the oblivion of life are expressed, as well as the social process for the reconstruction of the conditions of life in the planet, for building a sustainable world order grounded on the ontology of life, where the diversification of life becomes territorialized in different life-worlds through a politics of difference and an ethics of otherness. This historical process embraces a creative dialogue among the diverse regional political ecologies of the world, encountered by their power strategies in knowledge and rejoined by their search of a sustainable future for the whole of humanity and life on Earth.

Bibliography

Acosta, A. (2010), "El *Buen Vivir* en el camino del post-desarrollo. Una lectura desde la Constitución de Montecristi", Fundación Friedrich Ebert, Policy paper 9, Quito.

Arnold, D. & Guha, R. (1995), *Nature, culture, imperialism: essays on the environmental history of South Asia*, Delhi: Oxford University Press.

Bateson, G. (1972), *Steps to an ecology of mind: collected essays in anthropology, psychiatry, evolution, and epistemology*, Chicago: University Of Chicago Press.

Berthelot, J.M. (1998), *L'intelligence du social*, Paris: PUF.

Blaikie, P.M. & H.C. Brookfield (Eds.) (1987), *Land degradation and society*, London: Methuen.
Blaser, M. (2009), "Political ontology", *Cultural studies*, 23 (5): 873–896.
Boege, E. (2008), *El patrimonio biocultural de los pueblos indígenas de México. Hacia la conservación in situ de la biodiversidad y agrobiodiversidad en los territorios indígenas*, México: INAH/Comisión Nacional para el Desarrollo de los Pueblos Indígenas.
———. (2018), "Hacia una antropología ambiental para la apropiación social del patrimonio biocultural de los pueblos indígenas", in Toledo, V.M. & Alarcón-Cháires, P. (Eds.), *Tópicos bioculturales: reflexiones sobre el concepto de bioculturalidad y la defensa del patrimonio biocultural de México*, México: UNAM.
Boff, L. (1996a), *Ecologia, mundialização, espiritualidade*, São Paulo: Ática.
———. (1996b), *Ecología: grito de la tierra, grito de los pobres*, Madrid: Trotta.
Bookchin, M. (1962), *Our synthetic environment*, New York: Knopf.
———. (1971/1990), *Post-scarcity anarchism*, Montreal/New York, Black Rose Books, Second edition, 1990.
———. (1982/1991), *The ecology of freedom. The emergence and dissolution of hierarchy*, Montreal/New York: Black Rose Books, Second edition, 1991.
———. (1990a), Remaking society: Pathways to a green future, Boston: South End Press.
Bourdieu, P. (1984), *Homo academicus*, Paris: Les Éditions de Minuit.
Bourdieu, P. & L. Wacquant (1992), *An invitation to reflexive sociology*, Chicago: The University of Chicago Press.
Bryant, R. (1992), "Political ecology: an emerging research agenda in Third-World studies", *Political geography* 11: 12–36.
Commoner, B. (1971), *The closing circle: nature, man, and technology*. New York: Knopf.
———. (1976), *The poverty of power: energy and the economic crisis*. New York: Random House.
De Castro, J. (1975), *The geopolitics of hunger*, New York: Monthly Review Press.
Deleuze, G. (1969), *Logique du sens*, Paris: Les Éditions de Minuit.
Derrida, J. (1976), *Of grammatology*, Baltimore & London: Johns Hopkins University Press.
———. (1978), *Writing and difference*, Chicago: The University of Chicago Press.
———. (1982), *Margins of philosophy*, Chicago: Chicago University Press.
Dunlap, R.E. & W.R. Catton (1979), "Environmental sociology", *Annual review of sociology*, 5: 243–273.

———. (1994), "Struggling with human exemptionalism: the rise, decline and revitalization of environmental sociology", *The American sociologist*, 25 (1): 5–30.

Dussel, E. (1998), *Ética de la liberación en la edad de la globalización y de la exclusión*, Madrid: Trotta.

Enzensberger, H.M. (1974), "A critique of political ecology", *New left review*, 84 (3–31).

Escobar, A. (1992), "Imagining a post-development era: critical thought, development and social movements", *Social Text* 31–32:20–56.

———. (2008), *Territories of difference. Place, movements, life, redes*, Durham/London: Duke University Press.

———. (2010), "Post-constructivist political ecologies", in Redclift, M. & Woodgate G. (Eds.), *International handbook of environmental sociology*, 2nd edition, Cheltenham/Northampton: Edward Elgar Publishing Limited, pp. 91–105.

———. (2013), "Territorios de diferencia: la ontología política de los derechos al territorio", Segundo Taller Internacional SOGIP, "Los pueblos indígenas y sus derechos a la tierra: política agraria y usos, conservación, e industrias extractivas", www.sogip.ehess.fr.

Escobar, A., Grueso, L. & C. Rosero (1998), "The process of black community organizing in the Pacific coast of Colombia", in Álvarez, S., E. Dagnino & A. Escobar (Eds.) *Cultures of politics/politics of cultures: revisioning Latin American social movements*, Boulder, Colorado: Westview Press, pp. 196–219.

Esteva, G. & Prakash, M.S. (1998), *Grassroots post-modernism: remaking the soil of cultures*, London: Zed Books.

Foucault, M. (1980), *Power/knowledge*, New York: Pantheon.

———. (1998), *La verdad y las formas jurídicas*, Barcelona: Gedisa.

Georgescu-Roegen, N. (1971), *The entropy law and the economic process*, Cambridge, Mass.: Harvard University Press.

González Casanova, P. (1965), "Internal colonialism and national development", in *Studies in comparative international development*, Vol. 1, Num. 4:27–37.

Gorz, A. (1975), *Écologie et politique*, Paris: Seuil.

———. (1989), *Critique of economic reason*, London/New York: Verso.

Guha, R. (1989), *The unquiet woods: ecological change and peasant resistance in the Himalaya*, Delhi: Oxford University Press.

Guha, R. & Gagdil, M. (1992), *This fissured land: an ecological history of India*, Delhi: Oxford University Press.

Guha, R. & Martínez-Alier, J. (1997), *Varieties of environmentalism. Essays North and South*, London: Earthscan.

Habermas, J. (1984), *Theory of communicative action, Vol. I: Reason and the rationalization of society*, Boston, Mass.: Beacon Press.
———. (1987), *Theory of communicative action, Vol. II: Live-world and system: a critique of functionalist reason*, Boston, Mass.: Beacon Press.
Hardin, G. (1968), "The tragedy of the commons", *Science* No. 162: 1243–1248.
Harvey, D. (2003), *The new imperialism*, Oxford: Oxford University Press.
Heidegger, M. (1927/1962), *Being and time*, trans. John Macquarrie & Edward Robinson, San Francisco: HarperCollins Publishers.
———. (1957/1969), *Identity and difference*, Chicago: Chicago University Press.
———. (1971/2001), *Poetry, language, thought*, New York: Perennial Classics.
Holling, C.S. (1973), "Resilience and stability of ecological systems", *Annual review of ecological systems*, 4: 1–23.
Houtart, F. (2010), *A agroenergia: solução para o clima ou saída da crise para o capital?*, Petrópolis: Vozes.
Huanacuni, F. (2010), *Vivir bien/Buen vivir: Filosofía, políticas, estrategias y experiencias regionales*, La Paz: Convenio Andrés Bello/Instituto Internacional de Integración.
Kropotkin, P. (1902/2005), *Mutual aid: a factor of evolution*, Boston: Extending Horizons Books.
Kauffman, S. (2003), *Investigaciones: complejidad, autoorganización y nuevas leyes para una biología general*, Barcelona: Tusquets Editores.
Lander, E. Ed. (2000), *La colonialidad del saber: eurocentrismo y ciencias sociales. Perspectivas latinoamericanas*, Buenos Aires: Clacso/Unesco.
Leff, E. (Ed.) (1986), *Los problemas del conocimiento y la perspectiva ambiental del desarrollo*, México: Siglo XXI Editores.
———. (1995), *Green production: towards an environmental rationality*, New York: Guilford.
———. (1998a), "Murray Bookchin and the end of dialectical naturalism", *Capitalism, nature, socialism*, Vol. 9 (4):67–93.
———. (2002a), "La geopolítica de la biodiversidad y del desarrollo sustentable: economización del mundo, racionalidad ambiental y reapropiación social de la naturaleza", in Ceceña, A.E. & Sader, E., *La guerra infinita: hegemonía y terror mundial*, Buenos Aires: CLACSO-ASDI, pp. 191–216.
———. (2003), "La ecología política en América Latina: un campo en construcción", *Polis*, Revista de la Universidad Bolivariana, II (5) 2003, Santiago de Chile, pp. 125–145.
———. (2004), *Racionalidad ambiental: la apropiación social de la naturaleza*. México: Siglo XXI Editores.
———. (2010), "Imaginarios sociales y sustentabilidad", in *Cultura y representaciones sociales*. Num. 9, México, pp. 42–121.

———. (2011), "Sustentabilidad y Racionalidad Ambiental: Hacia 'otro' Programa de Sociología Ambiental", *Revista mexicana de sociología*, 73, núm. 1 (enero-marzo 2011): 5–46.

———. (2012), "Latin American environmental thinking: a heritage of knowledge for sustainability", *Environmental ethics*, Volume 34:4, Winter, 2012, pp. 431–450.

———. (2014a), *La apuesta por la vida: imaginarios sociales e imaginación sociológica en los territorios del Sur*, México: Siglo XXI Editores.

———. (2014b), "Los Derechos del Ser Colectivo y la Reapropiación Social de la Naturaleza", en Pereira de Cunha, Belinda; Agustín, Sergio; Araújo da Costa, Nálbia Roberta (Org.), *Saberes Ambientais, Sustentabilidade e Olhar Jurídico: visitando a obra de Enrique Leff*, Joao Pessoa e Caxias do Sul, Brasil.

———. (2018), *El fuego de la vida: Heidegger ante la cuestión ambiental*, Mexico: Siglo XXI Editores.

———. (2019), "Negentropic production", in Kothari, Ashish, Demaria, F., Acosta, A., Salleh, A. and Escobar, A. Eds., *Pluriverse: a postdevelopment lexicon*. London: Zed Books.

Levinas, E. (1977/1997), *Totalidad e infinito: ensayo sobre la exterioridad*, Salamanca: Sígueme, 4th edition [*Totality and infinity: an essay on exteriority*, Pittsburgh: Duquesne University Press, 1969].

———. (1999), *De otro modo que ser, o más allá de la esencia*, Salamanca: Ediciones Sígueme [Otherwise than Being or beyond essence, Springer Science + Public Media, B.V., 1991].

Lévi-Strauss, C. (1955), *Tristes tropiques*, Paris: Librairie Plon.

Martínez-Alier, J. (1995a), "Distributional issues in ecological economics", *Review of social economy*, Vol. LIII (4): 511–528.

Marx, K. (1965), *Œuvres, Économie I*, Paris: Gallimard.

Mignolo, W. (2000), *Local histories/global designs: coloniality, subaltern knowledges, and border thinking*, Princeton: Princeton University Press.

———. (2011), *Modernity and decoloniality*, Oxford bibliographies.

Mignolo, W. & A. Escobar, eds. (2009), *Globalization and the decolonial option*, London: Routledge.

Morin, E. (1980), *La méthode. La vie de la vie*, Paris, Editions du Seuil.

Murra, J. (1956), *The economic organization of the Inca state*, Chicago: University of Chicago.

Nietzsche, F. (1968), *The will to power*, New York: Vintage Books.

O'Connor, J. (1998), *Natural causes: essays in ecological Marxism*, New York: Guilford.

Orstrom, E. (1990), *Governing the commons. The evolution of institutions for collective action*, Cambridge: Cambridge University Press.

Peet, R. & Watts, M. (1993), "Introduction: development theory and environment in an age of market triumphalism", *Economic geography*, Vol. 69 (3): 227–253.
Porto-Gonçalves, C.W. (2001a), *Geo-grafías. movimientos sociales, nuevas territorialidades y sustentabilidad*, México: Siglo XXI Editores.
———. (2002), "O latifúndio genético e a r-existência indígeno-camponesa", *Geographia*, Ano 4, No. 8, Niterói: Universidade Federal Fluminense, pp. 7–30.
Porto-Gonçalves, C.W. & Betancourt, M. (2013), "Encrucijada latinoamericana en Bolivia: el conflicto del TIPNIS y sus implicaciones civilizatorias", mimeo.
Prigogine, I. (1955), *Thermodynamics of irreversible processes*, New York: John Wiley & Sons.
———. (1981), *From being to becoming: time and complexity in the physical sciences*, New York: W.H. Freeman & Co.
Prigogine, I. & Nicolis, G. (1977), *Self-organization in non-equilibrium systems*, New York: John Wiley & Sons.
Prigogine, I. & I. Stengers (1984), *Order out of chaos*, New York: Bentam Books.
Quijano, A. (2008), "Coloniality of power, eurocentrism, and social classification", in Moraña, M., Dussel, E. and Jáuregui, C.A. (Eds.), *Coloniality at large. Latin America and the postcolonial debate*, Durham, NC: Duke University Press, pp. 181–224.
Quintero, P. (Ed.) (2014), *Crisis civilizatoria, desarrollo y bien vivir*, Buenos Aires: Ediciones del Signo.
Rappaport, R.A. (1968/1984), *Pigs for the ancestors*, New Haven: Yale University Press.
———. (1971), "The flow of energy in an agricultural society", *Scientific American*, 224 (3):116–132.
Redclift, M. (1987), *Sustainable development: exploring the contradictions*, London: Methuen.
Ribeiro, D. (1973), "Etnicidade, indigenato e campesinato", *Revista de Cultura Vozes*, Ano 73, Vol. LXXIII, outubro, No. 8, p. 5–18, Rio de Janeiro.
Rivera Cusicanqui, S.; Domingues, J.; Escobar, A. & Leff, E. (2016), "Debate sobre el colonialismo intelectual y los dilemas de la teoría social latinoamericana", en *Cuestiones de Sociología,* 14, Universidad de La Plata, Argentina, www.cuestionessociologia.fahce.unlp.edu.ar/article/view/CSn14a09.
Robbins, P. (2012), *Political ecology: a critical introduction*, Chichester, UK: John Wiley & Sons, 2nd edition.
Schneider, E. & D. Sagan (2005), *Into the cool: energy flow, thermodynamics and life*, Chicago: Chicago University Press.
Schrödinger, E. (1944/1969) *¿What is life? The physical aspect of the living cell*, London/New York: Cambridge University Press.

Shiva, V. (1988), *Staying alive: women, ecology and development*, London: Zed.
Sousa Santos, B. (2008), *Conocer desde el Sur. Para una cultura política emancipatoria*, Buenos Aires: CLACSO/CIDES-UMSA/Plural Editores.
Stavenhagen, R. (1965), "Classes, colonialism, and acculturation. Essay on the system of inter-ethnic relations in Mesoamerica", *Studies in Comparative International Development*, 1 (6): 53–77.
Steward, J. (1972), *Theory of cultural change: the methodology of multilinear evolution*, Urbana: University of Illinois Press.
Thompson, E.P. (1971), "The moral economy of the English crowd in the 18th century", in *Past and present*, 50:76–136.
———. (1991), *Customs in common: studies in traditional popular culture*, London: Merlin Press.
Urry, J. (2003), *Global complexity*, Cambridge: Polity Press.
———. (2005), "The complexity turn", in *Theory, culture & society*, Vol. 22 (5): 1–14.
Vayda, P. (ed). (1969), *Environment and cultural behavior*, Garden City, NY: The Natural History Press.
———. (1983), "Progressive contextualization: methods for research in human ecology", *Human ecology* 11: 265–81.
Vernadsky, V.I. (1998), *The biosphere*, New York: Copernicus.
Walker, P.A. (2005), "Political ecology: where is the ecology?", *Progress in human geography* 29 (1):73–82.
Watts, M. (2008), "On the poverty of theory: natural hazards research in context", in Braun, B., Environment: critical essays in human geography, London: Routledge.
———. (2015), "Now and then: the origins of political ecology and the rebirth of adaptation as a form of thought", in Perrault, T., J. McCarthy & G. Bridge, *Handbook of political ecology*, London: Routledge.
Weber, M. (1922/2014), *Economía y sociedad: esbozo de sociología comprensiva*, México: Fondo de Cultura Económica.
Wilson, E.O. (1975), *Sociobiology, the new synthesis*, Cambridge, Mass. & London: The Belknap Press of Harvard University Press.
Wolf, E. (1972a), "Ownership and political ecology", *Anthropological quarterly* 45 (3): 201–205.
Wolf, E. & A. Palerm (1972a), "Potencial ecológico y desarrollo cultural en Mesoamérica", in *Agricultura y civilización en Mesoamérica*, No. 32, México: SepSetentas.
Zimmerman, M.E. (1994), *Contesting earth's future. Radical ecology and postmodernity*, Berkeley: University of California Press.

12

The Social *Enownment* of Nature, the Reinvention of Territories and the Construction of an Environmental Rationality

Carlos Walter Porto Gonçalves and Enrique Leff

Emergence of the Environmental Crisis: The Political-Epistemic Debate

The environmental crisis irrupted in the 1960s together alongside other public debates and social issues, and emancipation processes such as the feminist, gender, and student movements. This was a time when ethnic–racial questions gained centrality during the decolonization of Asia and Africa, and the Civil Rights Movement in the United States. Notwithstanding the critical nature of these issues for a more democratic world, the environmental question represented an urgent crisis, a *crisis of civilization*. The progression of modernity had reached a limit, and new

Published originally in Leff, E., Floriani, D., and L.H. de Oliveira Cunha (eds), dossier temático Pensamento Ambiental Latino-americano: movimentos sociais e territórios de vida, *Revista Desenvolvimento e Meio Ambiente* No. 35, Programa de Pós-Graduação de Meio Ambiente e Desenvolvimento, Universidade Federal do Paraná, Curitiba-Paraná-Brasil. That article derived in turn from a larger text: Leff, E., Argueta, A, Boege, E., and Porto-Gonçalves, C.W. (2002), Más allá del desarrollo sostenible. La construcción de una racionalidad ambiental para la sustentabilidad. Una visión desde América Latina, published originally in Leff, E., Ezcurra, E., Pisanty, I., and Romero, P. (eds), *La Transición Hacia el Desarrollo Sustentable. Perspectivas de América Latina y el Caribe*, México: INE-SEMARNAT/PNUMA/UAM, pp. 479–578.

E. Leff, *Political Ecology*, https://doi.org/10.1007/978-3-030-63325-7_12

civilization horizons had to be opened to gain a new perspective for the sustainability of life on planet Earth. This turning point in the course of history called for an epistemological merging of Nature and culture, which had been mutually alienated by the ontological differences and Cartesian dualism that were the foundations of modernity.

The ecological movement emerged in this context and debated the environmental crisis that had been triggered by economic growth and technological progress. This debate ranged from a critique of the arms race and the consumer society, to the world of waste and pollution brought about by capital accumulation and industrialization. New epistemic horizons were opened, including perspectives that span the sciences and methods of complexity, post-structuralism and the philosophy of postmodernity. Hermeneutics, deconstructionism, and constructivism were also associated with the search of new ways of thinking and constructing knowledge, and sought to orient emancipation processes from oppressive social structures and an objectified reality, from patriarchy and gerontocracy, capitalism and socialism, scientism and technology to a cosmopolitan dialogue of knowledge/*savoirs*.

From then on, the idea of setting limits to human intervention in Nature gained force. After the detonation of the atomic bomb in Hiroshima and Nagasaki, science lost its apparent neutral status, and its effects on society began to be questioned. Science was not necessarily at the service of life or of human emancipation, as assumed by the Enlightenment. However, the crisis of science did not arise primarily from its internal theoretical inquiries, but rather from damage resulting from its applications. The social repercussions of scientific rationality implied the critique of its metaphysical–ontological–epistemological foundations, mainly with regards to its hegemonic positivist and structural-functionalist derivations. This critique came to question the effects on society and on the environment of the power strategies embedded in knowledge (Foucault 1980). The crisis of scientific reason was an opportunity for other epistemological approaches to emerge and for other schemes of intelligibility and matrixes of rationality to be recognized, including the emancipation of subjugated knowledge from epistemological colonialism. Eurocentric thinking had ignored other worldviews and disqualified other human experiences and practices, including other forms of cognition, of knowledge, of understanding, and of constructing life-worlds of people, the place of humanity, and the destinies of life on planet Earth.

From Vance Packard's *The Waste Makers* (1960), to Rachel Carson's *Silent Spring* (1962), Paul Ehrlich's *The Population Bomb* (1968), Nicholas Georgescu-Roegen's *The Economic Process and the Entropy Law* (1971), the MIT/Club of Rome study on *The Limits to Growth* (Meadows et al. 1972), Hans Magnus Enzensberger's *Critique of Political Ecology* (1974), and Celso Furtado's *O Mito do Desenvolvimento* (*The Myth of Development*) (1974), the environmental debate acquired explicit political dimensions. In this intellectual context, the United Nations convened the first World Conference on Human Environment, held in Stockholm, in 1972, spreading the news of the environmental crisis and opening the environmental question to the international geopolitical agenda.

The environmental crisis came to question the civilization project that had been based on the anthropocentric ideal of man's *domination of Nature* originated in the Judeo-Christian onto-theological understanding of the divine mandate of God. Instituted later through the history of metaphysics in the rationality of modernity, its ethic, epistemic, scientific, technical, and political dimensions conflated in the centrality of the dominating hegemony of techno-economic rationality over Nature and social life. The separation between human and natural sciences, beyond the specialization within each one of these fields, corresponds to the separation of peasants and original peoples from their ecological conditions of existence, subjugating and banishing communities from their life-territories.

With the environmental crisis, the inquiry on Nature became an epistemological, political, and ethical debate on the conditions for the sustainability of life on the planet. Nature was subdued to modern scientific and technological development, mainly after economic rationality was instituted as the reason-of-being-in-the-world, and individual behaviors were trapped by the principles of rational choice. The search for fundamental essential unities of different ontological orders became an epistemic obsession of modern science: the atom in physics; the molecule in chemistry, the cell in biology, the individual in the social sciences. The belief in the scientific world that the mysteries of Nature are revealed in mathematical language was reflected in the mundane world in the unitary logic of market values. Thus, the hegemonic paradigms of modernity were instituted in the imaginaries and life-worlds of peoples and individuals. With the capitalist mode of production, the modern economy abandoned the physiocratic principle that affirmed that Nature was the source of wealth through the reproduction of seeds. With this obliviousness to Nature, economic rationality externalized Nature and abandoned its inquiry of the ecological conditions necessary for the sustainability of an economic process. The economic process became a mechanistic process of production mobilized by "productive factors" and fueled by fossil materials (carbon and oil) that irreversibly and in complex ways generated ecological decay, environmental degradation, climate change, and global warming.

To be sure, the economy cannot produce Nature: energy contained in a molecule of carbon or the atom, even if put into production by the modern economy, was not produced by humans; no country, no society, no people have produced iron, oil, air, or water. However, the degradation of matter and energy into soil, water, and air pollution were considered mere "externalities" of the economic process. The environmental crisis came to question the fallacies of economic rationality: the alienation of the economy from being Nature-based and reliance on the epistemological conception of ideas (*res cogitans*) existing outside Nature (*res extensa*).

Within this environmental crisis, sustainability has emerged as a goal and condition for global ecological balance and human survival. However, sustainability is a polysemic concept that cannot be universally and unanimously defined. Disputing its meanings and what strategies to engage for a social construction of a sustainable future, is at the very core of political ecology. Among the diverse approaches to sustainability in environmental and ecological economics, two radically different approaches are distinguished in political ecology: one being configured under the hegemonic economic rationality; the other based on an environmental rationality founded on the ecological conditions and cultural identities of the peoples of the Earth, based on the negentropic potentialities of the biosphere, the social reappropriation of the bio-cultural patrimony of the peoples and the reinvention and reconstruction of their life-territories

(Leff 2004, 2014a). Thus, people's emancipation strategies for sustainability contest the dominant geopolitics of sustainable development.

The Geopolitics of Sustainable Development

Political ecology is the disciplinary and political field of encountering different modes of understanding and constructing life-worlds: of confronting rationalities in the social appropriation of Nature, the enacting of the metabolism of the biosphere, and the construction of a sustainable future. This historical objective demands the deconstruction of theories and practices built on the foundations of scientific, economic, technological, and political rationality inscribed in scientific paradigms, instituted in national and international policies and rooted in peoples' life-worlds, to establish new socio-environmental relations. This deconstruction is not only operated by political ecology as a theory but above all, by emancipation practices of those peoples engaged in struggles for the reappropriation of Nature and the reinvention of their territorialities (Porto-Gonçalves 2006).

In this sense, political ecology transcends the purpose to make the economy more ecological by assigning market values to Nature and economic instruments to environmental management. Beyond the debates between "fictitious economy" and "real economy," between "speculative capital" and "productive capital," political ecology recaptures the economy based on the productive processes of Nature, inscribed in cultural meanings and embedded in the ecological conditions of the life-territories of the peoples. As pointed out by ecological economics, the economy should be understood and treated as a subsystem of a larger finite system, the biosphere, which implies the impossibility of perpetual growth. The distributive inequalities in economic wealth, ecological potentialities, and social costs, as well as the ineluctable environmental degradation generated by the established economic rationality, is an evidence of the false principle of the self-regulative mechanisms of the economy and the fallacy of the power of techno-science to dematerialize production. As stated by Abramovay,

> The difference in emissions of greenhouse gases between individuals, based on their economic wealth and consumption patterns [...] shows that the 500 million richest people of the world (7% of world population) generate half of the greenhouse gases dumped to the atmosphere [...] Even though today one unit of monetary value might be produced with 30% fewer materials than 30 years ago, there was an increase of 50% in the demand of these materials in the same period. Considering that the average consumption of natural resources by the American people is 88 kilograms per day, and that of the African people of south Sahara is only 10 kilograms daily (Friends of Earth et al. 2009), a generalization of consumption patterns to those of the wealthiest countries would lead to an increase of the pressure over the sustainability of ecosystems and their capacity to offer the essential environmental services. It is a false illusion to think that the reduction of present social inequalities can be compatible with the generalization of the consumption patterns of that 7% of the world population that is responsible for half of the emissions of greenhouse effect gases. (Abramovay 2010)

After the Stockholm Conference in 1972, a worldwide debate on the limits of the present trends of human intervention on Nature was launched; environmental movements oriented the debate on the economic and political interests involved in the social appropriation of Nature. Nature was resignified by the transformations of social and power relations induced by a new cycle of economic growth and technological development through biotechnology, genetic engineering, microelectronics, informatics, nanotechnology, and robotics (Porto-Gonçalves 2006). Biodiversity and germplasm became sources of capital accumulation for transnational corporations through new strategies for the appropriation of Nature, such as ethnic-biopiracy and the attempt to legitimize capital's intellectual property rights over natural processes. The principle of private property was extended not only to plants and animals but also to microscopic cells and the nanoscopic levels of genes. While until then all cultures had established their relations with Nature at a macroscopic level of the organism, nanotechnology expands the frontier of capital to the interior of matter and the living cell. The gene is only observable and modified with instruments of microscopic research. Thus, the locus for the production of knowledge and meaning of Nature shifted from the life-worlds and livelihoods of the people to the laboratories of biotechnology as centers of bio-power for the expansion of capital (Porto-Gonçalves 2007). At another level, the increasing emission of greenhouse effect gases resulting from the industrial metabolism mobilized by fossil fuels has transformed the composition of the atmosphere, generating intensified complex meteorological processes and triggering climate change. Their regulation has entered the financial circuits with the pretension of protecting the environment through the economic coding and valuing of carbon bonds as the basis for negotiations of global warming (Cornetta 2011).

The scientific controversies on biodiversity and climate change have been trapped in the power strategies of the discourse and geopolitics of "sustainable development" (Leff 2002a; Porto-Gonçalves 2006; Bartra 2008). The environmental question has been captured by the logic of the market and its financial strategies, as well as by normal science, ignoring the power relations that cut across the *geopolitics of economic globalization, biodiversity conservation, and sustainable development* that extends, intensifies, and complexifies previous processes of destructive appropriation of natural resources. The geopolitics of sustainable development is configured in the context of the deployment of techno-economic globalization. Thus, together with the historical forms of exploitation of Nature that characterized the "pillage of the Third World" (Jalée 1968), global capital promotes today a "conservationist" exploitation of Nature. Biodiversity appears not only as a multiplicity of life forms but as natural reserves—the habitat of biological and cultural diversity—valued for their genetic wealth, as eco-touristic resources or for its sink-function for collecting carbon. If in modern-colonial times, sugar cane, cotton, banana, or coffee monocultures were established in Latin America, the economic value of biodiversity is leading to a new type of landlordism: of conservation areas and genetic *latifundia* (Porto-Gonçalves 2002). Large areas are being transformed into conservation units or converted to modified genetic crops, ignoring the fact that those territories that remained at the margins of the market are the areas that harbor the most significant natural wealth in water and biodiversity of the planet, having been historically inhabited by traditional populations—indigenous, peasant, and maroon peoples—who have preserved these territories as a natural and cultural patrimony.

Political ecology becomes the field where the controversies between the economization of Nature and the ecologization of the economy are deployed; where different, and often opposing, strategies for the appropriation of Nature are confronted and collide. The discourse of sustainable development is a strategy for the economic appropriation of Nature that "denaturalizes" Nature through technology, inducing a process of *transgenesis* that invades and transmutes life, searching to normalize and to legitimize the commodification of Nature.

The merchandising of Nature deepens the differences between rich and poor countries under the principles of "sustainable development." Economic-ecologic globalization justifies the comparative advantage held by the more industrialized and polluting countries over the developing countries that are being induced to value their capacities economically to capture the excess of carbon dioxide and other greenhouse-effect gases dumped into the atmosphere by the rich countries and to offer the genetic resources from their reserves of biodiversity. The differences between central and peripheral countries are not only produced by the pillage and overexploitation of their resources but are now being masked by new functions assigned to their territories through economic strategies for the appropriation of environmental goods and services. This is no longer a race for development based on comparative advantages in the endowment and accumulation of productive factors, to bridgeh the technological gap and attain a more equitable world. Rather than valuing biodiversity as a potential for alternative development, it is reduced to a means to mitigate the ever-increasing ecological footprint of the developed countries and to continue extracting the materials needed for their unsustainable growth—oil, minerals, cellulose and foodstuffs—as well as for the capital accumulation of the emergent economies of countries like China, India, and Brazil.

For some governments and authors, this exchange of Nature for technology represents a win–win strategy, both in economic terms and for ecological conservation. That is the fallacy promoted by the power devices and discursive strategies of the geopolitics of "sustainable development," such as the "Program for Reducing Emissions from Deforestation and Forest Degradation" (REDD), the "Green Economy" and the "Clean Development Mechanism" (CDM). The REDD program pretends to reduce the negative contribution from deforestation and forest degradation to greenhouse effect emissions. It intends to refunctionalize the integration of the Third World territories in the global economy, to sustain the unsustainable growth of the more developed economies, unable to "dematerialize" their economies and to restrain their increasing emissions and ecological footprint through "green" technologies. Thus, the CDM pretends to preserve biodiversity, while inducing artificial forest plantations to increase the capacity to capture the excess emissions of industrialized countries and to produce natural commodities, like cellulose and other forest goods, as well as developing new pharmaceutical and food products derived from the richness in biodiversity of tropical ecosystems.

In this sense, an economic role is assigned to forests and biodiversity for their capacity to capture carbon and to balance the emissions of greenhouse-effect gases with the aim of mitigating global warming. This redefinition of unequal exchange in the integration of Third World countries and tropical regions into the global economy functions as a subvention to continuous unsustainable economic growth, offering limited and dubious benefits to tropical countries and the overall sustainability of the planet. In exchange for the artificiality of ecosystems in the North, to the unstoppable progress of industry and a highly capitalized and technologized agriculture, some exceptional territories are granted the luxury of maintaining a "natural economy," to continue living out of the generosity of Mother Earth by valuing the comparative advantages offered to them by the geographic localization of their territories.

This role imposed on Nature and culture by the geopolitics of sustainable development implies, over the purpose of reducing emissions, a reduction of the natural and cultural potentials for the construction of alternative sustainable economies and other possible worlds. From these considerations, the indigenous peoples represented in the First International Forum of Indigenous Peoples on Climate Change, held in Lyon, France in September 2000, rejected the inclusion of carbon sinks under the Clean Development Mechanism (CDM) because

> It reduces our sacred land and territories to mere carbon sequestration that is contrary to our worldviews and philosophy of life. Sinks in the CDM would constitute a worldwide strategy for expropriating our lands and ter-

> ritories and violating our fundamental rights that would culminate in a new form of colonialism. Sinks in the CDM would not help to reduce GHG emissions instead it would provide industrialized countries with a ploy to avoid reducing their emissions at source [...] the CDM pose the threat of invasion and loss of our land and territories by establishing new regimes for protected areas and privatization. We emphatically oppose the inclusion of sinks, plantations, nuclear power, mega-hydroelectric, and coal. Furthermore, we oppose the development of a carbon market that would broaden the scope of globalization. (International Indian Treaty Council 2000)

Contesting the capitalistic strategies for the appropriation of Nature, traditional peoples are developing new strategies to reappropriate their bio-cultural patrimony, to reinvent their modes of production and their ways of inhabiting their life territories.

Territories, Territorialities and Territorialization

The conscription of the Latin American and Caribbean peoples in the globalized world in the recent period of neoliberal hegemony triggered a theoretical-political debate in the environmental field where the concept of territory and its correlates, territoriality, and territorialization, have become the cornerstone of socio-environmental conflicts in the field of political ecology (Porto-Gonçalves 2001b; Haesbaert 2011; Leff 2014a). From the Barbados Declaration (1971) to Convention 169 of the International Labour Organization (1989) and the UN Declaration of the Rights of Indigenous Peoples (2007), the territorial rights of original peoples, peasants, and Afro-Americans were forged, recognizing other forms of living, appropriation of space and inhabiting the world. The historical struggles of these peoples and ethnic groups are configuring a new geopolitics drawn with/against the neoliberal policies after the 1990s, when critical political reforms in different countries—especially the national constitutions and legislative bodies of Bolivia and Ecuador—recognized the rights of Nature and original peoples, peasants, and maroon populations to their environments. These democratic changes triggered processes of cultural emancipation of traditional peoples for the reappropriation of their ancestral territories, encountering the interests and reaction of dominant social groups. Territorial struggles are becoming an outstanding issue of political ecology in Latin America, revealing the conflicts between the domineering economic and political power groups—global market, international consortia, and nation states—from the resistance/*rexistence* strategies for cultural reappropriation of Nature and the construction of new territorialities of indigenous, peasant, and maroon peoples arising in the perspective of an environmental rationality.

The *rights to difference* derived from the ontological difference in post-modern philosophy (Derrida 1978) reflected in a new politics of identity. Beyond its eagerness to deconstruct social theories and practices centered on class struggles, the politics of difference renewed the Latin American debates on the class/ethnic imbrications pointed out since 1920 by José Carlos Mariategui (1971), and more recently the race/class distinction developed by Anibal Quijano (2008). The theoretical debate on the politics of difference has given ground to new juridical-political procedures and new forms of territorialization of indigenous peoples and rural populations.

The "territorial question" is being debated and inscribed in a tense political process as the confrontation of power strategies for the appropriation of Nature and the construction of new life-territories. Two combined processes of land expropriation occurred in the constitution of the colonial-modern world system: (1) The conquest/invasion of *Abya Yala*/America that from its very start generated territorial conflicts with the deterritorialization of African peoples via the traffic of slaves and the colonization of the original peoples of Latin America; (2) The deterritorialization of the peasantry in Europe during the seventeenth and eighteenth centuries through the enclosures of their communal lands, in order to establish private properties for the development of capitalistic agriculture. Peasant's property forms traditionally were family-based under common use of land, forests, and water resources. This is the case of traditional peasant forms of property in different Third World regions, such as the *ayllus* Quechua/Aymara; the Russian *mir* (Zasulich/Marx); the Mexican *ejidos* with their *milpas* and family orchards; the territoriality of *Seringueiros* in the Brazilian Amazon (Porto-Gonçalves 2004), the

Retireiros of Araguaia river and their *varjões* (common lands); the *faxinais* in South Brazil or the *fundos de pasto* (pasture lands) in the north-east of Brazil (Campos 2000).

Nature and culture that had been politicized by the agrarian revolutions in the twentieth century seeking the social distribution of land acquired new meanings with the shift from the struggle for land to the claim for territory confronting the national state as a territory inhabited by multiple territorialities. The struggle over territories involves a theoretical-political debate, as assumed by indigenous leaders such as the Quechua-Ecuatorian agronomist Luis Macas, ex-president of the Coordination of Indigenous Nationalities of Ecuador—CONAIE—when he stated: "*our struggle is epistemic and political.*"

The concept of territory no longer designates the physical-geographical base for the exercise of state's sovereignty as consecrated by international law and political science, shifting to a process of appropriation and control of geographical space, of its peoples and resources, revealing the power relations that traverse the field of political ecology. The territory is being "reinvented" as a space–place for the cultural reappropriation of Nature (Porto-Gonçalves 2004; Raffestin 1980; Haesbaert 2011; Sack 1985; Harvey 2003). At the basis of this conceptual shift are the struggles of peasants, indigenous peoples, and maroons for their life-territories. Political ecology is thus conceptually constituted and deployed in practice by a politics within the triad territory–territoriality–territorialization. As cultural rights legitimate these processes of reterritorialization, the material/symbolic appropriation of Nature involves a dispute over the meanings assigned to Nature within the cosmogonies, imaginaries, and socioeconomic practices of cultural organizations. Thus, Nature and culture are politicized. The territory redefined as the *summa* of Nature–culture power relations becomes one of the identity traits of Latin American political ecology.

The condition for the reproduction of capitalist production relations is the permanent separation of peoples and ethnic groups from their material/natural conditions of existence, with the destruction of community modes of production and world lives, with individualized men and women in one side, and Nature without people in the other, converted into mere forces of production and commodities. The environment, reduced to a generic Nature–society relation under the geopolitics of sustainable development, appears as a political, ecological question: the struggles are then for the *social reappropriation of Nature*.

By the end of the 1980s and through the 1990s, the environmental question acquired new political projection with the emergence of new cultural-environmental peasant and indigenous movements, like that of *Seringueiros* in the Brazilian Amazonia, the afro-Colombians from the tropical forests of the Colombian South Pacific and the Zapatista indigenous movement in Mexico. The Sandinista–Miskitos *affair* in Nicaragua in 1979–1989 played an essential role in this breakthrough in historical materialism when the Miskito indigenous people opposed the Marxist Sandinista vision of progress guided by the development of productive forces, confronting the traditional conception and hegemonic vision within the left. A similar conflict had irrupted recently in Bolivia with the indigenous people that claim their rights to their biologically diverse ecological territory (TIPNIS) when the government intended to construct a road across the reserve to promote the economic development of the region.

Existential Emancipation, Cultural *Rexistence* and Reinvention of Life-Territories

The resignification of Nature after the 1960s involved new protagonists in the field of political ecology, peoples whose culture is interwoven with the ecological web of life in the territories they inhabit: indigenous peoples, Afro-Americans, peasants, riverside dwellers. New collective identities have been emerging from different ethnic conditions and cultural relations with Nature, involving different social practices and modes of being of rural and forest dwellers: *Seringueiros*, *castanheiros*, fishermen, *babassu* coconut-breaker women.

These social actors emerge from their resistance to being absorbed (deterritorialized) by globalization and their claims to reconstruct their life-territories, their cultural identities, and their autonomous, sustainable life-worlds. In this perspective, these resistance processes turn to be movements of *rexistence*. These populations do not only resist against dispossession and deterritorialization: they redefine their forms of existence through emancipation movements, by reinventing their identities, their modes of thinking, their modes of production and their livelihoods. After 500 years of modern colonization of Latin America and the Caribbean, despite the oppression to which they were submitted, these populations have not only persisted: they have reaffirmed themselves by reinventing their cultural beings that co-evolved with the Nature they inhabited. By claiming their rights to their territory, indigenous peoples are resignifying the ecological-cultural space that they inhabited, reinventing their biocultural patrimony and reterritorializing their productive and social practices.

An analysis of the different morpho-climatic domains of the Latin American landscapes reveals that in those regions, still covered by dense tropical forests—like the Amazon region with its eight million square kilometers, as well as the vast areas of the Atlantic Forest, that covers almost the totality of the Brazilian Atlantic Coast—there was previously no forest but herbaceous and shrub vegetation, known in Brazil as *cerrados* and *caatinga*. In these areas, the dominant weather 18,000 and 12,000 years ago, was much dryer and thus limited the formation of dense forests that were restricted to some niches and refuges (Ab'Saber 1970). Original peoples like the Tupi, the Guarani, and the Aruaque, among others, inhabited these ecosystems while they were being formed, thus *co-evolving* with Nature while transforming their life-territories (Posey 2004).

In the mountain region of the Tehuantepec Isthmus in Southern Mexico, in the states of Chiapas, Oaxaca, and Veracruz, the territories of Uxpanapa and Chimalapas represent areas particularly rich in biodiversity within the complex and dynamic refuges of flora characteristic of this region. Inventories of biodiversity of the tropical forest of Uxpanapa identified 924 plants, 150 birds, 34 mammals, seven turtles and 13 fish species in three ecosystems. In one *ejido* alone, that of Agustín Melgar in Uxpanapa, 168 useful species were recognized in the primary forest; 155 in the secondary forest and 33 in rivers, totaling 356 useful species for foodstuff, medical remedies, construction materials, woods, forages, skins, fibers, gums, waxes, poisons, dyes, and flavoring substances. The biodiversity being appropriated by indigenous communities in its growing areas and family orchards adds to 783 identified useful species (Toledo et al. 1978).

Central America and the Amazon are some of the most extensive biologically diverse areas on the planet. The Amazon forest that covers extended areas of Brazil, Colombia, Peru, Ecuador, Bolivia, Venezuela, Surinam, Guyana, and French Guyana covers a continuous expanse of eight million square kilometers of tropical forest containing between 500 and 700 tons of biomass per hectare. This mega-ecosystem is an immense "green ocean" responsible for the evapotranspiration of water that maintains a dynamic equilibrium of the planet's hydrology, an environmental service for the benefit of humanity (Uhl et al. 1991).

These regions contain a rich patrimony of biological diversity, of territories and landscapes interwoven with the different peoples that have inhabited them through history in the co-evolution of culture with Nature. These peoples hold an enormous legacy of knowledge welded in their traditional practices for living sustainably within their ecological conditions. Many of the principal crops that feed humanity were domesticated by Amerindian peoples: potato (*Solanum tuberosum*), originated in Peru, where more than 7,000 cultivars are known; sweet potato (*Ipomoea batatas*); bitter cassava (*Manihot esculenta*) and sweet cassava (*Manihot dulcis*); maize (*Zea mays*), a worldwide staple for human and animal nourishment; tomato (*Lycopersicum esculentum*); beans and peanuts (*Arachis hypogaea*); fruits such as cacao (*Theobroma cacao*), pineapple (*Ananas sativus*), *caju* or cashew nut (*Anacardium* occidentale), papaya (*Carica papaya*), *íngas* (*Inga* spp.), almonds like the *Pará chestnut* (*Bertholletia excelsa*); stimulating plants like *guaraná* (*Paullinia cupana*), mate herb (*Ilex paraguariensis*) and tobacco (*Nicotiana tabacum*); medicinal plants like *ipecacuana* (*Cephalis ipecacuanha*) from which emetine hydrochloride is extracted; copaiba (*Copaifera*) used against urinary tract disorders; quinine (*Cinchona officinale*), used against malaria; plants for industrial uses such as rubber (*Hevea brasiliensis*), that have not been totally substituted by synthetic rubber for uses like chirurgical gloves and high-quality condoms; carnauba plant (*Copernicia* sp.) from which wax is extracted; *timbó* (*Theprosia* sp.) that contains rotenone—an ingredient used as insecticide in sanitary medicine and agriculture; as well as other plants, wild or cultivated, used by indigenous peoples such as different kinds of corron (*Gossypium* spp.); *carúa* (*Neoglaziovia varietata*) a kind of bromeliad used to make yarn and fabric, and *piaçaba* (*Leopoldinia piasaba*) used to make brooms, mats, and baskets (Ribeiro 1992).

Throughout their long history of coexistence and co-evolution within their complex ecosystems, indigenous peoples have developed a complex knowledge of the ecology of these species derived from cultural modes of material and symbolic appropriation of Nature inscribed within their cosmogonies and imaginaries, embodied in their production practices, and embedded in their territories. The historic-cultural transformations under way are the result of the struggles of these indigenous peoples to defend their collective rights to preserve their cultural ways for reconstructing their territorialities in the face of domineering strategies for the appropriation of Nature geared by techno-economic rationality and the geopolitics of sustainable development.

The Rexistence of Indigenous Peoples: Tradition and Modernity

In the aftermath of the environmental crisis, Nature is being revalued because of its ecological regulatory functions to mitigate climate change and for its economic potential. The preservation of biodiversity emerges as a priority in the strategies of sustainable development not only for its intrinsic value in conservation areas but as territories defined as carbon sinks and for its biotechnological potential. The greatest wealth of biodiversity in the planet is found in the regions inhabited by indigenous and peasant peoples who, in recent years, are re-emerging in the political arena claiming their cultural rights for the reappropriation of their patrimony of natural resources. Following their historical resistance to modern colonization, struggles for new perspectives of emancipation and the construction of sustainability are emerging derived from their confrontation with the strategies for the appropriation/transformation of Nature opened by the expansion of the global economy revitalized by the biotechnological revolution and its counterpart, the legitimization of indigenous peoples' rights to their bio-cultural patrimony and their ancestral territories. The new geopolitics of biodiversity conservation, and sustainable development deploys its power strategies in the field of political ecology confronted by the indigenous and peasant peoples' cultural rights. What is at stake in these conflicts derived from the clash of alternative paths towards sustainability is not the distribution of benefits from the techno-economic reappropriation of Nature, but rather the emancipation and *rexistence* processes of traditional populations, which are the profound motivation of the socio-environmental movements in South and Latin America.

The Rexistence of Ecosystem's Peoples: The Case of the *Seringueiros*

Chico Mendes (1944–1988) became the leader of a new peasants' socio-environmental movement fighting against the hegemonic economic rationality for the exploitation of Nature. He conducted the social process to institute the extractive reserves as a new "agrarian reform," countering the overexploitation and the strategy of "sustainable development" in the Brazilian Amazon region (Porto-Gonçalves 2001b, 2004). *Seringueiros* are the self-identified inhabitants of the rich ecosystems. These people, attracted by the rubber fever of the second half of the nineteenth century from different parts of Brazil, underwent a long struggle over the twentieth century to settle definitively in those territories, to reappropriate its natural resources and reinvent their identities and livelihoods. At first, they were prevented by their *patrões* to practice subsistence agriculture, forcing them to dedicate all their labor time to the extraction of latex. Thus, they remained dependent on the owners of the rubber forests, the *seringals* as an agrarian proletariat. The rubber cycle was dependent on this double flux of labor that provided food to the *seringals* for the production of marketable rubber. Only after the crisis in the rubber market in the second decade of the twentieth century were the owners obliged to allow the *Seringueiro* to practice communal agriculture in order to maintain them in the forest. That was the source of the rich experience of these populations in combining agriculture and extractive activities.

The *Seringueiros* emerged in the political scene in the state of Acre, in the south-west of the Brazilian Amazon region in the 1970s. Their first actions of resistance were to prevent landowners from felling the forest to plant grass. Under the political direction of the National Confederation of Agricultural Workers (CONTAG), they considered themselves *posseiros*, squatters having possession of their common land for the extraction of rubber. By the early 1980s, the *Seringueiros* had exchanged their old subsistence family parcels (*colocaçoes*) for individual land plots. From then on, and under the political leadership of the Union of Rural Workers of Xapuri and Chico Mendes, and in opposition to the CONTAG in Acre, the *Seringueiros* started elaborating an original political proposal that combined their claim for land with the defense of their livelihood. This was the struggle to establish *Seringueiro territories*.

In 1985 the Seringueiros founded the *National Council of Seringueiros* (CNS) that maintained a necessary political link with the Union of Rural Workers to avoid their decoupling from their social and territorial base while broadening their struggle for land to the defense of the living forest and their lifestyles. Their proposal to establish *Extractive Reserves* is the crowning of their *Seringueiro* identity. This conservationist–productive strategy has become a living laboratory in exploring alternative sustainable strategies based on *savoirs* embodied in their cultural practices and knitted in their territories by their coexistence with the tropical forest. Behind the concept of Extractive Reserves was the idea of the Indigenous Reserves that established the first state guardianship over these communities and their different cultural traits. In the Extractive Reserves, communal land is the property of the Union with usufruct rights by families through their organized entities—unions, cooperatives, neighborhood associations—that elaborated the management plan. The *Extractive Reserves* combine the usufruct of each family with communal property under governmental legal guardianship to guarantee the sustainable use of natural resources while offering institutional conditions to transit towards a self-managed society.

Through the leadership of Chico Mendes, the *Seringueiros* extended their influence by creating the Alliance of Peoples of the Forest and are settled in 4 million hectares of land decreed as Extractive Reserves, diversifying the production and trade of the forest products and establishing cooperatives to defend their prices against the unequal exchange of their products. They have thus implemented a sustainable productive strategy. Their productive practices are founded in principles of sustainable eco-technological-cultural productivity and *environmental rationality* (Leff 1995). They take advantage of the enormous capacity of the Amazon ecosystems to produce biomass, an average of 500–700 tons per hectare. As informed by Manuela Carneiro da Cunha and Mauro Almeida from data offered by Susana Hecht and Steve Schwartzman (1988),

> Recent comparative studies quantified the difference in costs and benefits between cattle raising, agriculture, and extractivism in the state of Acre, taking into account the cost of soil recovery and excluding the global effects of burning or losses of germplasm. Without discounting the price of soil recovery for a 15 years project, extractivism generates average annual profits five times higher than agriculture and 15 times above livestock. If we add the cost for soil recovery [...] in 20 years, the results are negative in the amount of US$ 28,000–55,000 for agriculture and US$ 60,000–100,000 for livestock. Extractivism shows gains of US$ 30,460–50,000. (Da Cunha and Almeida 2000, p. 332)

Thus, agro-extractivism appears as the best sustainable production strategy for the Amazon region for its efficient negentropic management of energy with practices embodied in the reinvented identities and territorialities of the *Seringueiros* and other peoples of the Amazon ecosystems.

The Rexistence of Black Populations

Afro-descendant populations are among the most critical cultural identities and political actors emerging in the field of political ecology. In Brazil, the Constitution of 1988 recognized the rights of Black communities to their territories. These rights are the result of struggles for *rexistence*, after these populations, trying to flee from the regimes of slavery to which they were subjected during the eighteenth and nineteenth centuries, were forced to find refuge in regions of more difficult access (mountains, rugged terrains, floodplains, and savannahs) that happen to be territories of rich biodiversity.

An emblematic case is that of the Black populations of the Colombian Pacific, whose rights to their territories (*palenques*) and their cultures were recognized by the Constitution of 1991 (Escobar 2008). The Process of Black Communities (PCN) of the Colombian Pacific emerged from a project for the conservation of biodiversity as a result of the environmental policies derived from the Rio-92 process. By claiming their rights to participate in this project, these communities initiated an emancipation process leading to the reconstruction of their identities and the struggle for their cultural rights for autonomy and reappropriation of their territory. Hernán Cortés, leader of the PCN movement, expresses the radical imaginaries of his people that are in the roots of their cultural identity, confronting the hegemonic powers of modernity in their struggles to legitimate their rights of existence, opening the perspectives to reappropriate their bio-cultural patrimony and to reconstruct their life-territories. His word intertwines the textures of interculturality and the hybridizing of being within the context of biodiversity, culture, and sustainability as the *enowning* of their historical time in their "ecologic-cultural" existential ontology:

> *Ancestral mandates* determine the relation between afro-descendant peoples and Nature. Those grasp criteria preserved from our African ancestors, others appropriated from indigenous cultures, and criteria that were defined in the process of social and cultural reconstruction in the territories where liberty had been conquered. Our death people are never gone, they remain in the trees, in the brooks, in the rivers, in fire, in rain, on the shore [...] The ancestral mandate: *we all are a big family*, grants us a profound respect to all other beings in Nature that as living beings, the trees, the earth, the animals, water *have rights*. The dynamics of population, mobility, territorial occupation, and use and management practices of biodiversity depend on the conception that the trilogy *territory, culture, biodiversity* is an integral, indivisible whole; the territory is defined as a space for being and biodiversity as that which allow us to remain [...] the Afro-descendant peoples assume Nature as a *bio-cultural system* where social organization, productive practices, religiosity, spirituality and the word [...] determine our *living well* (*Vivir bien*). (Cortés, in Leff 2002, pp. 217–218)

These emergent "hybrid" identities are constructed not only as strategies of resistance in opposition to other identities and hegemonic powers; they are not mere fragmented political identities; they are the renewal of cultural beings constituted as a "*we* for living" in the shared *commons* of Nature: in new life-territories. The struggles for emancipation are struggles for the *rexistence* of cultural beings, for their reappropriation of Nature and the coexistence of diverse cultural beings in a global world ruled by a politics of difference and ethics of otherness. These are not only claims to improve economic and ecological distribution but disputes of meaning for the construction of alternative life-worlds: of other possible worlds emerging from the imaginaries—through the words and practices—of the people, as the reappropriation of their bio-cultural patrimony.

Thus, the emancipation process of black Afro-descendants, as that of indigenous and peasant peoples, is legitimizing the right to cultural difference and collective identities; to "other" knowledge, *savoirs* and practices that confront the truth of positivistic science and modern rationality. These struggles for environmental justice are decolonizing positive law and the power–knowledge devices that have been legalized and institutionalized by the hegemonic dominant order, emancipating new cultural beings and social actors for the construction of a sustainable future (Leff 2012).

The Rexistence of Babassu Coconut-Breaker Women

Babassu (*Orbygnia phalerata*) is a palm that flourishes, in association with other types of vegetation, in flood-plains near the valleys of rivers and small hills of the hot and humid tropics in the states of Pará, Maranhão, Tocantins, Mato Grosso and, outside of conventional Amazonia, in Piauí. Its habitat encompasses 14.5 million hectares of land altogether. Traditionally this was not a natural resource of commercial interest. On the contrary, it was intimately linked to cultural practices for the reproduction of peasant families, particularly in the valley of Mearim in Maranhão. *Babassu* was a resource for free exploitation. Its utilization fulfilled the need for landlords to reproduce the peasant's labor force.

From 1980 on, a robust social movement started to develop through associations that lead to the constitution in the 1990s of the Interstate Movement of *babassu* coconut-breaker women (MIQCB) that includes nearly 300,000 families of agro-extractivist workers.

In recent years, with capitalist expansion and greater access to these lands, many landowners started to prohibit the extraction of *babassu* arguing invasion of land, thus affecting cultural practices consecrated by traditional customs. Conflicts emerged from the different interests related to the exploitation of the palm. In their struggle for land and the defense of *babassu* groves, the *babassu* growers claim their natural means of production and their mode of living to be a way of overcoming poverty and maintaining sustainable livelihoods. These conflicts ended up aligning social groups in encountered positions, leading to the affirmation of collective identities that, given the importance of female and infant labor, propelled the coconut-breaker women into the political-cultural scene in a struggle against the devastation of *babassu* groves (*babaçuais*) and to put an end to the prohibition of collecting it, that is, to "*free babassu*." In the II Interstate Meeting of *Babassu* Coconut-Breakers, held in Teresina, PI, in 1993, these women demanded:

> 1. Disappropriation of all areas in conflict in the region of *babaçuais*; 2. Giving free access to *babassu* palms for women and children dedicated to its extraction, including private properties that do not fulfil a social function; 3. An end to the cutting of *babassu* palms; 4. An end all violence against rural workers in *babassu* grove zones; 5. Support for the development of cooperatives; 6. Proceeding to immediate implementation of settlement actions in already disappropriated areas and extractive reserves; 7. Compliance with the Statute of Children and Adolescents in Rural Zones; 8. Measures to ensure compliance with the Decree of Extractive Reserves. (Porto-Gonçalves 2001b)

Thus, the Law for Free *Babassu* (*Lei do Babaçu Livre*) was promoted, having been approved in different municipalities, as in the Lago do Junco, Lago dos Rodrigues, Esperantinópolis, São Luís Gonzaga and Imperatriz, in Maranhão; Axixá, Praia Norte e Buriti, in the State of Tocantins; and in São Domingos do Araguaia, in the State of Para. By this law, everybody can enter any property to collect coconuts for their needs, without interfering with the activities of the owner. The law prohibits the felling of palm trees and the use of chemical products. As a result, the income of these populations has increased as they diversified the self-use and commercialization of such *babassu* products as oil, soap and fleshy mesocarp for flour.

Fishing Reserves of Amazon River Dwellers

Fishing is one of the leading traditional activities developed in the largest hydrographical basin in the world: the Amazon; it is a practice inscribed in a polyvalent mode of living of ae population; it is an essential source of food. The riverside *caboclo* is one of the most typical Amazonian figures. Their current practices reflect the diverse cultures of indigenous peoples, Portuguese immigrants, migrants from the northeast and Black populations that converge and hybridize in their identities and practices. By dwelling in the flood-plains in the margins of the rivers, they have developed a deep, complex environmental knowledge from their coexistence with rivers and forests. These Amazonians have been characterized as *polyvalent fishermen* to differentiate them from the *univalent fishermen* that live from commercial fishing. In their vision and practices of Nature, forests and rivers are interconnected and interdependent; their modes of production and living are intertwined with the Amazon ecosystems in their agricultural, extractive, and fishing practices.

These *polyvalent* fishermen have survived colonial domination through their environmentally sound productive rationalities. Generally, they live in small villages and places located on the river margins—*igarapés, furos,* and *paranás*—managed with their traditional techniques through a rich tradition in the construction of boats and houses adapted to their ecosystems. They divide their time into cyclical activities related to terrestrial ecosystems such as small hatcheries, agricultural tillage, hunting, collecting, and extracting during the year; they work their own or rented land, raise cattle, cultivate *juta,* hunt, collect seeds, fruits, resins, and wild fibers; they produce coal, they mill cassava flour and fish, first for self-consumption and then for commercialization. The riverside populations of extractive fishermen-farmers have managed their fragile, complex, and productive ecosystems for years without any support from public policies.

These riverside communities are now involved in territorial conflicts confronting the present process of appropriation of natural resources by capitalistic enterprises, resulting in overfishing that threatens the sustainability of their livelihoods. As Lourdes Furtado explains,

> In the view of the inhabitants of the lake banks that depend on their ichthyological resources, this pressure over resources produces an impact as fish become scarce and the fishermen have difficulties in getting a good catch, sufficient to fulfil their needs for food and for selling a small surplus to acquire other goods; this threatens the sustainability of the stock of fishing resources and the food chain that allows the renewal of the necessary resources for life [...] generating expectations and internal tensions that many times lead to conflicts that manifest in different actions, from simple warnings to the "invader", the removal and burning of *malhadeira* nets, taking away fishing boats and canoes, up to the prohibition of fishing in the lakes. (Lourdes Furtado 2002, p. 38)

Furthermore, David Mcgrath points out the fact that:

> As pressure on varzea fisheries has increased, *riberinho* communities have attempted to assert control over local *varzea* (floodplain) lakes and exclude fishermen from outside the community. In a number of cases, *riberinho* communities have closed lakes to outsiders and established informal lake reserves under local community management. These lake reserves are a promising strategy for managing lake fisheries on a sustainable basis. (McGrath et al. 1993:167)

Important parallelisms and differences exist between the Lake Reserve and Extractive Reserves:

> The Lake Reserve is a form of land use very similar to the Extractive Reserve. Both of them are efforts of traditional populations to guarantee their access to resources that are the basis of their local family economies, preserving their livelihoods. If both of them search to preserve natural ecosystems and are based in traditional forms of occupation, there also exist

> essential differences between them due to the characteristics of the primary resources. In the case of the Lake Reserve the mobility of the fishing resources makes it unviable to create individual territories. All fishermen exploit the same population of fish, and in general, each fisher's production affects the productivity of others. Even though in the land around the lake there are individual owners, the lake is considered a "common good" for collective management, involving all fishermen of the community. In this context, the economic viability of the reserve depends not only on the rules established but also on the quality of the community organization, especially on the participation of the fishermen in the fishing agreements defined by the community. (ibid., p. 39)

From the dialogue of knowledge established between those populations of fishermen and conventional technicians, new proposals emerge that seek to overcome the colonial-modern paradigm that separates Nature and culture. A good example is that of the "ban-wage." The ban or prohibition is an instrument of protection of a species when limitations to fishing are set during the periods of reproduction. With the application of the ban-wage, fishers receive a minimum salary throughout that period, while at the same time they learn and develop a series of activities to diversify their sources of food.

The ban-wage proposal represents an innovation in the economic and juridical system to overcome the dichotomy between Nature and society. Thus, labor and Nature are revalued to avoid the overexploitation of labor and Nature by the traditional theory of value and the prevalent economic system that pays only for seasonal labor time. The ban-wage considers the time necessary for the reproduction of natural resources and the worker. By internalizing the ecological and cultural conditions for the sustainable productivity of biomass in a determinate ecosystem, that is, by considering the ecological and cultural value of the process of production, and not only its present market price, society as a community assigns itself the responsibility to preserve Nature—the river, the shore, the lake, the fish—as well as the fishermen's culture.

This reordering of production leads to revaluation of the fishermen's traditional knowledge, but moreover comes to consider him (or her) a citizen of rights—and the bearer of such rights—including the rights to inhabit their territories and to construct their economy according to their cultural values. Thus, environmental rationality opens new perspectives for the construction of sustainable societies beyond the adjustments of market prices or Marxist value theory (see Chaps. 6 and 7), through the innovative actions of emergent social actors that institute new productive–extractive practices that are culturally embodied and ecologically embedded in their territories.

Experiences of Community Forestry in Mexico

The experiences of community forestry in Mexico offer another excellent example of cultural management of biologically diverse territories and forest products by peasant and indigenous communities oriented towards the social construction of sustainable local economies (Merino and Segura 2002). Unlike the Chilean experience of large monoculture forest plantations—that is developing in vast extensions in Brazil for tradable forest products like cellulose—community forestry is based in the management of the biodiversity of natural forests. Forest coverage in Mexico is 127.6 million hectares, of which 63.5 million hectares are forests and rainforests. Mexico is the country with the largest certified area of communal land under social forestry management. Eighty per cent of these forests are under social ownership (*ejidos* or communal property). Following David Bray "A study of the ten most important forest states with 1,730 community forestry enterprises (75% of the total), showed that 163 communities (10 percent of the ten states) had achieved levels of collective action and forest industrial integration" (Bray 2008). In the State of Oaxaca alone, about 150 forest communities practice community forestry management in 650,000 hectares (Boege 2008). Until 2005, there were 26 certified communities and *ejidos* covering 587,143 hectares (Alatorre 2003). In general, these forests are located in mountain landscapes that contain in their different ecological levels some of the most abundant terrestrial biodiversity; thus, the management of the common forest goods implies the conservation of genetic, species, and ecosystems diversity, of their ecological structure and environmental services.[1]

[1] For an extensive bibliography on community forestry in Mexico, see Cossio et al. (2006).

Similar to the experience of extractive reserves in Brazil, social forestry in Mexico is the result of an intense struggle for the reappropriation of natural resources, previously in the hands of state or private concessionaires. The taking of lands, erection of sawmills, and introduction of transport units, leading to legal struggles against concessionaires, is part the process that led to the reinvention of the identity of the community for innovative management and new institutional frameworks to construct culturally and ecologically based sustainable forestry practices.

The more successful experiences on sustainable forest management are those where communities have undergone a process of reappropriation and reconstruction of their traditional knowledge and practices. In Quintana Roo, for example, tappers (gatherers of *chicle* resin from sapodilla, *Manilkara zapota*) who came in the first half of the nineteenth century from the state of Veracruz learned from the Mayas the names of local vegetation, wildlife behavior, traditional medicine, interpretations of the myths about annual rain cycles, and the complex classification and uses of soils. After the cancelation of forest concessions at the beginning of the decade in 1890, the local peasantry appropriated the techniques that for 25 years had been in use by a large state enterprise to harvest 600,000 cubic meters of cedar and mahogany in the natural forest.

Forest inventories are an essential requirement to maintain a constant harvest of commercial wood, without undermining the resource. The first forest inventories generated a collective intellectual appropriation process. The forest technicians who worked in the zone were trainees in forestry of temperate areas. Maya knowledge and wisdom of the tropical forest corresponded to another productive logic. The community assembly decided that the majority of its associates should participate in the inventories, opening paths, measuring, classifying and learning sampling techniques. Thus a horizontal exchange of knowledge took place, including the names of trees and soils following the Maya knowledge and the assimilation by the community of the resources from its territory.

Peasant forestry enterprises constantly innovate their production strategies in a dialogue of knowledge between researchers, technicians, and the local people involved. This peasant forestry strategy involves a process of reappropriation and blending of local knowledge with other technical knowledge. In this context, a constant negotiation takes place between the technical knowledge of forest engineers and traditional practices of forest regulation.

With the experiences of communal forestry in Mexico, as in other cases in Petén, Guatemala, the principles of community autonomy are being validated as a basis for an alternative project for peasant and indigenous peoples' communities and *ejidos*. These experiences contribute to defining alternative paths for the construction of local economies based on the self-management of sustainable territories by indigenous and peasant communities. They are significant laboratories to test and construct new forms of cultural and social occupation of the territory, respecting the potentials of biodiversity and cultural diversity. These new strategies of territorial ordering and natural forest management from multiple cultural and productive approaches are incorporating new agro-ecological and agroforestry approaches, multiple use of ecological levels and soil-management techniques, the conservation of endangered species and sustainable management of wildlife, with new rules for the collective access and use of shared resources for the sustainable management of complex forest and biologically diverse ecosystems based on the principles of environmental rationality.

Social Actors for the Construction of Sustainable Territories

The community agroforestry experiences in Mexico, as well as the experiences of the Extractive Reserves in Brazil, are examples of the invention of new productive rationalities for the sustainable appropriation of Nature in the conflictive terrain of political ecology, where local peoples' territorialities are confronted by the techno-economic power strategies of global capital. Thus, the indigenous peasant population of Los Chimalapas, located between the States of Oaxaca and Chiapas, has been struggling to gain control over an area of 600,000 hectares of tropical rainforest in southern Mexico, to create a *Peasant Ecological Reserve* (Reserva Ecológica Campesina de Los Chimalapas) (García 2014). The invention of this concept of Peasant Ecological Reserve parallels that of the Extractive Reserves of the *Seringueiros* in Brazil. Both of them stress the ecological (biodiversity) and sociological (peasantry) basis of these strategies that define the innovative and radical quality of these alternative paradigms and common practices of sustainability. Both the Peasant Reserves of Biodiversity and the Extractive Reserves go beyond the concept of Environmental Conservation Units, where the population is excluded. Here the local population becomes the main protagonist in the management of natural resources through existential experiences and the innovative processes that enact the cultural *enownment* of Nature in the ontological turn of political ecology.

Today, in the context of economic globalization and the geopolitics of sustainable development, new territorialities are being configured: it is no longer a struggle of national states to expand its limits, and not only the tensions emerging from the emancipation of indigenous peoples and other cultural groups within national states for the constitution of plurinational states. The displacement of peoples and the entropic degradation of their territories by the expansion of capital globalization, as well as emerging democratization processes in the modern world order are triggering emancipation processes oriented towards the construction of sustainable territories based on legitimization of cultural and environmental rights that offer the conditions for other social actors to enter the political arena, encountering the National State and its internal colonialism as well as the hegemonic World Economic System.

In the field of political ecology, a diversity of new political subjects are emerging. New voices are expressing environmental demands, such as those in the People's Summit of Climate Change and the Rights of Mother Earth celebrated in Cochabamba, Bolivia, in April 2010. It was precisely in Cochabamba where previously in the year 2000, the Water War (Guerra del Agua) involved peasants, indigenous peoples, environmentalists, and urban movements, to chase away Bechtel, the multinational enterprise involved in the privatization of water, to be followed later by the Gas War in 2003, and the raising to power of Evo Morales, the first president elected as the result of an indigenous and peasant movement, in 2005.

Peasant and indigenous peoples movements—the "*Campesinos*" (Bartra 2008) and the "*indigenato*" (Ribeiro 1980)—are undertaking a relevant role, especially after 1992, to reverse, from a long history of colonization and exclusion, of cultural subjugation, the deterritorialization and destruction of their patrimony of natural resources. Their emancipation from that dominating process involves the politicization of their ancestral territories. Their demands for territorialization go beyond traditional struggles for land. These struggles incorporate new rights to reappropriate their bio-cultural patrimony and to reconstruct their territories in the perspectives opened by sustainability for the destiny of humanity facing the environmental crisis and of life in the planet; against the processes of deterritorialization/expropriation and for the defense of their cultures and cultural diversity. Their struggles involve the defense of their natural conditions of existence under which they develop their cultural values and the meanings of their social practices (Boege 2008, 2018). Thus, the triad territory–territoriality–territorialization emerges in the core of political ecology.

In the crossroads of the contradictions of the colonial-modern world system and the resulting environmental crisis, the different social classes and ethnic groups revive their historic resistance to the expropriation of their territories, their lands and ecological conditions of existence, reconfiguring the ontological–epistemic–political debate on the increasing environmental risks and the dilemmas facing sustainability of life in the planet (Leff 2018). These emergent social movements introduce social and cultural questions into the field of political ecology. The slogan of Chico Mendes "There can be no defense of the forest without the peoples of the forest" synthesizes the social struggles in different places in the world: the Chipko movement in India, the conflict in Los Chimalapas in the tropical forest in between the states of Oaxaca and Chiapas in Mexico, the Process of Black Communities of the afro-Colombians of the Pacific, the Articulation of Peoples of the Savannahs (*cerrados*) in Brazil; the fishing agreements (*acordos de pesca*), the "closed season wages" (*salário defeso*), the marine reserves, and other forms of territorial configurations proposed by socio-environmental movements. In different geographic contexts culture is being politicized by the territorial question through struggles over the appropriation of Nature (Porto-Gonçalves 2004).

Even in countries where indigenous populations are relatively small in demographic terms, they deploy a broad cultural and linguistic diversity, as is the case of Brazil, where they represent only 0.4% of the total population, but speak 188 different languages, five times more than in Bolivia where they speak 32 languages, and the population corresponds to 62% of the total population. This opens a new ethical-political dilemma. Beyond the rights of existence of all peoples and their cultures, these populations occupy areas rich in biodiversity and ecological potentialities that are being threatened by regional integration projects such as the Initiative for the Integration of the Regional Infrastructure of South America (IIRSA) launched in 2000 with the participation of the 12 countries of South America which form the Union of South American Nations, supported by the Corporación Andina de Fomento (CAF), the Inter-American Development Bank (IDB) and the River Plate Basin Financial Development Fund (Fonplata) to link Southern economies through new transportation, energy, and telecommunications projects (to integrate highway networks, river ways, hydroelectric dams, and telecommunications links throughout the continent) to allow greater trade for the South American community of nations; or the contested and failed Mesoamerican Integration and Development Project—the Puebla–Panama Plan—launched in 2001 to promote the regional integration and development of the nine southern states of Mexico with all of Central America and Colombia.

These projects are opening these areas to regional infrastructure projects as well as to transnational enterprises for the exploitation of minerals, land, and water resources for agro-business and new commodity exportation. These territories have become strategic areas for alternative and confronting rationalities and interests, fronts of dispute for the appropriation of natural resources.

Indigenous peoples and peasants are taking new stands in the political ecology debate, escaping the political traps of multicultural postmodernity that keeps each cultural group apart, stimulating essentialism and xenophobia. Their struggles focus on decolonization, interculturality, and dialogue of knowledge. They construct their new rights for cultural being that entail their reidentification with Nature, the reappropriation of their history and their patrimony of natural resources, reinventing their life territories and reimagining their future. In reconstructing their territories, they redefine their notions of time and space, different from those of the hegemonic Eurocentric ontological thinking. Thus, peoples like the Aymara and Quechuas have their cosmogonies of the *Pacha*, a universe ordered in spatial-temporal categories characteristic of the Andean cultural rationality (Esterman 2006).

The emancipation of peoples in the reappropriation of Nature is starting to be reflected in significant transformations of the state at the national level. Thus Bolivia has been re-established as a plurinational state. Ecuador was the first country to introduce the rights of Nature into its constitution, followed by Bolivia. Also in Ecuador, following the initiative of the indigenous–peasant-environmental movement, the government of Rafael Correa tried to negotiate a project to maintain underground the oil in the National Yasuni Park, in the Andean-Amazonian foothills, by collecting half of the value that could be obtained with its exploitation. Breaking apart the compensation mechanisms such as the REDD and other economic transaction instruments in the CDM—that assign to poor countries the role of absorbing the surplus emissions of greenhouse effect gases while the rich countries keep expanding their ecological footprint—here the intention was to stop extracting oil—with the risks of polluting the natural ecosystems—to the benefit of the local indigenous peoples and of the planet as a whole, and to apply this revenue to clean energy and socio-environmental sustainable projects (Vogel 2009). The failure of this project due to the lack of commitment from the "International Community" and the inconsistency of ex-president Correa and his successor Lenin Moreno to comply with its principles, as well as those of the government of Bolivia regarding the Tipnis conflict, prove the difficulties of these "progressivist" governments to de-link from the logic and interests of the dominant economic order and to construct effective alliances with the peoples of the Earth in the construction of a sustainable planet.

Notwithstanding the lack of solidarity and consistency from governments in opening up paths to the construction of a sustainable and diverse world order, these emergent social movements are achieving the legitimization of new cultural and environmental rights, challenging the juridical system for the construction of the collective rights of the peoples of the Earth to the commons of the planet. This no longer implies broadening the scope of the hegemonic law system based on the principles of individual rights and private property as a means to deal and solve the "tragedy of the commons" (Hardin 1968), denying the customs in common (Thompson 1991) and customary rights of communities. These peoples claim intellectual property rights of another order (Posey and Dutfield 1996; Posey 2004; Leff 2014b). The original, traditional and communal knowledge of these communities is a common and collective patrimony that demands new juridical rationality which recognizes their communal and communitarian character and avoids reducing them to the principles of positivistic science and law that are contrary to their cultural values and foundations.

These socio-environmental struggles are redefining the power relations in the field of political ecology for the social reappropriation of Nature. As they gain legitimacy, they are incorporated into legal arrangements and public policies. There have been critical formal developments in recognition of the emergent indigenous peoples' rights in countries like Brazil, with its Constitution of 1988, Colombia, with Laws 70 and 121 in its Constitution of 1991, and the Organic Law of Indigenous Peoples in Venezuela. However, there have been enormous difficulties with their implementation and enforcement not only due to the factual powers that dispute their territorialities—as in the case of the Colombian Pacific—but also to the persistence of a nationalist development ideology deeply embedded in the power structures of the world-system and the national state, that continues to ignore the cultural rights claimed by the original peoples to live and produce within Nature, supported by the negentropic potentials of their territories, and according to their traditional knowledge and practices. The legitimation of the emergent environmental rights of the peoples of the Earth is leading to increasing conflicts arising from the tension between the interests derived from global economic rationality and the construction of environmental rationality in the transition to sustainability.

In Brazil, because of the defection of the state in recent years to neoliberal hegemony, the struggles of the *Seringueiros* that established new relations between the state and civil society have been under great stress to complete their purpose and extend to other communities. The Extractive Reserve was the first modern proposal to break from the paradigm of management of Nature reduced to conservation units that separate Nature and culture. In the Extractive Reserves, as previously in all traditional practices and cultural co-evolution with Nature, populations' knowledge is a *sine qua non* of conservation. Thus, a new paradigm was territorialized based on the culture of the people and in the access to land as a territory. The same principle applies to the demarcation of afro-Brazilian' *quilombos*, whose vast territories were reinvented by the *Maroons* to free themselves from slavery after having been ignored by the dominating culture and kept willingly invisible as a survival strategy. Today, these territories represent new opportunities for those people to reinvent and regain control over their modes of existence while becoming an obstacle to new fronts of expansion of capital.

The underlying question that surfaces amid these territorial struggles is a substantive issue that has remained invisible in the mainstream environmental debate. What is at stake is the *social reappropriation of Nature* and the fight to build sustainable societies founded in the diverse ecological and cultural conditions of the peoples of the Earth, the construction of *territories of difference* (Escobar 2008). The exploitation of Nature is not only a condition for the domination of some men over others, of rich countries over poor countries, of hegemonic rationality over subjugated cultures. Human beings, as well as life on the planet, have become hostages of a struggle to appropriate limited Nature by confronting rationalities. The techno-economic domination of life is expressed in conflicts of territorialities where alternative cultural and civilizing projects equip themselves for the appropriation of the natural conditions of human existence. These are the profound implications of the present social struggles for *reappropriation of Nature*, triggered today by the confrontation of a globalized unsustainable economic world-system and the emancipation of populations with strong territorial bonds, of many original and indigenous peoples that have managed to survive; of peasants, maroons, and landless peoples who, having been deterritorialized, are claiming their rights to reterritorialization and *rexistence*.

The territorial struggles for sustainability in the field of political ecology go beyond the old debate on development/underdevelopment trapped in the ideal of unlimited growth and progress. The theoretical–political debate on the destiny of humanity and life on the planet is now posed in terms of the alternatives for a sustainable future: from the doubtful possibilities of techno-economic solutions to the entropic death of the planet triggered by unsustainable economic growth, to an environmental rationality based on the negentropic conditions for life in this living planet, guided by new horizons of meaning, as those expressed today by the "living well" (*Suma Kawsay* or *Suma Qamaña*) of the people of the Earth (Huanacuni 2010).

These socio-environmental questions emerge today in the shaky ground of economic globalization, environmental crisis and climate change where the Latin American geography is confronting one of the most violent expropriation processes in the whole history. The prevailing economic developmental regime promotes the construction of highways, hydroelectric dams, thermoelectric and polluting extractive mineral enterprises over fragile ecosystems and traditional territories; the expansion of the agricultural frontier to new transgenic *latifundia* and monoculture forest plantations; deforestation and the reduction of the rich ecological and biodiversity potentials of Latin America to function as mere carbon sinks, with the purpose of absorbing the surplus emissions of greenhouse gases of the more polluting industrialized countries and emergent economies through the strategies of a "green economy." These are the dominant transformation processes generated by the compliance of governments with the new geopolitics of "sustainable development." Thus, Latin American and other Third World countries face a new wave of territorial conflicts, more complex and at a larger scale than the ones that were triggered by the "green revolution" and the construction of hydroelectric dams that displaced peoples from their territories and altered the ecological processes of the region profoundly.

The renewal of exploitative colonial-modern rationality today, driven by the growing capitalization of Nature is generating new territorial conflicts. An excellent example of the political tensions and the confronting processes of territorialization between capitalistic and environmental rationalities is the case of the recent struggle over TIPNIS—Tierra Indígena del Parque Nacional Isiboro Securé—located in the Andean–Amazonian foothills of Bolivia. The defense of this "cultural ecosystem" has mobilized indigenous peoples against the construction of a highway under the IIRSA project planned to cut through the middle of their territories to open a way to the Pacific for Brazil. Likewise, the invasion of mining enterprises in Latin America is generating all sorts of resistance movements. One example is that of the National Confederation of those

Affected by Mining (Confederación Nacional de los Afectados por la Minería, CONACAMI) in 2009 in Peru that was involved in one of the most violent conflicts of recent years in Bagua—where dozens of people, including military, died in the confrontation of indigenous peoples with mining industries that sought to expand over the Andean-Amazonian confines where these people live—followed by a more recent conflict in Cajamarca, in 2011. In Brazil, tense relations prevail with indigenous and peasant movements that are resisting the construction of dams in Belo Monte in the Xingu river, as well as in Jirau and Santo Antonio in the Madeira river, and the São Francisco river.

The Social *Enownment* of Nature and the Construction of Environmental Rationality

Recent experiences of socio-environmental movements in Latin America, as elsewhere in Asia and Africa, are showing the capacity of the local people not only to resist the hegemonic economic, instrumental, and utilitarian reasoning by which capitalism penetrates their geographic-socio-cultural life territories, but also to create alternative visions and new paths to sustainability from their cultural rationalities, reinventing their identities, their productive practices, and their livelihoods. They provide evidence that the traditional populations are holders of knowledge and practices interwoven with the ecological conditions for sustainable management of their environments. These successful experiences are legitimating new human rights to the territory and opening innovative perspectives for sustainability driven by social movements for a cultural reappropriation of Nature.

While the globalization process is penetrating every territory and ecosystem, every culture and individual, with its unsustainable techno-economic rationality, these new socio-environmental movements are constructing an alternative globalization from the potentialities of their ecosystems, their cultural identities, and their local autonomies, sustained and articulated by alternative *environmental rationality*. Beyond the purpose of facing the environmental crisis by assigning an economic value to Nature and culture, environmental rationality guides the construction of a sustainable global civilization by integrating a diversity of organizational processes oriented by the negentropic principle of life and cultural signification of Nature. This process encompasses the reconfiguration of cultural identities and the emergence of new social actors capable of innovating new productive processes based on the ecological potentials of Nature and the cultural creativity of the peoples of the Earth.

Traditional knowledge, oppressed and dominated by scientific, economic, and technological rationality of modernity, is being reconstructed in new local settings, hybridizing in conflicting ways traditional practices with modern scientific and technological knowledge. Traditional societies and local economies do not only produce use and exchange values; they also generate "meaningful use values" that reflect the complex relation of the natural and the symbolic order in socioeconomic and political relations of production. Under this rationale, Nature is not submitted to the strategies of sustainable development and guided by the dominant economic rationale. Environmental rationality deconstructs economic rationality by constructing an eco-technological-cultural paradigm of production founded on the principle of *negentropic productivity* (Leff 1995, 2019). The conditions of life and the inventions of diverse cultures, registered in the imaginaries and practices of the peoples re-emerge today under processes of resignification, reaffirmation, and updating of cultural identities in the reterritorialization of life-worlds.

While modern rationality tends to dissolve geographic referents and cultural meanings, space and place are being "reinvented" from the core of the emergent cultural identities to embody and root the conditions for sustainable societies in new life-territories. This change of rationality goes beyond the objective of grounding global economic rationality locally to establish a balance between ecological conservation and economic growth. The construction of sustainability rooted in principles of environmental rationality is the embodiment and grounding of new material forces, ecological conditions, cultural values, and symbolic meanings. The place becomes the *locus* for rooting diversity, where Nature and culture coexist in the complexity of natural processes and diverse cultural beings that construct their territorialities in different space and time settings. The concreteness of sustainability is defined in the encounter and convergence of different matrixes of cultural rationalities; in a *dialogue of knowledge* that constitutes different cultural beings in their relation to their imaginaries, their *savoirs*, and practices, in their inhabited environments (Leff 2004, 2014a, 2018).

Anchored and driven by these socio-environmental movements, crossed by tense conflicts in the complex matrix of rationality of modernity—the hegemonic homogeneity of instrumental economic rationality—the social construction of environmental rationality is emerging from the confluence of new ideas, projects, and rights, where old concepts—territory, autonomy, self-management, local knowledge—are being resignified, reconfiguring new territorial identities and new productive strategies. Thus, new socio-environmental movements are emerging in the field of political ecology searching to construct a sustainable future with social and environmental justice, with cultural and territorial diversity, with other horizons of meaning. From peoples' imaginaries of sustainability, cultural rights for the reappropriation of Nature are being created and legitimized, reconfiguring a new political agenda based in the cultural and natural patrimony of the peoples, in the reinvention of their identities and their life territories.

Final Remarks

The environmental crisis irrupted in the world in the 1960s, raising new epistemic and political challenges. Among the diverse issues debated in the conflictive field of political ecology, a radical question characterizes the socio-environmental movements in Latin America: the definition of new paths for constructing sustainable societies based on their ecological potentials and cultural identities. It is a struggle for an alternative social rationality through the recognition of sustainable societies' rights of existence, a reappropriation of their bio-cultural patrimony and the reconstruction of their life-territories.

Through conquest and colonization, the original peoples experienced a long history of exploitation and deterritorialization. After 500 years of resistance, they have started a process of emancipation and decolonization, a struggle for the reappropriation of their historical patrimony of natural resources and reinvention of their cultural identities. Countering the strategies of "sustainable development," these emergent social movements are opening new horizons of meaning for human history based on their social imaginaries for sustainability. The principles of these new cultural and environmental rights embrace political ethics that involves a critique of current trends of development, political domination, and exploitation of Nature. It is a claim to the rights to *rexistence*, to build their sustainable life-worlds based on their worldviews and forms of cognition, their cultural ways of inhabiting the planet and their own territories, establishing new relations with Nature and with other human beings: a spiritual and material balance with the cosmos, their ecological environment, and their social relations. These emergent social actors—indigenous peoples, peasants, and afro-descendants—place cultural autonomy and diversity at the center of the political ecology debate.

This response to the environmental question emerging from the emancipation process of the peoples of the Earth is molding a new *environmental rationality*, mobilizing their imaginaries and practices for the social construction of sustainable societies grounded in the conditions of life on the planet. A New World can still be possible, based on the ontology of diversity, a politics of difference and an ethic of otherness, where many different worlds can live in harmony, reopening the courses of history and the destinies of life to a multi-colored horizon through a dialogue of knowledge. A *wager of life* underlies and identifies Latin American political ecology.

Bibliography

Abramovay, R. (2010), "Reduzir a desigualdade entre os indivíduos para combater o aquecimento global", *Boletim da Sociedade Brasileira de Economia Ecológica*, special edition N° 23/24 January–August 2010, pp. 12–15.

Ab'Saber, A.N. (1970), "Províncias geológicas e domínios morfoclimáticos no Brasil", *Geomorfologia* 20:1–25, São Paulo.

Alatorre, E. (2003), *El proceso de certificación forestal en México*. Dirección de Certificación Forestal. Consejo Civil Mexicano para la Silvicultura Sostenible en México.

Barbados Declaration (1971), International Work Group for Indigenous Affairs/ World Council of Churches.

Bartra, A. (2008), *El Hombre de hierro. Los límites sociales y naturales del capital*, Mexico City: Universidad Autónoma de la Ciudad de México.

Boege, E. (2008), *El patrimonio biocultural de los pueblos indígenas de México. Hacia la conservación in situ de la biodiversidad y agrobiodiversidad en los territorios indígenas*, México: INAH/Comisión Nacional para el Desarrollo de los Pueblos Indígenas.

———. (2018), "Hacia una antropología ambiental para la apropiación social del patrimonio biocultural de los pueblos indígenas", in Toledo, V.M. & Alarcón-Cháires, P. (Eds.), *Tópicos bioculturales: reflexiones sobre el concepto de bioculturalidad y la defensa del patrimonio biocultural de México*, México: UNAM.

Bray, D. (2008), "Collective Action, Common Property Forests, Communities, and Markets", Commons forum Commentary.

Campos, N. (2000), *Terras de uso comum*, PhD thesis, Programa de Pós-graduação em Geografia da Universidade de São Paulo, São Paulo.

Cornetta, A. (2011), *Mecanismo de Desenvolvimento Limpo e Conflitos Locais*, XIII - Encontro de Geógrafos de América Latina 2011. San José, Costa Rica.

Cossio, R.E., Bray, D.B., Bult, S. & Merino, L. (2006), "Bibliografía anotada del manejo comunitario de los bosques en México, con bibliografía adicional sobre los bosques de México en general (Working Paper No. 2)". *LACC Working Paper Series*, http://digitalcommons.fiu.edu/laccwps/11.

Da Cunha, M.C. & Almeida, M. (2000), *Indigenous people, traditional people and conservation in the Amazon, Daedalus* Vol. 129 (2): 315–338.

Derrida, J. (1978), *Writing and difference*, Chicago: The University of Chicago Press.

Escobar, A. (2008), *Territories of difference. Place, movements, life, redes*, Durham/ London: Duke University Press.

Esterman, J. (2006), *La filosofía andina*, Quito: Ed. Abya Yala.

Foucault, M. (1980), *Power/knowledge*, New York: Pantheon.

Furtado, L. (2002), *Pescadores do Rio Amazonas. Um estudo antropológico da pesca ribeirinha numa área amazônica*, MCT-CNPq-Museu Emilio Goeldi.

García, M.Á. (2014), "La reserva ecológica campesina de Los Chimalapas", en Betancourt, A. et. al., *Del monólogo a la polifonía: proyectos supranacionales y saberes indígenas en la gestión de la áreas naturales protegidas (1990–2010)*, México: UNAM.

Haesbaert, R. (2011), *El mito de la desterritorialización*, México: Siglo XXI Editores.

Hardin, G. (1968), "The tragedy of the commons", *Science* No. 162: 1243–1248.

Harvey, D. (2003), *The new imperialism*, Oxford: Oxford University Press.

Hecht, S.B. & Schwartzman, S. (1988), *The good, the bad and the ugly: extraction, colonist agriculture and livestock in comparative economic perspective*, Los Angeles: Westing Papers, UCLA.

Huanacuni, F. (2010), *Vivir bien/Buen vivir: Filosofía, políticas, estrategias y experiencias regionales*, La Paz: Convenio Andrés Bello/Instituto Internacional de Integración.

International Indian Treaty Council (2000), http://www.treatycouncil.org/.

Jalée, P. (1968), *Le pillage du Tiers Monde*, Paris: François Maspero.

Leff, E. (1995), *Green production: towards an environmental rationality*, New York: Guilford.

———. (2001), *Epistemología ambiental*, Sao Paulo: Cortez Editora.

———. (2004), *Racionalidad ambiental: la apropiación social de la naturaleza*. México: Siglo XXI Editores.

———. (2012), "Latin American environmental thinking: a heritage of knowledge for sustainability", *Environmental ethics*, Volume 34:4, Winter, 2012, pp. 431–450.

———. (2014a), *La apuesta por la vida: imaginarios sociales e imaginación sociológica en los territorios del Sur*, México: Siglo XXI Editores.

———. (2014b), "Los Derechos del Ser Colectivo y la Reapropiación Social de la Naturaleza", en Pereira de Cunha, Belinda; Agustín, Sergio; Araújo da Costa, Nálbia Roberta (Org.), *Saberes Ambientais, Sustentabilidade e Olhar Jurídico: visitando a obra de Enrique Leff*, Joao Pessoa e Caxias do Sul, Brasil.

———. (2018), *El fuego de la vida: Heidegger ante la cuestión ambiental*, Mexico: Siglo XXI Editores.

———. (2019), "Negentropic production", in Kothari, Ashish, Demaria, F., Acosta, A., Salleh, A. and Escobar, A. Eds., *Pluriverse: a postdevelopment lexicon*. London: Zed Books.

Leff, E. (Ed.) (2002), *Ética, vida, sustentabilidad*, México: PNUMA, Serie Pensamiento Ambiental Latinoamericano No. 5.

Mariategui, J.C. (1971), *Seven interpretive essays on Peruvian reality*. Texas Pan American Series, Austin: University of Texas Press.

Meadows, D.H., D.L. Meadows, J. Randers & W.W. Behrens III (1972), *The limits to growth*, New York: Universe Books.

Merino, L. & G. Segura (2002), *El manejo de los recursos forestales en México, 1992–2002. Procesos, tendencias y políticas públicas*, in Leff, E., E. Ezcurra, I. Pisanty & P. Romero (Ed), *La Transición Hacia el Desarrollo Sustentable. Perspectivas de América Latina y el Caribe*, México: PNUMA/INE-SEMARNAT/UAM, pp. 237–256.

McGrath, D.G., F. de Castro, C.R. Futema, B.D. de Amaral, and J.C. de Araujo (1993), "Fisheries and the evolution of resource management on the lower Amazonian floodplain. Human Ecology" 21(2): 167–195.
Porto-Gonçalves, C.W. (2001b), *Amazônia, Amazônias*, São Paulo: Contexto.
———. (2002), "O latifúndio genético e a r-existência indígeno-camponesa", *Geographia*, Ano 4, No. 8, Niterói: Universidade Federal Fluminense, pp. 7–30.
———. (2004), *Geografando nos varadouros do mundo: da territorialidade seringalista à territorialidade seringueira*, Brasilia: IBAMA.
———. (2006), *A Globalização da natureza e a natureza da globalização*, Rio de Janeiro: Civilização Brasileira.
———. (2007), *Em defesa dos organismos laboratorialmente modificados: em busca da precisão conceitual*, http://www.geopolitica.ws/article/em-defesa-dosorganismos-laboratorialmente-modificados/.
Posey, D.A. (2004), *Indigenous knowledge and ethics: a Darrell Posey reader*, New York: Routledge.
Posey, D.A., & Dutfield, G. (1996), *Beyond intellectual property: toward traditional resource rights for indigenous peoples and local communities*, Ottawa: IDRC.
Quijano, A. (2008), "Coloniality of power, eurocentrism, and social classification", in Moraña, M., Dussel, E. and Jáuregui, C.A. (Eds.), *Coloniality at large. Latin America and the postcolonial debate*, Durham, NC: Duke University Press, pp. 181–224.
Raffestin, C. (1980), *Pour une géographie du pouvoir*, Paris: Librairies techniques.
Ribeiro, D. (1980), *Indigenato e campesinato*, *Revista de Cultura VOZES*, Ano 73, Vol. LXXIII, Outubro de 1980, n° 8, p. 5–10, Rio de Janeiro.
Ribeiro, B.G. (1992), *Amazônia urgente: Cinco séculos de história e ecologia*, Belo Horizonte: Ed. Itatiaia/INEP-MEC/CNPq/Vitae.
Sack, R. (1985), *Human territoriality: its theory and history*. Cambridge: Cambridge University Press.
Thompson, E.P. (1991), *Customs in common: studies in traditional popular culture*, London: Merlin Press.
Toledo, V.M., Caballero, J. & A. Argueta (1978), "El uso múltiple de la selva basado en el conocimiento tradicional", *Biótica* 3: 85–101.
Uhl, C.; Nepstad, D.; Silva, J.M.C. & Vieira, I.C.G. (1991), "Restauração da floresta em pastagens degradadas", *Ciência Hoje* 13:23–31.
Vogel, J.H. (2009), *The economics of the Yasuní initiative: climate change as if thermodynamics mattered, London/New York/Delhi:* Anthem Press.

Epilogue

We have come to the end of this book, a gathering of reflections cultivated over forty years, about the forging of the field of political ecology. Beyond displaying a map of the diversity of socio-environmental conflicts that exist on the planet's territories (and of which we have an abundance of case studies carried out by academics and activists throughout Latin America, the global South and the entire world), my intention has been none other than to deconstruct the theoretical discourses that rendered political ecology invisible during the years when the socio-environmental field was under construction. My objective has been to unravel the discursive fabric of the theoretical and institutional language that bounded the discipline, in order to name and free up the emerging processes that mark the boundaries of conceptual and territorial disputes in the processes of appropriation of Nature; by means of theoretical and practical strategies, that is strategies of resistance, the goal has been to unbind processes of re-existence for the peoples of the Earth. This is the task of deconstructing capital and territorializing life on the planet.

My goal over the years has been none other than to weld the field of Latin American political ecology with the conceptual core of *environmental rationality*, to envision a historical process towards the emancipation of life based on three principles: an ontology of diversity, a politics of

E. Leff, *Political Ecology*, https://doi.org/10.1007/978-3-030-63325-7

difference, and an ethics of otherness. Beyond the purpose of deconstructing the legacy of philosophical thought and modern science as the historical causes of the environmental crisis, the reflection of political ecology sought to make visible the process of dispossession of peoples and the ecological degradation of their life-territories. Not only has political ecology established a new interdisciplinary field of social sciences; its historical transcendence is a social movement engaged in the sustainability of life, a process of resistance and re-existence that is mobilizing a dialogue of knowledges/ *savoirs,* which are in turn capable of reorienting the civilizing course of humanity. This turn in history stems from the conditions of life and the rights to exist of all peoples on Earth through their diverse social imaginaries and practices, opening new horizons for the sustainability of life on the planet.

Earth Day, established half a century ago, gave way to a critical reflection that came to dispute the perceived normality of the historical process: one driven by a social order marked by economic growth, and by the technological intervention of biological evolution that in turn has degraded the conditions of life on the planet. In an attempt to solve the climate emergency, alarm signals were launched from the Earth summits by using a catastrophic discourse, with apocalyptic tones and lethal signs. But the official discourse on sustainable development has not managed to unravel the complexity of the civilizing crisis that humanity is going through: a systemic crisis that is economic and ecological, environmental and epidemiological, ontological and existential.

Humanity woke up in 2020 infected with a lethal virus. These events have moved our uncertain quests towards rethinking and guiding socio-environmental conflicts for the sustainability of life. The environmental crisis had confronted that normality of the economic regime, which induced the greatest abnormality in the order of life: a precipitation towards the entropic death of the planet. If throughout this book we sought to expose the discursive strategies of the *geopolitics of sustainable development* and the *green economy* that have simulated and co-opted the senses of sustainability, the COVID-19 pandemic has sharpened the inquiry into the environmental crisis and into strategies that will enable a sustainable world.

The COVID-19 pandemic has radicalized reflections that began with the environmental crisis and should guide a civilizing transformation towards the sustainability of life. In this book we named its fundamental principles: (1) An *ontology of diversity*, which affirms the essence of life in its diversifying process of becoming, an ontology that speaks of the complexity in the creative evolution of life; (2) A *politics of difference*, as the expression of the conflict involved in the construction of a sustainable world "made of many worlds," in living together in a world inhabited by different ways of being-in-the-world, with different visions, meanings, and interests; (3) An *ethics of otherness*, that is the coexistence of different life-worlds, irreducible to one identity or one rationality for governing the different rights of existence and ways of living-in-the-world.

Today, the outbreak of the pandemic has combined the environmental and epidemiological crises as the most radical symptom of the crisis of capitalism, a crisis of the techno-economic rationality that governs the world and has disrupted the metabolism of life on the planet. This rationality has intervened in the biosphere through extractive processes that exposed the insatiability of capital that feeds on limited Nature. As such, extractivist capital has become a fundamental axis of critical reflection for the political ecology of Latin America, for revealing the effects of environmental crisis on the lives of peoples and individuals, and on the vital senses and the existential rights of people to live in a sustainable world.

The pandemic has revealed and exacerbated a malaise in the present stage of extractivist capital. There have certainly been other times of virus shedding, housed in cells since the origins of life throughout human history. But a critical reflection leads to questioning whether the COVID-19 event is indeed natural. Capital had been expanding and engulfing Nature for years, stirring up the molecules and viruses embedded in life cycles of ecosystems, expropriating nature in every territory by means of its accumulation process that dispossess the biocultural heritage of the peoples of Earth. Capital had already invaded the whole biosphere and penetrated planet's geological layers through fracking, tearing off Earth's web of life and stealing its last breath. Extractivist capitalism has now invaded life's cells, recombining and mutating its genes without ethical restraint, in the way that incest prohibition took care of human life throughout history. Globalization has altered the metabolism of life in the whole biosphere, recombining and promoting mutations of the viruses that inhabit its organisms. Capital has spread the pathogenic virus by transporting it through trade and tourism as capital appreciation agencies. The coronavirus has reached all corners of the world, an indication of capitalism's agency in the globalization process. This coronavirus is the messenger of a *promiscuous extractivist capitalism*, the kind that "liberates" viruses from the biosphere, and turns them into agents deadly to human life.

As I am writing this epilogue, today we have already reached over two million fatalities from COVID-19, and with the availability of the vaccine, governments are preparing to return to a "new normal". But what would that new normal life be? How many human lives will it cost to reinstate this "normal" economic life that ceased to be normal long ago, and will inevitably repeat new cycles of abnormal capital accumulation? The pandemic questions the non-normality of the planetary state and of human life, both driven by an ontological regime rooted in techno-economic rationality. This regime mobilizes the metabolism of the biosphere to the downfall of life on the planet. It elicits questions about the value and meaning we ascribe to life itself.

The climate emergency and the COVID-19 pandemic expose a confrontation between the will to preserve life and capital's domination; that is, a confrontation between the techno-economic regime that has objectified Nature and has subjected life to capital, to be ultimately appropriated and exploited until both biodiversity and human life become extinct. Capital is the sovereign regime that governs the world and degrades life on the planet. Life has been the forgotten one in history, due to a dominating Reason. For the first time in history, humanity faces an imperative to take charge of its human condition, to take responsibility for its will to dominate Nature, to deconstruct its unsustainable Reason, to rediscover its place in the sustainability of life on the planet. Humanity has been summoned to the titanic challenge of reinventing its ways of inhabiting the Earth within life's conditions.

The abnormality of the environmental crisis, like that of the COVID-19 pandemic, involves questions about what possible knowledge can humanity embody when considering the becoming of life and its unforeseeable events: Does humanity have the ability to direct the creative evolution of life, instead of simply responding to critical events under the rationality that has caused and triggered the crisis? Can we embody an ethics of life at the core of environmental rationality? Beyond scientific knowledge, the essential questions interrogate whether philosophy can illuminate the paths of history, whether the philosopher can become a thinker of his own time, and whether, rather than dating his thoughts in history, the contemporary thinker can understand and influence the events of his/her own time. Time steeped in history, in a historical grounding that not only imprints a direction and inertia to the historical events that have deviated the courses of life, also limits what can be thinkable and put into practice in a particular historic moment. Because, inevitably, human thought tends to follow rather than anticipate the events of history. Because beyond the resistance to historical transcendence of the sovereign ontological regime that rules the world, the imperative of survival impinges and hampers the possibilities of enacting a politics of difference and an ethics of responsibility of life and for the life of others.

The COVID-19 pandemic burst into a world that was already at ground zero, in a countdown to stabilizing the planet's climate risk. The Paris Agreements during COP24 warned humanity that there was only a decade left to recover and balance life's *normality*. The coronavirus has denounced the abnormality of this imperative call. The pandemic is not a common outbreak of Nature, but rather an event caused by capital's intervention on life. As governments begin to implement plans to return to a *new normal life*, who can actually define or even understand this "new normalcy"? Albert Einstein had already warned us that we cannot solve problems with the same way of thinking that created them.[1] The exceptional nature of this epidemiological emergency cannot be dissociated from the socio-environmental crisis that has been spreading on the planet, that has degraded the life-support system of the Earth, interrogating our modes of understanding and inhabiting the biosphere under the conditions for life on the planet.

The virus has challenged humanity's responsibility in the face of life's destiny: in light of the thermodynamic and ecological conditions of the biosphere on which life depends; in the face of the unconscious drives

[1] See Calaprice, A. (ed.). 2005. *The ultimate quotable Einstein*, Princeton University Press, p. 476; see also "A new type of thinking is essential if mankind is to survive and move toward higher levels," from "Atomic education urged by Einstein," *New York Times*, May 25, 1946.

and the *will to power* that predisposes human actions to dominate Nature. The environmental crisis prompted a reflection towards deconstructing rationality that has normalized the unequal conditions of existence of life. Critical environmentalism is shifting its critique of unsustainable reason to bring into play the body's sensitivity. Since Nietzsche, Merleau-Ponty and Levinas, since Freud and Lacan, bioethics has activated the knowledge of the body in the understanding of life; it has moved from the unconscious drives to dominate life, Nature and the Other, to the embracing of desires and aspirations for emancipating life. In the material body, life's impulses, sensations, perceptions, emotions, feelings, and intuitions come alive, the kind that cannot be illuminated by consciousness; the kind that cannot be conjugated in language, and that cannot acquire transparency through Reason.[2]

The viral crisis has come to alert human consciousness to the fragility and vulnerability of life: to protect our vital sensitivity. Today we know that we put our lives at risk by nearness to the other: by shaking hands, hugging, and kissing. The virus sounded once again an alarm about ways we express the joys of life that was set off by the risks of liberating sexual desire, inherent to human nature, with the HIV virus. The coronavirus brought back this predicament to the main scene and promoted the use of devices as a normality to condone and preserve life. Today, the COVID-19 pandemic expands the use of filters and masks that protect body's contact with life: whenever we breathe, smell, touch, look, eat, embrace, love.

"A caress does not know what it is searching for," said Emmanuel Levinas, but the body's sensitivity is no longer on the skin's surface or in the proximity of another human body. The *epiphany of the face,* as an ethical principle of radical responsibility for the other, should show us the way to collectively live with the Other, that *other* whom I cannot reduce to my own self, to my way of being, to my way of thinking. The ethics of human coexistence leads us to accept the *other* as Other. Today, human contact is not only distorted when facing the world "through the looking glass," as in *Alice in Wonderland*, or in Freud's vision of the *future of an illusion*. If safe coexistence had already led us to look away in public in

---

[2] Cf. Leff, E. (2020), *El conflicto de la vida*. Mexico City: Siglo XXI Editores.

order to avoid sexual harassment, today the pandemic forces us to look at each other through monitor screens and cell phones, since the risk of contagion prevents us from facing the other person "in person"; the virtual gaze of Facebook has interposed itself into looking into the face of the Other.

The ethics of life faces this enigma: How to love life without looking into the eyes of the *other*? How to love the other when the *other*'s gaze has been perverted and contaminated? Without the naked gaze of the *other*, love dissolves. Life dies when poetry lies in blood due to human insensitivity to life itself, as it was the case during the Holocaust or today with genocide, terrorism, organized crime, and the entropic death of the planet as extreme consequences of the oblivion of life.

The pandemic forces us to confront the nature of human desire. Beyond the existential condition that emerges from Aristotle's axiom "all men are mortal" and the dictum "being-towards-death," which Heidegger placed at the center of existential ontology, the coronavirus has brought a real risk to life. However, there is no guarantee that the virus will generate a "species consciousness," or a new way of understanding the conditions of life on the planet. Will to power is resilient and ingrained in our institutional, rational, and instinctive structure. It is the logic of force majeure that governs the world and has confined the social body to the insensitivity of life. If the environmental crisis is a symptom of the siege brought forth by Reason, the pandemic enables a perception of life that is closer to the anguish that precedes death; it brings light to our vulnerability and it renders the fragility of life tangible. The virus has taken away our sleep; but its greatest danger is that it will bring us all back to the plague of oblivion, that oblivion into which fell the *Macondo* of Nobel prize-winner Gabriel Garcia Márquez, a waking up without signs of life.

The pandemic has imposed the temporary confinement of bodies, within the mental prison of established reason. The tragedy of the current pandemic is the viral epistemic genocide of which COVID-19 is a carrier. We bet on the intuition of the body, but its capacities for perception have been hampered and our sensitivity toward life has been degraded. The turns of phrases and the language we use fail us in our instinct to hold onto life, to enunciate pain and frustration; but above all our logic curtails the desire to live and to know and to embody life conditions for the sustainability and meaning of life. The imperative need to care for life has imposed a distance to its bare contact; the unconscious anguish that restricts closeness has also become the prohibition of vital body contact. Existential meanings and body senses have been disrupted; life's destinies have been diverted. We will have to reinvent the words needed to speak-out and to acknowledge life; we will need to address the new knowledge of a life that has been infected with a deadly virus; in order to absorb and be absorbed by life itself; in order to renew life's paths towards a sustainable future.

From the dark background of confinement, uncertain questions strike us: Will we manage to deconstruct the dominant form of rationality and access other ways of understanding life, while simultaneously sharpening the intelligence of reason and the body's sensitivity? Will we be able to adjust our ways of life to the conditions of the biosphere, leaving oil underground, avoiding further pollution of the atmosphere, allowing viruses to regain their place in the biosphere, and cease the disturbance of agents turned aggressors to human life? Will we learn to look into the Other and caress life in a sustainable way? Will we learn to think about and understand the conditions of life and how to live with those conditions in order to territorialize life on Earth? Or will the post-pandemic be another step in the direction of unsustainability, towards the insensitivity of life and the destination of the entropic death of the planet?

The only normal for the future announced by the pandemic is a state of exception in which the hope of life must be renewed. The renewed civilizing transition towards sustainability cannot be the re-adapting to a life under a permanent threat of death. Before we plunge into the abyss of the environmental crisis, of the ecological catastrophe and the climatic emergency, life requires a new pact with Nature, a looking-up at the stars in order to harmoniously re-signify the senses with the music of the cosmos, and the song of the Earth. COVID-19's threat to life must shake humanity to comply with life and to venture towards other possible worlds. This is the radical reflection brought forth by the pandemic, the challenge that humanity faces to recall its nature, to become *in* Nature: to learn how to inhabit the planet in the conditions of life in this living planet.

Mexico City Enrique Leff
January 15, 2021

Bibliography

Abramovay, R. (2010), "Reduzir a desigualdade entre os indivíduos para combater o aquecimento global", *Boletim da Sociedade Brasileira de Economia Ecológica*, special edition N° 23/24 January–August 2010, pp. 12–15.

Ab'Saber, A.N. (1970), "Províncias geológicas e domínios morfoclimáticos no Brasil", *Geomorfologia* 20:1–25, São Paulo.

Acosta, A. (2010), "El *Buen Vivir* en el camino del post-desarrollo. Una lectura desde la Constitución de Montecristi", Fundación Friedrich Ebert, Policy paper 9, Quito.

Adams, R.N. (1975), *Energy and structure: a theory of social power*, Texas University Press, Austin.

Alatorre, E. (2003), *El proceso de certificación forestal en México*. Dirección de Certificación Forestal. Consejo Civil Mexicano para la Silvicultura Sostenible en México.

Alcorn, J. (1993), "Los procesos como recursos. La ideología agrícola tradicional del manejo de los recursos entre los boras y huastecos y sus implicaciones para la investigación", in Leff & Carabias (Eds.), *Cultura y manejo de los recursos naturales*, México: UNAM, Miguel Ángel Porrúa, Vol. 2, pp. 329–365.

Alimonda, H. (Ed.) (2002), *Ecología política. Naturaleza, sociedad y utopía*, Buenos Aires: CLACSO.

———. (Ed.) (2006), *Los tormentos de la materia: aportes para una ecología política latino-americana*, Buenos Aires: CLACSO.

Althusser, L. (1969), *For Marx*, London, Allen Lane.

E. Leff, *Political Ecology*, https://doi.org/10.1007/978-3-030-63325-7

———. (1970), "Idéologie et appareils idéologiques d'Etat," *La pensée*, 151.

Althusser, L., Balibar, E., Establet, R., Macherey, P. & Rancière, J. (1965/2016), *Reading Capital*, translated by Ben Brewster and David Fernbach, London: Verso.

Altvater, E. (1993), *The future of the market*, New York: Verso.

Alvarez-Buylla, Elena y Alma Piñeyro (Coords.) (2013), *El maíz ante los transgénicos. Un análisis integral sobre el caso de México*, México: CEIICH-UNAM/UCCS.

Amin, S. (1976), *Unequal development: an essay on the social formations of peripheral capitalism*, New York: Monthly Review Press.

Anand, A. (1983), "Saving trees, saving lives: Third world women and the issue of survival", in L. Caldecott & S. Leland (Eds.), *Reclaim the Earth*, London: The Women's Press.

Argueta, A. et al. (1994), *Atlas de las plantas de la medicina tradicional mexicana*, 3 volumes, México: Instituto Nacional Indigenista.

Arnold, D. & Guha, R. (1995), *Nature, culture, imperialism: essays on the environmental history of South Asia*, Delhi: Oxford University Press.

Bachelard, G. (1938/1972), *La formation de l'esprit scientifique*, Paris, Librairie Philosophique J. Vrin.

———. (1972), *L'engagement rationaliste*, París, Presses Universitaires de France.

Barbados Declaration (1971), International Work Group for Indigenous Affairs/ World Council of Churches.

Bartra, A. (2008), *El Hombre de hierro. Los límites sociales y naturales del capital*, Mexico City: Universidad Autónoma de la Ciudad de México.

Bataille, G. (1957/1997), *El erotismo*, México: Tusquets. [1962. *Death and sensuality; a study of eroticism and the taboo*, New York: Walker and Company].

Bateson, G. (1972), *Steps to an ecology of mind: collected essays in anthropology, psychiatry, evolution, and epistemology*, Chicago: University Of Chicago Press.

———. (1979), *Mind and nature: a necessary unity (advances in systems theory, complexity, and the human sciences)*, Hampton Press.

Baudrillard, J. (1974), *Crítica de la economía política del signo*, Mexico, Siglo XXI Editores.

———. (1976), *L'échange symbolique et la mort*, Paris: Gallimard.

———. (1983), *Les stratégies fatales*, Paris: Éditions Grasset & Frasquelle.

———. (1990), *Seduction*, New York: St. Martin's Press.

———. (1993), *The transparency of evil*, London, Verso.

de Beauvoir, S. (1968), *The second sex*, London: Jonathan Cape.

Beck, U., Giddens, A. & S. Lash (1994), *Reflexive modernization: politics, tradition and aesthetics in the modern social order*, Stanford: Stanford University Press.

Benjamin, W. (1936), "The work of art in the age of mechanical reproduction", and (1940), "Theses on the philosophy of history", in Arendt, H. (ed.) (1968), *Illuminations*, London: Fontana.

Benton, T. (Ed.) (1996), *The greening of Marxism*, New York: Guilford Press.

Bergson, H. (1907/1998), *Creative evolution*, Mineola, New York: Dover Publications.

Berthelot, J.M. (1998), *L'intelligence du social*, Paris: PUF.

Biersack, A. & Greenberg, J.B. (Eds.) (2006), *Reimagining political ecology*, Durham/London, Duke University Press.

Blaikie, P. (1985), *The political economy of soil erosion in developing countries*, London: Longman.

Blaikie, P.M. & H.C. Brookfield (Eds.) (1987), *Land degradation and society*, London: Methuen.

Blaser, M. (2009), "Political ontology", *Cultural studies*, 23 (5): 873–896.

Boege, E. (1988), *Los mazatecos ante la nación*, México: Siglo XXI Editores.

———. (2008), *El patrimonio biocultural de los pueblos indígenas de México. Hacia la conservación in situ de la biodiversidad y agrobiodiversidad en los territorios indígenas*, México: INAH/Comisión Nacional para el Desarrollo de los Pueblos Indígenas.

———. (2018), "Hacia una antropología ambiental para la apropiación social del patrimonio biocultural de los pueblos indígenas", in Toledo, V.M. & Alarcón-Cháires, P. (Eds.), *Tópicos bioculturales: reflexiones sobre el concepto de bioculturalidad y la defensa del patrimonio biocultural de México*, México: UNAM.

Boff, L. (1996a), *Ecologia, mundialização, espiritualidade*, São Paulo: Ática.

———. (1996b), *Ecología: grito de la tierra, grito de los pobres*, Madrid: Trotta.

Boltvinik, J. (2010), "Poverty and persistence of the peasantry", Background paper for the International Workshop on this Issue, CROP/El Colegio de México.

Boltvinik, J. & Archer Mann, S. (Ed.), *Peasant's Poverty and Persistence in the 21st Century. Theories, realities and policies*, COLMEX/CROP/Zed Books.

Bonfil Batalla, G. (1987), *El México profundo, una civilización negada*, México: Grijalbo.

Bookchin, M. (1962), *Our synthetic environment*, New York: Knopf.

———. (1964), "Ecology and revolutionary thought", *Comment* (republished in the British monthly *Anarchy* in 1965).

———. (1971/1990), *Post-scarcity anarchism*, Montreal/New York, Black Rose Books, Second edition, 1990.

———. (1982/1991), *The ecology of freedom. The emergence and dissolution of hierarchy*, Montreal/New York: Black Rose Books, Second edition, 1991.

———. (1990a), *Remaking society: Pathways to a green future*, Boston: South End Press.

———. (1990b), *The philosophy of social ecology. Essays on dialectical naturalism*, Montreal/New York: Black Rose Books.

Borrero, J.M. (2002), *La imaginación abolicionista. Ensayos de ecología política*, Cali: PNUMA/CELA/Hivos.

Boulding, K.E. (1966), "The economics of the coming spaceship earth", in Jarrett, H. (ed.), *Environmental quality in a growing economy*, Baltimore: Johns Hopkins University Press, 1966: 3–14.

Bourdieu, P. (1984a), *Distinction: social critique of the judgment of taste*, Boston: Harvard University Press.

———. (1984b), *Homo academicus*, Paris: Les Éditions de Minuit.

Bourdieu, P. & L. Wacquant (1992), *An invitation to reflexive sociology*, Chicago: The University of Chicago Press.

Brand, Ulrich/Görg, Christoph (2008), Sustainability and Globalisation: A Theoretical Perspective, in Park, Jacob/Conca, Ken/Finger, Mathias (eds.), *The Crisis of Global Environmental Governance. Towards a new political economy of sustainability*. London and New York: Routledge, 13–33.

Bryant, R. (1992), "Political ecology: an emerging research agenda in Third-World studies", *Political geography* 11: 12–36.

Burgoa, M. & Flores Galindo, A. (1982), "La utopía andina", *Allpanchis*, Vol. 20.

Burkett, P. (1999), *Marx and nature: a red and green perspective*, New York, St. Matyin's Press.

Campos, N. (2000), *Terras de uso comum*, PhD thesis, Programa de Pós-graduação em Geografia da Universidade de São Paulo, São Paulo.

Canguilhem, G. (1966/1971), *Lo normal y lo patológico*. México: Siglo XXI Editores.

———. (1971), *La connaissance de la vie*, Paris: J. Vrin.

———. (1977), *Idéologie et rationalité dans l'histoire des sciences de la vie*. Paris: J. Vrin.

Capra, F. (1996), *The web of life*, New York: Anchor Books.

Cardona, G.R. (1986), *La floresta di piume. Manuale di etnoscienzia*, Roma: Editore Laterza.

Cardoso, F.H. & E. Faletto (1979), *Dependency and development in Latin America*, University of California Press.

Carlson, R. (1962), *The silent Spring*, Boston: Houghton Mifflin.

Castoriadis, C. (1998), *The imaginary institution of society*, Cambridge: MIT Press.

de Castro, J. (1946/1975), *The geopolitics of hunger*, New York: Monthly Review Press.

———. (2003), "Subdesenvolvimento: causa primeira da poluição", in: *Fome: um tema proibido—últimos escritos de Josué de Castro*, Rio de Janeiro: Civilização Brasileira, pp. 135–137.

Ceceña, A.E. & E. Sader, *La guerra infinita. Hegemonía y terror mundial*, CLACSO-ASDI, Buenos Aires.

CEPAL/PNUMA (1983), *Sobrevivencia campesina en ecosistemas de altura*, 2 Vols., Santiago de Chile.

CEPAUR (1989), "Human scale development", in *Development dialogue*, Num. 1, Motala: CEPAUR/Dag Hammarskjold Foundation.

Césaire, A. (1955), *Discourse on colonialism*, New York: Monthly Review.

Chardin, P. Teilhard de, (1961), *The phenomenon of man*, San Francisco: Harper & Row Publishers.

Chayanov, A. (1974), *La organización de la unidad económica campesina*, Buenos Aires: Nueva Visión.

Clark, J. (2008), "Domesticating the dialectic: a critique of Bookchin's neo-Aristotelian metaphysics", *CNS*, Vol. 19, No. 1, March, p. 95.

Commoner, B. (1971), *The closing circle: nature, man, and technology*. New York: Knopf.

———. (1976), *The poverty of power: energy and the economic crisis*. New York: Random House.

Condarco, R. & Murra, J. (1987), *La teoría de la complementariedad vertical eco-simbiótica*, La Paz: Hisbol.

Cornetta, A. (2011), *Mecanismo de desenvolvimento limpo e conflitos locais*, XIII Encontro de Geógrafos de América Latina 2011. San José, Costa Rica.

Cosby, A.W. (1986), *Ecological imperialism. The biological expansion of Europe, 900–1900*, New York: Cambridge University Press.

Cossio, R.E., Bray, D.B., Bult, S. & Merino, L. (2006), "Bibliografía anotada del manejo comunitario de los bosques en México, con bibliografía adicional sobre los bosques de México en general (Working Paper No. 2)". *LACC Working Paper Series*, http://digitalcommons.fiu.edu/laccwps/11.

Da Cunha, M.C. & Almeida, M. (2000), *Indigenous people, traditional people and conservation in the Amazon, Daedalus* Vol. 129 (2): 315–338.

Daly, H. E. (1991), *Steady-state economics*, Washington, D. C., Island.

Debeir, J.C., Deléage, J.P. & D. Hémery (1986), *Les servitudes de la puissance. Une histoire de l'énergie*, Paris: Flammarion.

De Gortari, E. (1963), *La ciencia en la historia de México*, México: Fondo de Cultura Económica.

De la Torre, C. & M. Burgoa (Comps.) (1986), *Andenes y camellones en el Perú andino*, Lima: Concytec.

Deleuze, G. (1969), *Logique du sens*, Paris: Les Éditions de Minuit.

———. (1994), *Difference and repetition*, New York: Columbia University Press.

Deleuze, G. & Guattari, F. (1983), *Anti-Oedipus*, Minneapolis: University of Minnesota Press.

———. (1987), *A thousand plateaus. Capitalism & schizophrenia*, Minneapolis/London: University of Minnesota Press.

Denevan, W.M. (1980), "Tipología de configuraciones agrícolas prehispánicas", in *América Indígena*, No. 40, pp. 610–652.

Derrida, J. (1976), *Of grammatology*, Baltimore & London: Johns Hopkins University Press.

———. (1978), *Writing and difference*, Chicago: The University of Chicago Press.

———. (1982), *Margins of philosophy*, Chicago: Chicago University Press.

———. (1983/2008), *Geschlecht: différence sexuelle, différence ontologique*, *Cahier de l'Herne* consacré à Heidegger, dirigé par Michel Haar, Paris, in *Psyche. Inventions of the Other*, Vol. II, Edited by Peggy Kamuf and Elizabeth Rothenberg, Stanford: Stanford University Press, pp. 7–26.

Descola, Ph. (1996), *La selva culta. Simbolismo y praxis en la ecología de los Achuar*, Quito: Ediciones Abya Yala.

———. (2005), *Par—delà nature et culture*, Paris: Gallimard.

Descola, Ph. & G. Pálsson (Eds.) (1996), *Nature and society: anthropological perspectives*, London: Routledge.

Díaz Polanco, H. (1991), *Autonomía regional. La autodeterminación de los pueblos indios*, México, Siglo XXI/UNAM.

Dos Santos, Th. (1978), *Imperialismo y dependencia*, México: Ed. Era.

Dunlap, R.E. & W.R. Catton (1979), "Environmental sociology", *Annual review of sociology*, 5: 243–273.

——— (1994), "Struggling with human exemptionalism: the rise, decline and revitalization of environmental sociology", *The American sociologist*, 25 (1): 5–30.

Dussel, E. (1998), *Ética de la liberación en la edad de la globalización y de la exclusión*, Madrid: Trotta.

Echeverría, B. (1998), *Valor de uso y utopía*, México: Siglo XXI Editores.

Ely, J. (1988), "Lukacs' construction of nature", in *Capitalism, nature, socialism*, No. 1.

Engels, F. (1968), *Dialectique de la nature*, Paris: Editions Sociales.

Enzensberger, H.M. (1974), "A critique of political ecology", *New left review*, 84 (3–31).

Erlich, P. (1968), *The population bomb*, New York: Ballantine Books.

Escobar, A. (1992), "Imagining a post-development era: critical thought, development and social movements", *Social Text* 31–32:20–56.

———. (1995), *Encountering development. The making and unmaking of the Third World*, Princeton University Press, Princeton.

———. (1997), "Cultural politics and biological diversity: State, capital and social movements in the Pacific coast of Colombia", in O. Starn & R. Fox (Eds), *Culture and social protest: Between resistance and revolution*, New Brunswick, Rutgers University Press.

———. (1999a), *El final del salvaje. Naturaleza, cultura y política en la antropología contemporánea*, Bogotá: CEREC/ICAN.

———. (1999b), "Gender, place and networks. A political ecology of cyberculture", in Harcourt, W. (ed.), *Women@Internet. Creating new cultures in cyberspace*, London: Zed Books, pp. 31–54.

———. (2006), "An ecology of difference: equality and conflict in a glocalized world", in *Focaal, European Journal of Anthropology* 47: 120–140.

———. (2008), *Territories of difference. Place, movements, life, redes*, Durham/London: Duke University Press.

———. (2010), "Post-constructivist political ecologies", in Redclift, M. & Woodgate G. (Eds.), *International handbook of environmental sociology*, 2nd edition, Cheltenham/Northampton: Edward Elgar Publishing Limited, pp. 91–105.

———. (2013), "Territorios de diferencia: la ontología política de los derechos al territorio", Segundo Taller Internacional SOGIP, "Los pueblos indígenas y sus derechos a la tierra: política agraria y usos, conservación, e industrias extractivas", www.sogip.ehess.fr.

Escobar, A., Grueso, L. & C. Rosero (1998), "The process of black community organizing in the Pacific coast of Colombia", in Álvarez, S., E. Dagnino & A. Escobar (Eds.) *Cultures of politics/politics of cultures: revisioning Latin American social movements*, Boulder, Colorado: Westview Press, pp. 196–219.

Esterman, J. (2006), *La filosofía andina*, Quito: Ed. Abya Yala.

Esteva, G. & Prakash, M.S. (1998), *Grassroots post-modernism: remaking the soil of cultures*, London: Zed Books.

Evans Pritchard, E. E. (1962), *Social anthropology and other essays*, New York: The Free Press.

Faber, D. (1988), "Dependent Development, Disarticulated Accumulation and Ecological Crisis in Central America," *CNS*, 1.

Fals Borda, O. (1981), "La ciencia del pueblo", in *Investigación participativa y praxis rural. Nuevos conceptos en educación y desarrollo comunal*, Lima, Perú Editorial Mosca Azul.

———. (1987), *Ciencia propia y colonialismo intelectual. Los nuevos rumbos.* 3a. edición, Bogotá: Carlos Valencia Editores.

Fanon, F. (2004), *The wretched of the earth*, New York: Grove.

Ferry, L. (1995), *The new ecological order*, Chicago: The University of Chicago Press.

Fossaert, R. (1977), *La société, Tome 2, Les structures économiques*, Paris, Seuil.

Foster, J.B. (2000), *Marx's ecology: materialism and nature*, New York: Monthly Review Press.

———. (2018), "Marx, value and nature", *Monthly review*, Vol. 70, No. 3 July-August.

Foucault, M. (1966), *Les mots et les choses*, París, Gallimard.

———. (1969), *L'archéologie du savoir*, Paris, Gallimard.

———. (1968/1979), *Microfísica del poder*, Madrid: Las Ediciones de La Piqueta, 2a Ed.

———. (1980), *Power/knowledge*, New York: Pantheon.

———. (1998), *La verdad y las formas jurídicas*, Barcelona: Gedisa.

Friends of the Earth (2006), *Overconsumption?: our use of the world´s natural resources*. In http://www.foeeurope.org/publications/2009/Overconsumption_Sep09.pdf.

Funtowicz, S. & J. Ravetz (1993), *Epistemologia política. Ciencia con la gente*, Buenos Aires, Centro Editor de América Latina.

——— (1994), "Emergent complex systems", *Futures*, 26 (6): 568–582.

Furtado, C. (1974), *O mito do desenvolvimento econômico*, São Paulo: Ed. Paz e Terra.

Furtado, L. (2002), *Pescadores do Rio Amazonas. Um estudo antropológico da pesca ribeirinha numa área amazônica*, MCT-CNPq-Museu Emilio Goeldi.

Gabor, D. (1964), *Inventing the future*, Harmondsworth: Penguin.

Gagdil, M. (1985), "Social restraints on resource utilization: the Indian experience", in McNeely & Pitt, op. cit.

Gagdil, M. & P. Iyer (1993), "La diversificación en el uso de los recursos de propiedad común en la sociedad india", in Leff & Carabias, Vol 2, op. cit., pp. 551–573.

Galeano, E. (1971), *Open veins of Latin America: five centuries of the pillage of a continent*, New York: Monthly Review Press.

García, M.Á. (2014), "La reserva ecológica campesina de Los Chimalapas", en Betancourt, A. et. al., *Del monólogo a la polifonía: proyectos supranacionales y saberes indígenas en la gestión de la áreas naturales protegidas (1990–2010)*, México: UNAM.

García, R. (1986), "Conceptos básicos para el estudio de sistemas complejos", in E. Leff (Coordinador) (1986), *Los problemas del conocimiento y la perspectiva ambiental del desarrollo*, Siglo XXI Editores, México.

———. (1994), "Interdisciplinariedad y Sistemas Complejos", in E. Leff (Coordinador), *Ciencias sociales y formación ambiental*, GEDISA/UNAM/PNUMA, Barcelona.

García Canclini, N. (2005), *Hybrid cultures: strategies for entering and leaving modernity*, Minnesota: University of Minnesota Press.

Georgescu-Roegen, N. (1971), *The entropy law and the economic process*, Cambridge, Mass.: Harvard University Press.

———. (1993a), "Looking back", in J.C. Dragan, E.K. Seifert, M.C. Demetrescu Eds. *Entropy and bioeconomics*, EABS, Nagard, Milano, pp. 11–21.

———. (1993b), "Thermodynamics and we, the humans", in *Entropy and bioeconomics*, op. cit., pp. 184–201.

Giampietro, M. (1993), "Escaping the Georgescu-Roegen Paradox on Development: Equilibrium and Non-Equilibrium Thermodynamics to Describe Technological Evolution", in *Entropy and Bioeconomics*, Op. cit., pp. 202–229.

Giddens, A., Beck, U. & S. Lash (1994), *Reflexive modernization. Politics, tradition and aesthetics in the modern social order*, Stanford: Stanford University Press.

Giménez, G. (1994), "Los movimientos sociales. Problemas teórico-metodológicos", *Revista mexicana de sociología*, Vol. LVI, Num. 2, pp. 3–14.

Gligo, N. & Morello, J. (1980), "Notas sobre la historia ecológica de América Latina", In Súnkel, O. & N. Gligo, *Estilos de desarrollo y medio ambiente en la América Latina*, México: Fondo de Cultura Económica.

Godelier, M. (1969), *Rationalité et irrationalité en économie*, Paris, Maspero.

———. (1974), *Economía, fetichismo y religión en las sociedades primitivas*, México, Siglo XXI Editores.

Godelier, M.M. (1984), *L'Idéel et le Matériel*, Paris, Fayard.

Goldmann, L. (1959), *Recherches dialectiques*, Paris, Gallimard.

Gómez, M. (Coord.) (1997), *Derecho indígena*, México: INI/AMNU.

González Casanova, P. (1965), "Internal colonialism and national development", in *Studies in comparative international development*, Vol. 1, Num. 4:27–37.

González Casanova, P. & M. Roitman (Coords.) (1996), *Democracia y Estado multiétnico en América Latina*, México: La Jornada Ediciones/CIICH-UNAM.

Gorz, A. (1975), *Écologie et politique*, Paris: Seuil.

———. (1977), *Écologie et liberté*, Paris: Galilée

———. (1989), *Critique of economic reason*, London/New York: Verso.

———. (2006), "Gorz. Ecologie une éthique de la libération", interview in *EcoRev*, January.

———. (2008), *Ecologica*, Paris: Galilée.

Gould, K.A., D. Pellow & A. Schnaiberg (2008), *The treadmill of production. Injustice and unsustainability in the global economy*, Boulder: Paradigm publishers.

Granou, A. (1973), "Le Capitalisme Face à la 'Non-Croissance'", *Les temps modernes*, 236, 1973.

Grinevald, J. (1993), "The biosphere and the noosphere revisited: biogeochemistry and bioeconomics", in *Entropy and bioeconomics*, op. cit., pp. 241–258.

Grünberg, G. (Coord.) (1995), *Articulación de la diversidad: pluralidad étnica, autonomías y democratización en América Latina*, Quito: Ediciones Abya-Yala.

Guha, R. (1989), *The unquiet woods: ecological change and peasant resistance in the Himalaya*, Delhi: Oxford University Press.

Guha, R. & Gagdil, M. (1992), *This fissured land: an ecological history of India*, Delhi: Oxford University Press.

Guha, R. & Martínez-Alier, J. (1997), *Varieties of environmentalism. Essays North and South*, London: Earthscan.

Gunder-Frank, A. (1966), *The development of underdevelopment*, New York: Monthly Review Press.

Günther, F. (1993), "Man in living systems", in *Entropy and bioeconomics*, op. cit., pp. 259–275.

Habermas, J. (1984), *Theory of communicative action, Vol. I: Reason and the rationalization of society*, Boston, Mass.: Beacon Press.

———. (1987a), *Theory of communicative action, Vol. II: Live-world and system: a critique of functionalist reason*, Boston, Mass.: Beacon Press.

———. (1987b), *The philosophical discourse of modernity*, Translated by Frederick Lawrence, Cambridge: Polity Press/Blackwell Publishers.

Haesbaert, R. (2004), O mito da desterritorialização: do "fim dos territórios" à multiterritorialidade, Rio de Janeiro: Bertrand Brasil.

———. (2013), A global sense of place and multi-territoriality: notes for dialogue from a 'peripheral' point of view", in Featherstone, D. & Painter, J., *Spatial Politics: Essays for Doreen Massey*, Chichester: John Wiley and Sons.

Hall, D.O. & Rosillo-Calle, F. (1999), "Biomass: A future renewable carbon feedstock for energy", in Parman,V. N., Tribusch, H., Bridgwater, A., & Hall, D.O., *Chemistry for the Energy Future*, Oxford: Oxford University Press, pp. 101–102, 109, 118.

Haraway, D. (1991), *Simians, cyborgs and women. The reinvention of nature*, New York, Routledge.

Hardin, G. (1968), "The tragedy of the commons", *Science* No. 162: 1243–1248.
Harvey, D. (2003), *The new imperialism*, Oxford: Oxford University Press.
Hecht, S.B. & Schwartzman, S. (1988), *The good, the bad and the ugly: extraction, colonist agriculture and livestock in comparative economic perspective*, Los Angeles: Westing Papers, UCLA.
Heidegger, M. (1927/1962), *Being and time*, trans. John Macquarrie & Edward Robinson, San Francisco: HarperCollins Publishers.
———. (1935/1977), "The question concerning technology", in *Heidegger. Basic writings*, trans. D. Farrell Krell, New York: HarperCollins, pp. 283–317.
———. (1946/1977), "Letter on humanism", in *Heidegger. Basic writings*, trans. D. Farrell Krell, New York: HarperCollins, pp. 189–242.
———. (1957/1969), *Identity and difference*, Chicago: Chicago University Press.
———. (1975), *La pregunta por la cosa*, Buenos Aires: Alfa Ed.
———. (1999), *Contributions to philosophy (from enowning)*, Translated by Parvis Amad & Kenneth Maly, Bloomimngton & Indianapolis: Indiana University Press.
———. (1971/2001), *Poetry, language, thought*, New York: Perennial Classics.
Heller, A. (1976), *The theory of need in Marx*. London: Allison and Busby.
Herrera, A.O. et al. (1976), *Catastrophe or new society: a Latin American model*, Ottawa: IDRC.
Hetman, F. (1973), *Society and the assessment of technology*, Paris: OCDE.
Hinterberger, F. & Seifert E. (1995), "Reducing material throughput: A contribution to the measurement of dematerialization and sustainable human development", in J. van der Straaten y A. Tylecote (eds.), *Environment, technology and economic growth: The challenge to sustainable development*, Aldershot, Edward Elgar.
Hobbelink, H. (1992), "La diversidad biológica y la biotecnología agrícola", *Ecología política*, No. 4, Barcelona, Icaria.
Holling, C.S. (1973), "Resilience and stability of ecological systems", *Annual review of ecological systems*, 4: 1–23.
Horkheimer, M. & Th. Adorno (1944/2002), *Dialectics of Enlightenment*, Stanford: Stanford University Press.
Houtart, F. (2010), *A agroenergia: solução para o clima ou saída da crise para o capital?*, Petrópolis: Vozes.
Huanacuni, F. (2010), *Vivir bien/Buen vivir: Filosofía, políticas, estrategias y experiencias regionales*, La Paz: Convenio Andrés Bello/Instituto Internacional de Integración.
Illich, I. (1973), *Tools for conviviality*, New York: Harper & Row.
Infgold, T. (2000), *The perception of the environment*, Londres, Routledge.

Ingelhart, R. (1991), *El cambio cultural en las sociedades industriales avanzadas*, Madrid: Siglo XXI Editores.

Instituto Indigenista Interamericano (1990), "Política indigenista 1991–1995", *América indígena*, Vol. 50, Num. 1.

International Indian Treaty Council (2000), http://www.treatycouncil.org/.

Janicaud, D. (1994), *Powers of the rational: science, technology and the future of thought*, Indiana: Indiana University Press.

Kahn, D. (2014), "The chain of the Past" (Daniel Kahn on Avrom Sutzkever), Azymptote Blog, April, 2014.

Kapp, W. (1983), "Social costs in economic development", in J.E. Ullmann (Ed), *Social costs, economic development and environmental disruption*, Lanham, Mass., University Press of America.

Kautsky, K. (1899/1998), *The agrarian question*, Swan: London.

Kosik, K. (1970), *La dialectique du concret*, Paris, François Maspero.

Kropotkin, P. (1902/2005), *Mutual aid: a factor of evolution*, Boston: Extending Horizons Books.

Kuhn, T. (1962), *The structure of scientific revolutions*, Chicago: The University of Chicago Press.

Lacan, J. (1971), "Subversión del sujeto y dialéctica del deseo en el inconsciente freudiano", *Escritos*, México, Siglo XXI Eds., Third edition, 1976.

———. (1974–75), *Séminaire XXII RSI (Réel, Symbolique, Imaginaire)*, mimeo.

———. (1976), "Subversión del sujeto y dialéctica del deseo en el inconsciente freudiano", Escritos, México, Siglo XXI.

———. (1998), *On feminine sexuality the limits of love and knowledge*: The Seminar of Jacques Lacan, Book XX *Encore* (edited by Jacques-Alain Miller), New York: Norton.

Laclau, E. (1996), *Emancipations*, London & New York: Verso.

Laclau, E. & Mouffe, C. (1985), *Hegemony and socialist strategy*, London: Verso.

Lander, E. Ed. (2000), *La colonialidad del saber: eurocentrismo y ciencias sociales. Perspectivas latinoamericanas*, Buenos Aires: Clacso/Unesco.

Latouche, S. (2009), *La apuesta por el decrecimiento: ¿cómo salir del imaginario dominante?*, Barcelona: Icaria.

Latour, B. (2004), *Politics of nature. How to bring the sciences into democracy*, Cambridge, Mass.: Harvard University Press.

Leff, E. (1975), "Hacia un proyecto de ecodesarrollo", *Comercio Exterior*, Vol. XXV, No. 1, México, pp. 88–94.

———. (1980), "La teoría del valor en Marx frente a la revolución científico-tecnológica", en Leff, E. (Editor), *Teoría del valor*, México: UNAM.

———. (1986a), *Ecología y capital: hacia una perspectiva ambiental del desarrollo*, México: UNAM
———. (1986b), "Eco-technological productivity: a conceptual basis for the integrated management of natural resources", *Social science information*, 25 (3): 681–702.
———. (1986c), "Ambiente y articulación de ciencias", in E. Leff, *Los problemas del conocimiento y la perspectiva ambiental del desarrollo*, Mexico City: Siglo XXI Editores.
———. (1992a), "Cultura democrática, gestión ambiental y desarrollo sustentable en América Latina", *Ecología política* No. 4, ICARIA, Barcelona, pp. 47–55.
———. (1992b), "A second contradiction of capitalism? Notes for the environmental transformation of historical materialism", *Capitalism, Nature, Socialism*, 3 (4):109–116, December, 1992.
———. (1993a), "Marxism and the environmental question: from critical theory of production to an environmental rationality for sustainable development", *Capitalism, nature, socialism*, Vol. 4 (1): 44–66, Santa Cruz, California.
———. (1993b), "La dimensión cultural del manejo integrado sustentable y sostenido de los recursos naturales", en Leff E. & J. Carabias (Eds), *Cultura y manejo sustentable de los recursos naturales*, México: CIIH-UNAM/Miguel Angel Porrua.
———. (1994), *Ecología y capital: racionalidad ambiental, democracia participativa y desarrollo sustentable*, México: Siglo XXI Editores.
———. (1995), *Green production: towards an environmental rationality*, New York: Guilford.
———. (1996), "Los nuevos actores sociales del ambientalismo en el medio rural", en Carton de Grammont H. & H. Tejera, *La sociedad rural frente al nuevo milenio*, Vol. 4 "Los nuevos actores sociales y los procesos políticos en el campo", México: UNAM/INAH/UAM/Plaza y Valdez Editores.
———. (1998a), "Murray Bookchin and the end of dialectical naturalism", *Capitalism, nature, socialism*, Vol. 9 (4):67–93.
———. (1998b), *Saber ambiental: Sustentabilidad, racionalidad, complejidad, poder*, México: Siglo XXI/UNAM/PNUMA (3rd revised and extended edition, 2002).
———. (1999), "On the social appropriation of nature", *Capitalism, nature, socialism*, Vol. 10 (3), No. 38, September, 1999, pp. 89–104.
———. (2000a), "Pensar la complejidad ambiental", en Leff, E. (Coord.), *La complejidad ambiental, Colección* "Aprender a aprender", México: Siglo XXI Editores (2nd revised edition, 2003).

———. (2000b), "The scientific-technologic revolution, the forces of nature and Marx's theory of value", in *Capitalism, nature, socialism*, Vol. 11 (4):109–130.

———. (2001), *Epistemología ambiental*, Sao Paulo: Cortez Editora.

———. (2002a), "La geopolítica de la biodiversidad y del desarrollo sustentable: economización del mundo, racionalidad ambiental y reapropiación social de la naturaleza", in Ceceña, A.E. & Sader, E., *La guerra infinita: hegemonía y terror mundial*, Buenos Aires: CLACSO-ASDI, pp. 191–216.

———. (2003), "La ecología política en América Latina: un campo en construcción", *Polis*, Revista de la Universidad Bolivariana, II (5) 2003, Santiago de Chile, pp. 125–145.

———. (2004), *Racionalidad ambiental: la apropiación social de la naturaleza.* México: Siglo XXI Editores.

———. (2006a), *Aventuras de la epistemología ambiental: de la articulación de las ciencias al diálogo de saberes*, México: Siglo XXI Editores.

———. (2006b), "Ética por la Vida. Elogio de la Voluntad de Poder" en *Polis*, Revista de la Universidad Bolivariana, vol. 5, núm. 13, Universidad de Los Lagos Santiago, Chile.

———. (2009), "De-growth or Deconstruction of the Economy: Towards a Sustainable World", in U. Brand, N. Bullard, E. Lander & T. Mueller, *Contours of Climate Justice: Ideas for Shaping New Climate and Energy Politics*, in *Critical Currents*, Dag Hammarsjöld Foundation, Occasional Paper Series, No. 6. Oct. 2009, Uppsala, pp. 101–107.

———. (2010), "Imaginarios sociales y sustentabilidad", in *Cultura y representaciones sociales.* Num. 9, México, pp. 42–121.

———. (2012), "Latin American environmental thinking: a heritage of knowledge for sustainability", *Environmental ethics*, Volume 34:4, Winter, 2012, pp. 431–450.

———. (2014a), *La apuesta por la vida: imaginarios sociales e imaginación sociológica en los territorios del Sur*, México: Siglo XXI Editores.

———. (2014b), "Los Derechos del Ser Colectivo y la Reapropiación Social de la Naturaleza", en Pereira de Cunha, Belinda; Agustín, Sergio; Araújo da Costa, Nálbia Roberta (Org.), *Saberes Ambientais, Sustentabilidade e Olhar Jurídico: visitando a obra de Enrique Leff*, Joao Pessoa e Caxias do Sul, Brasil.

———. (2015a), "Political ecology: a Latin American perspective" in Leff, E., Floriani, D., & L.H. de Oliveira Cunha (Eds.), dossier temático "Pensamento Ambiental Latino-americano: movimentos sociais e territórios de vida", *Revista Desenvolvimento e Meio Ambiente* 35, Programa de Pós-Graduação de Meio Ambiente e Desenvolvimento, Universidade Federal do Paraná (Curitiba-Paraná-Brasil).

———. (2015b), "Encountering political ecology: epistemology and emancipation", in Bryant, Raymond & Soyeun Kim (Eds.), *The International Handbook of Political Ecology*, Cheltenham, UK: Edward Elgar, pp. 44–56.
———. (2017), "Power-knowledge relations in the field of political ecology", in *Ambiente & Sociedade*, Vol. 20 (3): 225–254.
———. (2018), *El fuego de la vida: Heidegger ante la cuestión ambiental*, Mexico: Siglo XXI Editores.
———. (2019), "Negentropic production", in Kothari, Ashish, Demaria, F., Acosta, A., Salleh, A. and Escobar, A. Eds., *Pluriverse: a postdevelopment lexicon*. London: Zed Books.
———. (2020a), *El conflicto de la vida*, Mexico: Siglo XXI Editores.
———. (2020b), "Heráclito: A Φύσις e o Desocultamento da Vida", *Desenvolvimento e meio ambiente*, 54.
———. (Ed.) (1986), *Los problemas del conocimiento y la perspectiva ambiental del desarrollo*, México: Siglo XXI Editores.
———. (Ed.) (2002b), *Ética, vida, sustentabilidad*, México: PNUMA, Serie Pensamiento Ambiental Latinoamericano No. 5.
Leff, E. & J. Carabias (Eds.), (1993), *Cultura y manejo de los recursos naturales*, México: UNAM, Miguel Ángel Porrúa, 2 Vols.
Leff, E.; Argueta, A; Boege, E. & Porto-Gonçalves, C.W. (2002), "Más allá del desarrollo sostenible. La construcción de una racionalidad ambiental para la sustentabilidad. Una visión desde América Latina", in Leff, E.; Ezcurra, E.; Pisanty, I. & Romero, P. (Eds.), *La transición hacia el desarrollo sustentable. Perspectivas de América Latina y el Caribe*, México: PNUMA/INE-SEMARNAT/UAM, pp. 479–578.
Leis, H. (2001), *La modernidad insustentable: las críticas del ambientalismo a la sociedad contemporánea*, Montevideo PNUMA/Nordan.
Leroi-Gourhan, A. (1964–1965), *Le geste et la parole*, 2 Vols., Paris: Albin Michel.
Levinas, E. (1977/1997), *Totalidad e infinito: ensayo sobre la exterioridad*, Salamanca: Sígueme, 4th edition [*Totality and infinity: an essay on exteriority*, Pittsburgh: Duquesne University Press, 1969].
———. (1993), *El tiempo y el otro*, Barcelona: Paidós [*Time and the Other*, Pittsburgh: Duquesne University Press, 1987].
———. (1999), *De otro modo que ser, o más allá de la esencia*, Salamanca: Ediciones Sígueme [Otherwise than Being or beyond essence, Springer Science + Public Media, B.V., 1991].
Lévi-Strauss, C. (1955), *Tristes tropiques*, Paris: Librairie Plon.
Light, A. (Ed.) (1998), *Social ecology after Bookchin*, New York/London: The Guilford Press.

Lipietz, A. (1999), ¿Qu'est-ce que l'écologie politique? La grande transformation du XX Siècle, Paris: La Découverte.

Lopez-Austin, A. & Lopez-Lujan, L. (2001), *Mexico's indigenous past*, Norman: University of Oklahoma Press.

Lotka, A. J. (1922), "Contribution to the energetics of evolution" *Proc. Natl. Acad. Sci. USA*, 8: pp. 147–51.

Lovelock, J. (1979), *Gaia. A new look at life on Earth*, Oxford: Oxford University Press.

Lozada, G.A. (1993), "Georgescu-Roegen's critique of statistical mechanics revisited", in *Entropy and bioeconomics*, op. cit., pp. 389–398.

Lukács, G. (1923/1971), *History and class consciousness*, Merlin Press.

Maier, C. (2006), "Transformations of territoriality 1600–2000", in G. Budde, S. Conrad & O. Janz, Eds., *Transnationale Geschichte: Themen, Tendenzen und Theorien*, Göttingen: Vandenhoeck & Ruprecht.

Mann, S.A. & J.A. Dickinson (1978), "Obstacles to the development of a capitalist agriculture", *Journal of peasant studies*, Vol. 5 (4): 466–481.

Marcuse, H. (1937), "Philosophy and critical theory", in *Negations*, USA/Great Britain/Australia, Penguin Books, 1968.

———. (1968), *L'homme unidimensionnel*, Paris: Ed. Minuit [*One-dimensional man*, Boston: Beacon Press, 1964].

———. (1972), *Counterrevolution and revolt*, Boston: Bacon Press.

———. (1992), "Ecology and the critique of modern society", *Capitalism, nature, socialism*, Vol. 3 (3):29–38.

———. (2005), *Heideggerian Marxism*, edited by Richard Wolin & John Abromeit, Lincoln & London: University of Nebraska Press.

Margulis, L. & Sagan, D. (1995), *What is life?*, Berkeley and Los Angeles: University of California Press.

Mariategui, J.C. (1971), *Seven interpretive essays on Peruvian reality*. Texas Pan American Series, Austin: University of Texas Press.

Marini, R.M. & dos Santos, T. (Coords.) (1999), *El pensamiento social latinoamericano en el siglo XX*, Caracas: UNESCO, 2 Vols.

Margalef, R. (1968), *Perspectives in ecological theory*, The University of Chicago Press, Chicago.

Martí, J. (1963), *Obras completas*, La Habana: Editorial Nacional de Cuba.

Martínez-Alier, J. (1987), *Ecological economics. Energy, environment and society*, Oxford/Cambridge: Basil Blackwell.

———. (1989), "Ecological economics and eco-socialism," *CNS*, 2.

———. (1994), "The Merchandizing of Biodiversity", *Etnoecologica*, 2 (3): 69–86, Mexico.

———. (1995a), "Distributional issues in ecological economics", *Review of social economy*, Vol. LIII (4): 511–528.

———. (1995b), "Political ecology, distributional conflicts and economic incommensurability", *New left review* I/211.

Marx, K. (1965), *Œuvres, Économie I*, Paris: Gallimard.

———. (1973), *Grundrisse*, Penguin Books, UK

———. (1976, 1978, 1981), *Capital* (three volumes) Penguin, UK.

Marx, K. & F. Engels (1975–2005), *Marx and Engels collected works*, New York: International Publishers.

Masuda, S, Shimada, I & Morris, C. (1985), *Andean ecology and civilization: an interdisciplinary perspective on Andean ecological complementarity*, Tokyo; University of Tokyo Press.

Mayumi, K. (1993), "Georgescu Roegen's 'fourth law of thermodynamics' and the flow-fund model", in *Entropy and bioeconomics*, op. cit., pp. 399–413.

Meadows, D.H., D.L. Meadows, J. Randers & W.W. Behrens III (1972), *The limits to growth*, New York: Universe Books.

Meillassoux, C. (1977a), *Terrains et Théories*, París, Editions Anthropos.

———. (1977b), *Mujeres, graneros y capitales*, México: Siglo XXI Editores.

Mellor, M. (1997), *Feminism & ecology*, Cambridge: Polito.

Merchant, C. (1992), *Radical ecology*, London: Roultledge.

Merino, L. & G. Segura (2002), *El manejo de los recursos forestales en México, 1992–2002. Procesos, tendencias y políticas públicas*, in Leff, E., E. Ezcurra, I. Pisanty & P. Romero (Ed), *La Transición Hacia el Desarrollo Sustentable. Perspectivas de América Latina y el Caribe*, México: PNUMA/INE-SEMARNAT/UAM, pp. 237–256.

Mignolo, W. (2000), *Local histories/global designs: coloniality, subaltern knowledges, and border thinking*, Princeton: Princeton University Press.

———. (2011), *Modernity and decoloniality*, Oxford bibliographies.

Mignolo, W. & A. Escobar, eds. (2009), *Globalization and the decolonial option*, London: Routledge.

Moguel, J., C. Botey & L. Hernández (1992), *Autonomía y nuevos sujetos sociales en el desarrollo rural*, México, Siglo XXI Editores.

Mol, A. & G. Spaargaren (2006), "Toward a sociology of environmental flows: a new agenda for twenty-first Century environmental sociology", in Spaargaren, G., Mol, A.P.J. y F.H. Buttel (2006), pp. 39–82.

Morales, H. (2011), *Otra historia de la sexualidad*, México: Palabra en Vuelo.

Morello, J. (1986), "Conceptos para un manejo integrado de los recursos naturales", en Leff, E. (Ed.), *Los problemas del conocimiento y la perspectiva ambiental del desarrollo*, México, Siglo XXI Editores.

———. (1990), "Insumos para la agenda ambiental latino-americana", in *Nuestra propia agenda*, BID/PNUD.

Morin, E. (1977), *La méthode. La nature de la nature*, Paris, Editions du Seuil.

———. (1980), *La méthode. La vie de la vie*, Paris, Editions du Seuil.

———. (1990), *Introduction à la pensée complexe*, Paris: Éditions du Seuil.

Moscovici, S. (1972), *La société contre nature*, Paris: Union générale d'édition Collection 10/18.

Mumford, L. (1970), *Pentagon of power: The myth of the machine*, Vol. II, New York: Harcourt.

Murra, J. (1956), *The economic organization of the Inca state*, Chicago: University of Chicago.

———. (1975), *Formaciones económicas y políticas del mundo andino*, Lima: IEP

Naess, A. (1989), *Ecology, community and lifestyle*, trans. D. Rothenberg, Cambridge: Cambridge University Press.

Naredo, J.M. (2006), *Raíces económicas del deterioro ecológico y social*, Madrid: Siglo XXI Editores.

Needham, J. (1954), *Science and civilization in China*, Cambridge, UK: Cambridge University Press.

Nietschmann, B. (1973), *Between land and water: the subsistence ecology of the Miskito Indians, Eastern Nicaragua*, New York: Seminar Press.

Nietzsche, F. (1968), *The will to power*, New York: Vintage Books.

Norgaard, R. (1984), "Coevolutionary development potential", *Land economics*, 60:160–173.

———. (1994), *Development betrayed*, Routledge: London.

O'Connor, J. (1988), "Capitalism, nature, socialism: a theoretical introduction", *Capitalism, nature, socialism*, 1 (1): 11–38.

———. (1989), "Combined and uneven development and ecological crisis: a theoretical introduction," *Race and class*, 30, 3, 1989.

———. (1998), *Natural causes: essays in ecological Marxism*, New York: Guilford.

O'Connor, M. (1991), "Entropy, structure and organizational change", *Ecological economics* 3:95–122.

———. (1993a), "On Steady State: A Valediction", in *Entropy and bioeconomics*, op. cit., pp. 414–457.

———. (1993b), "On the misadventures of capitalist nature", *Capitalism, nature, socialism*, 4 (3): 7–40.

O'Connor, M. Ed. (1994), *Is capitalism sustainable?*, New York: Guilford.

Orstrom, E. (1990), *Governing the commons. The evolution of institutions for collective action*, Cambridge: Cambridge University Press.

Packard, V. (1960), *The waste makers*, Brooklyn: Ig Publishing.

Park, Jacob/ Finger, Matthias/Conca, Ken (2008), The Death of Rio Environmentalism, in Conca, Ken/Finger, Matthias/ Park, Jacob (eds.), *The Crisis of Global Environmental Governance: Towards a New Political Economy of Sustainability*. London: Routledge.

Parra, M. (1993) "La producción silvo-agropecuaria de los indígenas de los Altos de Chiapas", en Leff & Carabias, Vol. 2, op. cit., pp. 445–487.

Passet, R. (1979), *L'économique et le vivant*, Paris: Payot.

———. (1985), "L'Économie: des choses mortes au vivant", *Encyclopaedia Universalis "Symposium"*, pp. 831–841.

Peet, R. & Watts, M. (1993), "Introduction: development theory and environment in an age of market triumphalism", *Economic geography*, Vol. 69 (3): 227–253.

———. Eds. (1996/2004), *Liberation ecologies: environment, development, social movements*, London: Routledge, 2nd edition.

Peet, R., Robbins, P. & Watts, M., Eds. (2010), *Global political ecology*, London: Routledge.

Pitt, D. (1985), "Towards ethnoconservation", in McNeely J. & D. Pitt, *Culture and Conservation: the human dimension in environmental planning*, London: IUCN/Croom Helm.

PNUMA (2002), *Manifesto for life. For an ethic for sustainability*, www.rolac.unep.mx.

Polanyi, K. (1944) *The great transformation: the political and economic origins of our time*, Boston: Beacon Press.

Popper, K. (1973), *La logique de la découverte scientifique*, Paris: Ed. Payot

Porto-Gonçalves, C.W. (2001a), *Geo-grafías. movimientos sociales, nuevas territorialidades y sustentabilidad*, México: Siglo XXI Editores.

———. (2001b), *Amazônia, Amazônias*, São Paulo: Contexto.

———. (2002), "O latifúndio genético e a r-existência indígeno-camponesa", *Geographia*, Ano 4, No. 8, Niterói: Universidade Federal Fluminense, pp. 7–30.

———. (2004), *Geografando nos varadouros do mundo: da territorialidade seringalista à territorialidade seringueira*, Brasilia: IBAMA.

———. (2006), *A Globalização da natureza e a natureza da globalização*, Rio de Janeiro: Civilização Brasileira.

———. (2007), *Em defesa dos organismos laboratorialmente modificados: em busca da precisão conceitual*, http://www.geopolitica.ws/article/em-defesa-dosorganismos-laboratorialmente-modificados/.

———. (2010), *Territorialidades y lucha por el territorio en América Latina. Geografía de los movimientos sociales en América Latina*, Caracas: IVIC.

Porto-Gonçalves, C.W. & Betancourt, M. (2013), "Encrucijada latinoamericana en Bolivia: el conflicto del TIPNIS y sus implicaciones civilizatorias", mimeo.

Porto-Gonçalves, C.W. & E. Leff (2015), "Political Ecology in Latin America: the social reappropriation of nature, in Leff, E., Floriani, D., & L.H. de Oliveira Cunha (Eds.), dossier temático "Pensamento Ambiental Latino-americano: movimentos sociais e territórios de vida", *Revista Desenvolvimento e Meio Ambiente* 35, Programa de Pós-Graduação de Meio Ambiente e Desenvolvimento, Universidade Federal do Paraná (Curitiba-Paraná-BRASIL).

Posey, D.A. (2004), *Indigenous knowledge and ethics: a Darrell Posey reader*, New York: Routledge.

Posey, D.A., & Dutfield, G. (1996), *Beyond intellectual property: toward traditional resource rights for indigenous peoples and local communities*, Ottawa: IDRC.

Prigogine, I. (1955), *Thermodynamics of irreversible processes*, New York: John Wiley & Sons.

———. (1961), *Introduction to thermodynamics of irreversible processes* (Second ed.). New York: Interscience.

———. (1981), *From being to becoming: time and complexity in the physical sciences*, New York: W.H. Freeman & Co.

Prigogine, I. & Nicolis, G. (1977), *Self-organization in non-equilibrium systems*, New York: John Wiley & Sons.

Prigogine, I. & I. Stengers (1984), *Order out of chaos*, New York: Bentam Books.

Przybylski, T. (1993), "Entropy as a measure of forest damages", in *Entropy and bioeconomics*, op. cit., pp. 476–481.

Puntí, Albert (1988), "Energy accounting: some new proposals", *Human ecology* 16 (1):79–86.

Quijano, A. (2008), "Coloniality of power, eurocentrism, and social classification", in Moraña, M., Dussel, E. and Jáuregui, C.A. (Eds.), *Coloniality at large. Latin America and the postcolonial debate*, Durham, NC: Duke University Press, pp. 181–224.

Quintero, P. (Ed.) (2014), *Crisis civilizatoria, desarrollo y bien vivir*, Buenos Aires: Ediciones del Signo.

Raffestin, C. (1980), *Pour une géographie du pouvoir*, Paris: Librairies techniques.

Rappaport, R.A. (1968/1984), *Pigs for the ancestors*, New Haven: Yale University Press.

———. (1971), "The flow of energy in an agricultural society", *Scientific American*, 224 (3):116–132.

Redclift, M. (1987), *Sustainable development: exploring the contradictions*, London: Methuen.

Redclift, M. & G. Woodgate (Eds.) (2010), *The international handbook of environmental sociology*, second edition, Cheltenham, UK/Northampton MA, USA: Edward Elgar.

Ribeiro, D. (1973), "Etnicidade, indigenato e campesinato", *Revista de Cultura Vozes*, Ano 73, Vol. LXXIII, outubro, No. 8, p. 5–18, Rio de Janeiro.

———. (1980), *Indigenato e campesinato, Revista de Cultura VOZES*, Ano 73, Vol. LXXIII, Outubro de 1980, n° 8, p. 5–10, Rio de Janeiro.

Ribeiro, B.G. (1992), *Amazônia urgente: Cinco séculos de história e ecologia*, Belo Horizonte: Ed. Itatiaia/INEP-MEC/CNPq/Vitae.

Richta, R. (1969), *La civilization au carrefour*, Paris: Ed. Anthropos.

Rivera Cusicanqui, S.; Domingues, J.; Escobar, A. & Leff, E. (2016), "Debate sobre el colonialismo intelectual y los dilemas de la teoría social latinoamericana", en *Cuestiones de Sociología,* 14, Universidad de La Plata, Argentina, www.cuestionessociologia.fahce.unlp.edu.ar/article/view/CSn14a09.

Robbins, P. (2012), *Political ecology: a critical introduction*, Chichester, UK: John Wiley & Sons, 2nd edition.

Rodin, L.E., N.I. Bazilevich e N.N. Rozov (1975), "Primary productivity of the main world ecosystem", *Ecology First International Congress*, The Hague, September 1974, in W.H. van Dobben e Lowe-McConell, *Unifying concepts in ecology*, The Hague, W. Jung B. V. Publishers and Wageningen: Centre for Agricultural Publishing and Documentation.

Romanini, C. (1976), *Ecotécnicas para el trópico húmedo*, México: Cecodes-Conacyt/PNUMA

Rorty, R. (1979), *Philosophy and the mirror of nature*, Princeton: Princeton University Press.

Saal, F. (1998), *Palabra de analista*, Mexico: Siglo XXI Editores.

Sachs, I. (1980), *Ecodesarrollo: desarrollo sin destrucción*, México: El Colegio de México.

Sack, R. (1985), *Human territoriality: its theory and history*. Cambridge: Cambridge University Press.

Safouan, M. (1981), "Is the Oedipus complex universal?", in *M/F*, No. 5/6, pp. 83–90.

Santos, M. (1996), *A natureza do espaço: técnica e tempo; razão e emoção*, São Paulo: Hucitec.

Sartre, J. P. (1960), *Critique de la raison dialectique*, París, Gallimard.

Scheer, H. (2000), *Economía solar global. Estrategias para la modernidad ecológica*, Barcelona, Galaxia Gutenberg.

Schmidt, A. (1971), *The concept of nature in Marx*, London: New Left Books.

Schmidt-Bleek, F. (2008), "Future. Beyond climate change", Position paper 08/01, Factor 10 Institute, Provence, France.

Schneider, E. & D. Sagan (2005), *Into the cool: energy flow, thermodynamics and life*, Chicago: Chicago University Press.

Schrödinger, E. (1944/1969) *¿What is life? The physical aspect of the living cell*, London/New York: Cambridge University Press.

Schumacher, E.F. (1973), *Small is beautiful. Economics as if people mattered*, London: Blond & Briggs.

Schumpeter, J. (2014/1942), *Capitalism, socialism and democracy* (2nd ed.) Floyd, Virginia: Impact Books.

Shiva, V. (1988), *Staying alive: women, ecology and development*, London: Zed.

———. (1993), *Monocultures of the mind*, London: Zed Press.

Sousa Santos, B. (2008), *Conocer desde el Sur. Para una cultura política emancipatoria*, Buenos Aires: CLACSO/CIDES-UMSA/Plural Editores.

Spaargaren, G., Mol, A.P.J. y F.H. Buttel (2006), Governing environmental flows. Global challenges to social theory, Cambridge, Mass: The MIT Press.

Stavenhagen, R. (1965), "Classes, colonialism, and acculturation. Essay on the system of inter-ethnic relations in Mesoamerica", *Studies in Comparative International Development*, 1 (6): 53–77.

Steiner, G. (1975/2001), *Después de Babel*, México: Fondo de Cultura Económica, 3a edición.

Steward, J. (1972), *Theory of cultural change: the methodology of multilinear evolution*, Urbana: University of Illinois Press.

Svampa, M. (2016), *Debates latinoamericanos. Indianismo, desarrollo, dependencia y populismo*, Buenos Aires y Bolivia: Edhasa/CEDIB.

Tapia, L. (2009), *La invención del núcleo comum—ciudadanía y gobierno multisocietal*, La Paz, Bolívia: Ed. Muela del Diablo.

Thompson, E.P. (1971), "The moral economy of the English crowd in the 18th century", in *Past and present*, 50:76–136.

———. (1991), *Customs in common: studies in traditional popular culture*, London: Merlin Press.

Throne, F. (1935), "Nature rambling: We fight for grass", *The science newsletter* 27, 717, Jan. 5: 14.

Toledo, V.M. (1980), "Ecología del modo campesino de producción", *Antropología y marxismo* No. 3, pp. 35–55.

Toledo, V.M., Caballero, J. & A. Argueta (1978), "El uso múltiple de la selva basado en el conocimiento tradicional", *Biótica* 3: 85–101.

Toledo, V.M. & A. Argueta (1993), "Naturaleza, producción y cultura en una región indígena de México: las lecciones de Pátzcuaro", in Leff & Carabias, Vol. 2, op. cit., pp. 413–443.

Torres, R. (1997), *Entre lo propio y lo ajeno. Derechos de los pueblos indígenas y propiedad intelectual,* Quito, COICA.

Tricart, J. (1978), "Vocations des terres, ressources ou contraintes et développement rural", *Hérodote,* 12: 65–75.

———. (1982), "Géographie/Écologie", *Hérodote,* 26.

Uhl, C.; Nepstad, D.; Silva, J.M.C. & Vieira, I.C.G. (1991), "Restauração da floresta em pastagens degradadas", *Ciência Hoje* 13:23–31.

Uribe, R. (org.) (1988), *Medio ambiente y comunidades indígenas del sureste: prácticas tradicionales de producción, rituales y manejo de recursos,* Villahermosa, Tabasco, México: UNESCO/Gob. de Tabasco.

Urry, J. (2003), *Global complexity,* Cambridge: Polity Press.

———. (2005), "The complexity turn", in *Theory, culture & society,* Vol. 22 (5): 1–14.

Varèse, S. & Martin, G. (1993), "Ecología y producción en dos áreas indígenas de México y Perú. Experiencias y propuestas para un desarrollo culturalmente sustentable", in Leff & Carabias, op. cit., pp. 717–740.

Vayda, P. (ed). (1969), *Environment and cultural behavior,* Garden City, NY: The Natural History Press.

———. (1983), "Progressive contextualization: methods for research in human ecology", *Human ecology* 11: 265–81.

Vernadsky, V.I. (1998), *The biosphere,* New York: Copernicus.

Vogel, J.H. (2009), *The economics of the Yasuní initiative: climate change as if thermodynamics mattered, London/New York/Delhi:* Anthem Press.

von Bertalanffy, L. (1968), *General systems theory: foundations, development, applications,* New York: George Braziller.

Walker, P.A. (2005), "Political ecology: where is the ecology?", *Progress in human geography* 29 (1):73–82.

Wallerstein, I. (1974), *The modern world-system, vol. I: capitalist agriculture and the origins of the European world-economy in the sixteenth century,* New York/ London: Academic Press.

———. (1980), *The modern world-system, vol. II: mercantilism and the consolidation of the European world-economy, 1600–1750,* New York: Academic Press.

———. (1989), *The modern world-system, vol. III: the second great expansion of the capitalist world-economy, 1730–1840,* San Diego: Academic Press.

———. (2011), *The modern world-system, vol. IV: centrist liberalism triumphant, 1789–1914,* Berkeley: University of California press.

Ward, B. (1966), *Spaceship earth*, New York: Columbia University Press.

Watts, M. (2008), "On the poverty of theory: natural hazards research in context", in Braun, B., Environment: critical essays in human geography, London: Routledge.

———. (2015), "Now and then: the origins of political ecology and the rebirth of adaptation as a form of thought", in Perrault, T., J. McCarthy & G. Bridge, *Handbook of political ecology*, London: Routledge.

WCED (1987), *Our common future*, Oxford: Oxford University Press.

Weber, M. (1922/2014), *Economía y sociedad: esbozo de sociología comprensiva*, México: Fondo de Cultura Económica.

Weiszsäcker, E.U. von, Lovins, A.B. & L.H. Lovins (1997), *Factor four. Doubling wealth halving resource use*, a report to the Club of Rome, London: Earthscan.

White, L. (1949), *The science of culture*. Nueva York: Grove Press.

Whiteside, K (2002), *Divided natures*, Cambridge: Massachusetts Institute of Technology.

Whittaker, R.H. (1975), *Communities and ecosystems*, Macmillan, New York/London.

Wilson, E.O. (1975), *Sociobiology, the new synthesis*, Cambridge, Mass. & London: The Belknap Press of Harvard University Press.

Wolf, E. (1972a), "Ownership and political ecology", *Anthropological quarterly* 45 (3): 201–205.

———. (1972b), *Las luchas campesinas del siglo XX*, México, Siglo XXI Editores.

Wolf, E. & A. Palerm (1972), "Potencial ecológico y desarrollo cultural en Mesoamérica", in *Agricultura y civilización en Mesoamérica*, No. 32, México: SepSetentas.

WRI/UNEP (1990), *World resources 1990–1991*, Oxford University Press, Oxford/New York.

Yearly, S. (1991), *The green case. A sociology of environmental issues, arguments and politics*, London: Harper Collins Academic.

Zimmerman, M.E. (1994), *Contesting earth's future. Radical ecology and postmodernity*, Berkeley: University of California Press.

Index[1]

[1] Note: Page numbers followed by 'n' refer to notes.

E. Leff, *Political Ecology*, https://doi.org/10.1007/978-3-030-63325-7

The manufacturer's authorised representative in the EU is Springer Nature Customer Service Centre GmbH, Europaplatz 3, 69115 Heidelberg, Germany. If you have any concerns regarding our products, please contact ProductSafety@springernature.com

Printed and bound by CPI Group (UK) Ltd, Croydon, CR0 4YY

30/09/2026

02225453-0008